GM Passes Ford, 1918–1938

GM Passes Ford, 1918–1938

Designing the General Motors Performance-Control System

Arthur J. Kuhn

THE PENNSYLVANIA STATE UNIVERSITY PRESS
UNIVERSITY PARK AND LONDON

Library of Congress Cataloging-in-Publication Data

Kuhn, Arthur J.
  GM passes Ford, 1918–1938.

  Bibliography: p.
  Includes indexes.
  1. General Motors Corporation—Management—History—
20th century.   2. Ford Motor Company—Management—
History—20th century.   3. Automobile industry and
trade—United States—Management—History—20th century.
I. Title.
HD9710.U54G4745   1986   338.7′6292′0973   85–31965
ISBN 0-271-00432-0

To Past Generations: The Designer's Educator

To Future Generations: The Designer's Client

Contents

Preface

This book presents a cybernetic and system-theory analysis of the General Motors Corporation (GM) and Ford Motor Company (FM) reversals between 1918 and 1938. During these years, GM's Alfred Sloan—aided by codesigners such as Donaldson Brown—conceived, implemented, and then refined a sophisticated performance-control system harmonious with propositions later articulated by cyberneticists and system theorists. Meanwhile, FM's Henry Ford eschewed such notions, failing, for instance, to match GM's extensive product variety or environmental monitoring.

By the mid-1920s GM had begun to narrow FM's once monumental sales and profit leads. GM first passed FM in 1927 and shortly thereafter outdistanced it consistently.

Later, as successive GM managements lapsed from the Sloan-Brown team's early policies, GM's sales and profits suffered as Ford's had earlier. For instance, GM's 1980 product line appeared as standardized and dated as FM's Model T had in 1927 when compared to the varied and sophisticated array of automobiles coming from a host of foreign competitors. And GM lost money in 1980 for the first time since the Sloan team took command in 1921.

The GM–FM analysis rests on a strong theoretical foundation and is intended to begin linking cybernetic and system theory with business practice. Ashby (1956:1) defined cybernetics "in a word, as the art of *steersmanship*." Over the past few decades much has been written about the potential contributions this paradigm offers organizational analysts and planners, yet few concrete linkages have been made, and much work remains before the potential can be realized. The basic theme of the present developmental effort is that policy decisions, taken in the ordinary course of business, comprise acts of organizational design and that these acts must be compatible with cybernetic and system propositions if the enterprise is to be successful. Thus cybernetic and system concepts have been combined to create a system-design-for-performance-control (SPDC) model that outlines a normative design process consisting of seven phases. Here, history is reinterpreted within this systematic framework so that the lessons learned may be applied to other cases whether historical or contemporary. I present a more extensive and general treatment of the SDPC model in *Organizational Cybernetics and Business Policy: System Design for Performance Control*.

Acknowledgments

I would like to acknowledge and express my appreciation to several individuals and institutions upon whose efforts and resources this work has been built.

Joseph Litterer first stimulated my interest in organization theory at the University of Illinois. At the University of California, Berkeley, John Wheeler, George Strauss, Charles Perrow, Karlene Roberts, Arthur Stinchcombe, Raymond Miles, and especially C. West Churchman helped me strengthen the foundation of organization and system theory on which this work rests.

Colleagues and students at Duke University also provided significant support. In particular, Richard Burton's continuing comments honed my thinking and sustained my interest. I owe my students the greatest appreciation, for they patiently listened while I developed the system-design-for-performance-control model.

Joseph Litterer, John Wheeler, C. West Churchman, Helmy Baligh, Richard Burton, and Alfred Chandler all read various portions of the manuscript and suggested helpful changes or offered welcome encouragement. Willis L. Parker's "comma chasing" proved invaluable in untangling many editorial snags. John M. Pickering and Chris W. Kentera of the Penn State Press offered useful suggestions for tightening my presentation.

And of course much appreciation goes to those many cyberneticians, system theorists, organizational writers, and business historians upon whose work this book has so heavily drawn. Five works have been particularly helpful: Allan Nevins and Frank E. Hill's three-volume history, *Ford: The Times, the Man, the Company; Ford: Expansion and Challenge, 1915-1933;* and *Ford: Decline and Rebirth, 1933-1962;* Alfred D. Chandler's classic *Strategy and Structure;* and

Chandler and Stephen Salsbury's *Pierre S. duPont and the Making of the Modern Corporation.*

The numerous writings of GM's early designers were invaluable in this research effort as well. In particular, Donaldson Brown's little-known book, *Some Reminiscences of an Industrialist,* calls for special mention as does Alfred Sloan's *Adventures of a White Collar Man.* Without doubt, Sloan's later work, *My Years with General Motors,* is most noteworthy. It should long remain a classic of American management literature for both practitioners and researchers.

Carol Van Steenberg, throughout this endeavor, earned my deepest gratitude. She listened and responded to formative ideas; she read and edited many draft pages; she asked difficult questions that uncovered inconsistencies or omissions; she initiated discussions that improved conceptual clarity; and she alerted me to many relevant writings.

PART I

The General Motors-Ford Motor Comparison

1

Introduction: "Steersmanship" in Action

This book analyzes General Motors Corporation's (GM's) upturn under Alfred Sloan and Ford Motor Company's (FM's) downturn under Henry Ford between 1918 and 1938. Although cybernetic principles and system concepts were not articulated until after World War II, the thesis proffered here is that GM passed FM by implicitly designing a performance-control system in accord with such theoretical propositions—an approach rejected at FM. Thus, a system-design-for-performance-control (SPDC) model, incorporating cybernetic and system concepts, drives the explanation.

On close reading, Alfred Sloan's writings reveal that his thinking on controlling GM's performance was uncannily parallel to the positions taken by cyberneticians and system-design theorists, notably Ashby and Churchman. Sloan (1927c), for example, using a ship/management analogy similarly employed by Churchman forty years later, likened GM's dangerous position in 1920

> to a big ship in the ocean. We were sailing along at full speed, the sun was shining, and so far as could be seen there was no cloud in the sky that would indicate an approaching storm. . . . yet . . . almost overnight, values commenced to fall. The liquidation from the inflated prices resulting from the war had set in. Practically all schedules or a large part of them were cancelled. Inventory commenced to roll in, and, before it was realized what was happening, this great ship of ours was in the midst of a terrific storm. (p. 3)

Churchman, in *The Systems Approach*, offers via the ship analogy a generalized procedure for minimizing such dangers.

> The captain of the ship as the manager generates the plans for the ship's operations and makes sure of the implementation of his plans. He institutes various kinds of information systems throughout the ship that inform him where a deviation from plan has occurred, and his task is to determine why the deviation has occurred, to evaluate the performance of the ship, and then finally, if necessary, to change his plan if the information indicates the advisability of doing so. This may be called the "cybernetic loop" of the management function, because it is what the steersman of a ship is supposed to accomplish. (p. 46)

The close correspondence between the Sloan and Churchman positions can be seen in Sloan's description of GM's new "Control Systems."

> Our first duty was to obtain a proper control over the operations of this big ship. We should not be satisfied to go along ... when times were good, with no thought of the future. We should first devise scientific means of administration and control ... to project ourselves as much as possible into the future and discount changing trends and influences and, second, ... we should be prepared at all times to alter the course of this ship of ours promptly and effectively should circumstances develop that required us to do so. This has all been accomplished. (p. 4)

In contrast, Henry Ford in the development of his company took what can only be described as an antidesign tack. Ironically, Ford articulated his approach with the help of an analogy between systems of management and the sailing of ships.

> The sea captain sails by his charts; he can do that because the sea route has been charted. But American business is now sailing over seas that have never been passed before. There are no management charts in the wheelhouse of progressive business; if there are, they are charts of areas long past. What American business is depending on now is the man on the bridge—the living manager—whose only charts are his foresight and his insight and his sense of the new trade winds that are beginning to blow. To suggest that you can make a manager by stuffing him full of a system of management which worked fairly well ten or twenty years ago is to suggest that business is a ferryboat plying in familiar waters between two commonplace ports. But business today is an explorer's ship; it is always meeting new conditions. No system of management can guide it; it depends on the man on the bridge. If we can only get clearly in mind the difference between a system of management and a manager, then we shall have come a long step toward facing business problems from the angle of mastery. (Ford and Crowther, 1930:24)

Churchman (1968) characterizes antiplanners as believing "that any attempt to lay out specific and 'rational' plans is either foolish or downright evil" (p. 14). That description fitted Henry Ford very well.

The widely differing design philosophies found within the two firms can be traced ultimately to the degree of systematic technical training and scientific education attained by their key participants. At GM the principals represented an emerging breed of engineering-oriented executives who were rationalizing the American system of corporate capitalism (see Noble's *America by Design*). Sloan and Donaldson Brown, his chief lieutenant, both had graduated as electrical engineers, Sloan from Massachusetts Institute of Technology and Brown from Virginia Polytechnic Institute. In contrast, Ford lacked the most rudimentary of engineering fundamentals. Many associates even questioned whether Ford could read a blueprint.

Nonetheless, Ford appreciated organization in his factories. In the early days, for instance, FM's engineering department kept extensive statistics. "The richness and accuracy of this information" enabled the company to "increase its production in the period between 1915 and 1923 with few serious problems" (Hounshell, 1984:272). So in terms of its productive machinery, FM grew immensely; by 1923 the company had become the envied giant of American industry. In 1924, however, the automobile market began to change. Over the next two years a metamorphosis occurred that rendered FM's Model T machinery and the accompanying historical information worthless. Even worse, Ford had long held that FM's non-factory organization should be minimized (Nevins and Hill, 1957:271). Without a central headquarters to guide it, FM foundered in 1927. Rather than learn from the problems associated with the Model T's demise, Ford became increasingly resentful of the business organization needed to run his company (Galbraith, 1967:90). Throughout the 1920s, 1930s, and into the 1940s, Ford's strong antidesign attitude toward systems of performance control made FM a giant ship without a helm.

The intriguing analogies offered by Sloan, Churchman, and Ford provide a convenient—if colloquial—introduction for the organizational investigation at hand. They juxtapose the conflicting orientations of Sloan and Ford, and more importantly, the similarity of the Sloan and Churchman orientations. The opposing orientations of Sloan and Ford thus become the broad independent variable of this work.

Another important distinction between GM under Sloan and FM under Ford is the former's soaring prosperity and the latter's spiraling decline—as measured by automotive market shares and capital investment returns. Typically, when comparing an organization with its own past and/or with another organization, the analyst searches for the dichotomy of "high versus low performance." A performance difference between two firms—or between two points in time—is important to both business planners and social scientists in that such variations (i.e., high versus low performance) are usually the ultimate concern, or dependent variable, in organizational design and research.

The GM–FM opportunity for comparative analysis is particularly attractive because it lends itself quite readily to the methodology Barton (1968) terms the "strategic paired comparison" (p. 337), which requires two organizations that are generally similar but differ strongly with respect to the explanatory independent and intermediate variables. The two automobile firms were amazingly alike—in terms of industry, technology, sales, location, maturity, etc.—yet strikingly different in their design/antidesign orientations and the resultant policy variations. The more similar two organizations "are with respect to crucial variables, in short, the better able is the investigator to isolate and analyze the influence of other variables that might account for the differences he wishes to explain" (Smelser, 1968:72).

The GM–FM differences are also important because GM and FM have long played dominant—though somewhat misinterpreted—roles in American society: the automobile business is America's largest industry and these are two of America's largest firms. Not surprisingly, both organizations possess extended and well-documented histories, several of which provide valuable insight into the GM designers' and FM antidesigners' thinking as they were written by the firms' participants themselves or by researchers who interviewed the participants and/or studied their firms' archives. In addition, the GM participants wrote numerous articles describing the performance-control systems they were developing, and Henry Ford granted many interviews to writers interested in describing his methods. This high visibility has influenced both managers and researchers; because of its well-publicized economic success, the influence of the Sloan regime at GM has truly been profound.

Sloan's managerial innovations of the 1920s and 1930s have a continuing relevance. Even firms in the modern high-technology electronic industry can draw from these early automotive experiences. For the automobile industry, the 1920s and 1930s marked the transition from an early entrepreneurial stage dominated by the Henry Fords to a corporate phase dominated by the Alfred Sloans. Many electronics firms in the 1980s face similar shifts (see, for example, Moritz's *The Little Kingdom: The Private Story of Apple Computer*). The market and technological uncertainty found in the electronics industry (as Kidder describes it in *The Soul of a New Machine*) also parallels that in the early years of the automobile industry, which "could only be characterized as severe" (Thomas, 1973:135). The product and its use evolved rapidly, the processes shifted dramatically, and the market changed abruptly. Even the societal/political realm changed sharply (in the 1930s) necessitating considerable adaptation. In essence, the early automobile business constituted America's first major high-tech industry.

In the automobile industry itself Sloan's techniques and procedures have a recurring relevance. During the late 1940s and early 1950s, for instance, the

Ford Motor Company adopted many of Sloan's methods when Henry Ford II succeeded his ailing grandfather.

More importantly, almost a half century after Sloan's major design work ended, researchers studying the domestic auto industry were finding that deteriorated management practices underlay the industry's malaise of the 1970s and early 1980s. (See, for example, Abernathy, Clark, and Kantrow's *Industrial Renaissance*, 1983:4-5; Lawrence and Dyer's *Renewing American Industry*, 1983:21; and Yates's *Decline and Fall of the American Automobile Industry*, 1983.)

Indeed, the managements of the domestic automobile firms seemed to be coming full circle, almost unwittingly readopting the old Sloan ways after having strayed far afield in the interim. At GM, for instance, successive managements had violated important policy guidelines set by Sloan and his colleagues in the 1920s and 1930s. GM's triumph over FM did not result simply from Sloan's wily switch to marketing unwanted frills and Ford's stolid adherence to producing basic transportation. Sloan did not vanquish Ford's popular Model T by cajoling consumers to buy upgraded automobiles and annual models. Rather he monitored changing tastes and pushed GM's divisions to keep their offerings synchronized with market demands. Moreover, Sloan's production experience—almost as vast as Ford's—made him strive for the largest possible production runs. "By getting the small profit more times a year, and piling up these small profits into a heap available for dividends and surplus at the end of the year," Sloan (1926) stressed, "the competent manufacturer can hold his aggregate earnings at a satisfactory point" (p. 993). Out of this competitive necessity arose GM's ever-growing array of interchangeable engine components, chassis parts, automotive accessories, and automobile bodies. One of Sloan's more significant achievements was balancing the market's demand for product variety and GM's need for product standardization. Ford, like Sloan's own successors at GM, failed at this balancing act and sacrificed product differentiation for production volume.

In the 1970s and early 1980s, GM's financially dominated corporate headquarters ignored the need to "preserve a distinction of appearance, so that one knows on sight a Chevrolet, a Pontiac, an Oldsmobile, a Buick or a Cadillac" (Sloan, 1964:265). Still worse, "some Chevrolets" started to look "more impressive than some Cadillacs. The economy luxury car and the luxurious economy car were beginning to be confused" (Boorstin, 1973:554). This trend culminated in 1981 when a once-proud Cadillac Division—originator of such automotive achievements as the exquisitely styled and impeccably engineered V-16 of 1929—introduced its lowly Cimarron model. "Even Pete Estes, a former GM president, complained that the Cimarron looked too much like a Chevrolet. Leather seats and automatic headlight dimmers were not enough to distinguish it from a basic J-car" (Iacocca, 1984:254). As a result of this and other blunders, GM's product line lacked variety, confused consumers, and hurt sales. Having needlessly muddled its product images to gain short-run cost savings, GM then rushed to restore each division's identity.

More than a year later (1984) GM's management was still struggling to end its plethora of look-alike models by initiating a massive internal reorganization. (See, for example, "Can GM Solve Its Identity Crisis?" *Business Week*, January 23, 1984:32-33.) Reorganization was necessary because GM had become far too centralized, given the top corporate executives' lack of automotive experience.

Indeed, centralization without the requisite corporate-level expertise was the primary cause of GM's otherwise perplexing difficulties in the late 1970s and early 1980s. Following Sloan's lead, his successors further concentrated decision making in GM's headquarters.

Though most writers note that the Sloan-Brown team instituted tight financial controls, few acknowledge GM's *early* centralization in nonfinancial areas such as purchasing, engineering, styling, manufacturing, and marketing. Nevertheless, a careful reading of *My Years with General Motors* reveals abundant evidence to dismiss Sloan's decentralization thesis as a myth—long unquestioned in a free-enterprise economy and a democratic society.

Sloan started this trend toward centralization simply because he had inherited from his predecessor, William Durant, a group of division managers almost totally ignorant of financial matters. Durant's "automobile men," for instance, often accumulated dangerously excessive inventories through inattention, poor forecasting, or even inflationary speculation. This and numerous other abuses were ended when the Sloan-Brown team imposed strict financial controls over divisional operations. Interestingly, when Lee Iacocca took over the ailing Chrysler Corporation in 1978, he also discovered excessive inventories, euphemistically termed a "sales bank," and "no overall system of financial controls" (Iacocca, 1984:154). "If the bean counters are too weak," as Iacocca puts it, "the company will spend itself into bankruptcy" (p. 43).

"But if they're too strong, the company won't meet the market or stay competitive" (pp. 43-44). Sloan and his financially oriented colleagues avoided the latter pitfall by teaching themselves all facets of the automobile business. They inspected plants; they talked with divisional engineers, production experts, and salesmen; they studied car designs and searched for improvements. Even Donaldson Brown, GM's top financial officer, toured the country visiting dealers with Sloan. Soon GM was delivering cars far better than anything offered by Durant's or even Ford's "automobile men."

But over the years of diminished competition the balance shifted increasingly to the financial side. By 1980, then, GM's top executives were almost exclusively "financial experts" without automotive engineering, production, and marketing experience. Finally aware of the long-term consequences of such limitations, Roger Smith broke from his own financial mold to attempt a GM revival.

On another front in the early 1980s, GM, like other domestic auto manufacturers, hurriedly copied "just-in-time" or *kanban* supply methods from Japanese auto producers—suddenly the new paragons of modern industrial efficiency. (See, for example, Burch, 1982:36.) But hand-to-mouth buying (Stillman, 1927:1) to pare costly in-process inventories was not really new, only temporarily forgotten. As early as 1924 Sloan taught GM's buyers that "the fundamental principles of inventory control cannot be too often studied or too intelligently applied. Our ratio of stock on hand to production is constantly going down, without interfering with continuity of manufacture" (p. 194). Sloan's early pupils at GM learned this lesson well. One Buick executive wrote: "The bolts which come out of the machine this morning are on the finished automobile which a dealer drives away on the highway this afternoon" (Durham, 1927:56). Similarly at FM in the early days, Ernest Kanzler—an extremely competent executive eventually fired by Henry Ford—had synchronized parts supplies with FM's finished-tractor production. "So exact were his schedules that supplies arrived practically as needed, and freight cars bringing in wheels, radiators, castings, etc. were utilized a few hours after their arrival to dispatch completed tractors" (Nevins and Hill, 1957:155).

The passage of time also led GM to change other early policies. When fighting Ford's static-model policy in the 1920s and 1930s, for instance, GM stressed the use of general-purpose machinery. The flexibility of this equipment, in contrast to FM's single-purpose machinery, allowed GM to change its products rapidly. As its domination increased and innovation decreased, however, GM adopted many of Ford's techniques. Indeed, the integrated transfer lines that produced V-8 engines through the 1950s, 1960s, and 1970s would have delighted Henry Ford as they minimized space and ran on endlessly for years. But, as Ford found earlier, such machinery inhibited innovation (Abernathy, 1978:99). Thus to meet the more varied and competitive conditions of the 1980s, the domestic automobile industry and GM in particular was reintroducing more flexible machinery—much of it now computer-controlled (Cole and Yakushiji, 1984:147).

To cite another example of the relevance of Sloan's teachings: American businessmen and academics alike have praised the Japanese for their "participatory managerial techniques," seemingly unaware that Sloan pioneered this practice with GM's management in the 1920s. Though increasingly autocratic, he always spent considerable time soliciting opinions from his managers and explaining his position before finalizing a policy decision. A very important Sloan achievement, then, was "one of intercommunication, getting all the facts before all the people concerned" ("Alfred P. Sloan, Jr.:

Chairman," *Fortune*, 1938:114). With complete candor and without any hesitation, Sloan when the occasion arose could say: "General Motors is a group organization. I have just returned from Detroit, where I consulted everybody who could possibly contribute to a certain decision which had to be reached" (Forbes and Foster, 1926:238). Sloan's "zeal" was "not so much to 'run' G.M. as to arrange and activate its committees"; and his "science" was "not only automobiles but management itself" ("General Motors," *Fortune*, 1938:178).

Although it might be assumed that Sloan's "science" of the 1920s and 1930s was a given in the late 1970s in the American automobile industry, the contrary was true. Lee Iacocca found Chrysler's internal communication network seriously deficient. "There was no real committee setup, no cement in the organizational chart, no system of meetings to get people talking to each other" (Iacocca, 1984:152). This deficiency extended right to the top: "Chrysler's board of directors had even less information than their counterparts at Ford— and that's saying a mouthful. There were no slides and no financial reviews" (p. 155). In stark contrast to this approach, Sloan kept GM's board fully informed about internal operations as well as market developments and economic conditions.

Environmental intelligence thus played a key role in GM's early management scheme. Besides tirelessly visiting GM's dealers and talking with their community leaders, Sloan generously supported the firm's Research Laboratories, Proving Ground, and Customer Research Staff. In addition, Sloan and other top GM executives toured Europe looking for technical developments and styling advances. As he said about the latter activity, "Naturally I tried to keep my eyes open for any new things that applied to the automobile industry" (testimony of Sloan, *U.S.* v. *DuPont*, 1956:1312).

To reverse GM's poor showing in the California market, which usually anticipates national trends, the firm's management of the 1980s established an Advanced Concepts Center in Southern California. This small design, research, and engineering facility was to reach its objective by monitoring technical, social, and aesthetic trends, especially on the West Coast.

Such an effort became necessary because Sloan's successors had grown insular. They failed to follow his dictum to always "get the facts." To them, success was natural; they were always right. Unlike Sloan and his cohorts, they had not seen GM when it competed against the then-dominant FM. Perhaps this limited sense of history explains their failure to monitor the market's shifts just as Henry Ford had done fifty years earlier. Whatever the cause, GM's modern managements overlooked the possibility that developing foreign manufacturers could profit from Sloan's strategy for passing the unchallengeable Ford back in the 1920s: anticipating a shift in consumer preferences and exploiting the flux at the competition's expense. Whereas Ford resisted the shift by clinging to his standardized, outmoded Model T, Sloan and his colleagues embraced the change by offering a varied array of updated makes and models. Fifty years later, however, the tables had turned. In 1980 GM suffered with the poorly differentiated, dated product line.

Finally, the early GM–FM histories help explain the societal/political problems encountered by the American auto manufacturers in the late 1960s, 1970s, and early 1980s. The GM history in the middle and late 1930s remains particularly important because the firm's enviable sales and profit records made it an inviting target for political leaders and labor organizers. Though reluctantly, Sloan responded to complaints against the business community far more responsibly than Ford did. Indeed, GM's early capitulation to organized labor in 1937 marked a new era of industrial relations in the United States. Unfortunately, the truce hammered out in the turbulent 1930s came at a high price for consumers, and it left unresolved serious antagonisms between management and labor that still haunt the American auto industry. Worse yet, it was shortly after GM's defeat by organized labor that Sloan turned the firm further inward by narrowing the membership of its top policy-setting committee and by ceasing the work of its short-lived Policy Group–Social and Economic Trends. The latter unit—chaired by Donaldson Brown—held the key to the long-run future since its charge was to monitor the impact of social and political conditions on GM's more immediate economic surroundings. These moves along with Sloan's tendency to centralize and integrate—which had accelerated under the cost-cutting exigencies of the depression—absorbed the corporate headquarters more and more in lower-level, shorter-term divisional affairs. So occupied, it neglected the firm's higher-level, longer-term social responsibilities.

GM's success was thus limited to the economic arena and was not extended to the broader societal milieu. And given the increasingly myopic focus of Sloan's successors on immediate sales and short-term profits, it was not at all surprising that GM's top executives failed to anticipate the major shift in what previously had been the firm's long-stable political environment.

Hence, an unprepared GM (as well as the other American auto firms) suddenly faced tough new safety, pollution, and energy legislation in the 1960s and 1970s. Only in 1970, after the legislative gears had been turning a full five years, would GM's Board of Directors create a Public Policy Committee to focus on national concerns. While certainly a commendable addition (Kanter, 1983:317), it was thirty years too late. GM's reputation had been sullied by such incidents as the Corvair tragedy and the accompanying Nader exposé.

In spite of their contemporary relevance, the GM case and the FM contrast between 1918 and 1938 have received little analytical attention enabling the lessons learned to be applied to other cases whether historical or contemporary. A generalizable framework is needed to move from Sloan's planning and Ford's antidesign predilections (the independent variable) to specific policy variations (the intermediate variables) that caused GM's improvement and FM's deterioration in market share and return rate (the dependent variables).

2

Contrasting the General Motors–Ford Motor Policies Via the System-Design-for-Performance-Control Model

SETTING THE STAGE

The SDPC model guides the selection of specific policy variations (the intermediate variables) from the vast GM–FM histories. This model is built upon several key cybernetic and system-theory concepts.

As applied to formal organizations, the system-design process begins with the conceptualization of an organization and its environment as a system. According to Ashby, "System . . . means, not a thing, but a list of variables" (1956:40) selected by the designers conducting the analysis. Ashby's set of variables yields "a *system of decisions*" (Ackoff, 1970:2) needed to specify all the variable values. And the human being, albeit the most important decision maker, is only one of several "actors" fixing the organization's and environment's performance-related variables at specific values.

In this context, a system's "largeness" refers to the number of distinctions made: either to the number of variables involved or to the number of states, i.e. values, that are available (Ashby, 1956:61). So as firms expand into different products, technologies, markets, and/or countries, their sizes increase simply because their managements face more decision variables. Compared with entering different but well-established industries, moving into new lines undergoing rapid technological and market development adds even greater variety for many decisions will demand frequent revisions. The need for incessant adjustments, however, moderates with experience and fewer changes are called for per decision period. Considering time, then, system size expands as the decision frequency increases and diminishes as it slackens.

Another of Ashby's major contributions to organization planners is his "law of Requisite Variety" (1956:206). That law, interpreted in light of Simon's (1965) limited rationality concept, leads to this precept: To keep environmental disturbances from damaging performance, the environment's variety must be exceeded (or equaled) by the firm's decision variety which, in turn, must be exceeded (or equaled) by its management's decision-making and information-handling capacity, or variety.

For an example of the first limit, consider a simplified situation drawn from the automobile industry. From an auto manufacturer's perspective, the economy, or environment, can generate roughly three environmental states: depression, normalcy, and boom. A manufacturer of a low-priced car risks doing comparatively poorly during booms, as did FM in the 1920s with its inexpensive, standardized Model T (and its minuscule Lincoln sales). A manufacturer of high-priced cars risks doing poorly during depressions, as did Packard in the 1930s. A firm that builds a whole range of mass-produced automobiles, however, holds a much safer position, for it possesses the requisite variety to counter the economy's variation. Only GM offered such a variegated product line during the 1920s and 1930s and thus matched the first limit of Ashby's law of Requisite Variety.

What about the second limit? Certainly the decision makers for manufacturers of either low-priced or high-priced cars are unlikely to face excess variety. Even with the price-diversified auto firm, however, management overload is not too great a possibility since an inexpensive auto is mass produced and marketed much as are its more expensive substitutes.

Still, a prudent group of designers may not be satisfied. To guard further against unwanted disturbances that could damage performance, they will want to take additional precautions.

By expanding the corporate offices, for example, they can increase management's ability to handle the firm's decision variety, which in turn increases its ability to counter environmental variety. The large divisionally organized corporation partially "solves the problem of supplying requisite capacity at the top . . . by creating a team of general executives and providing them with an elite staff" (Williamson, 1970:124). Such an expanded corporate group can enhance the firm's variety or flexibility, say, by preventing the otherwise decentralized divisions from accumulating unmovable inventories. Hence cash remains available to counter the effects of an economic downturn.

The firm's decision problem also can be simplified by partitioning its controlled variables so that a large number of small management teams can be arrayed to handle the many day-to-day decisions. Theoretically, maximum simplicity results when the "parts are all identical, mere replicates of one another, and between [which] the couplings are of zero degree" (Ashby, 1956:65). The divisionally organized firm often approximates these conditions since the divisions usually resemble one another functionally and interact minimally. Organization designers can also hold steady with "*walls of constancy*" (Ashby, 1966:165) any residual interactions of significance. "If a high proportion of [such] variables go constant," i.e., become parameters for the decision period involved, the firm's decision problem may be cut into component parts "that are quite independent of one another" (p. 169).

Moreover, when the firm's many components are similar but separate, as GM's divisions were in the 1920s, management's capacity can be amplified (Ashby, 1956; 1966). One need only train a host of decision makers *en masse*.

The firm's decision variety can be increased further by periodically reconnecting the segmented managerial units. Such an intermittent communication network ensures that the firm's separate components are operating correctly on their own and coordinating their efforts as a whole. More importantly, these low-frequency information linkages make "*possible a greater repertoire of behaviours*," i.e., combinations of variable states, yet avoid the risk of lengthening adaptation time (Ashby, 1966:223). Thus a firm that pools and disseminates the experience of its decision makers via recurring meetings can expect to perform better than a more fragmented competitor. A divisionally organized firm, for instance, might recoup lost economies of scale by forming an interdivisional purchasing committee to buy in bulk.

To synchronize the firm's decisions with external conditions (such as consumer tastes and demand levels), environmental information can be collected and disseminated. Feedback information reduces the contingencies the firm must consider "by reinserting into it the results of its past performance" (Wiener, 1954:61). Feedback, moreover, can correct for performance deviations of unknown cause. Corporate-level executives, for instance, often exhort divisional subordinates to increase efforts when performance dips below expectations for unknown reasons.

Though relatively simple and inexpensive, the feedback-correction process may prevent timely adaptation to environmental shifts. This lag often can be eliminated by correlating known environmental fluctuations (e.g., the economy's recurring boom-bust cycle) with past performance deviations. Then, by predicting forthcoming environmental disturbances and by anticipating future performance feedback (i.e., feedforward information) corrective action can be taken before performance deteriorates (Ashby, 1956:201).

Finally, longer-range feedback, "working intermittently and at a much slower order of speed" (Ashby, 1966:98), can be used to guard against disturbances beyond the corrective capacity of the firm's primary feedback (and feedforward) channels. Under these circumstances, longer-term parameters must be shifted to new values that restore sales and profit performance — e.g., as occurred in the middle 1980s after Ford Motor and others succeeded in limiting the new Japanese competition by lobbying against the United States' long-held policy of free trade. In sum, fast-acting feedback provides stability against frequent impulsive disturbances emanating from the economic arena while slow-acting feedback provides stability against infrequent stepwise shifts coming from the societal/political realm and beyond the control of the first-order regulators. Ashby terms such a double feedback system "ultrastable" (p. 98).

While Ashby's cybernetic concepts lay an important analytical footing, considerable effort remains to use them in historical analysis. Churchman's (1971) work helps relate Ashby's concepts to the GM-FM case histories. Specifically, Churchman introduces the major players and props in the SDPC story: clients, designers, and decision makers; system, components, and environment; system and component performance measures; and resources. The GM counterparts for Churchman's story characters and stage props are:

CHURCHMAN	GM
Clients	Stockholders
Designers	Board of directors
Decision makers	Corporate and divisional managers
System	Firm, government, competitors, labor, suppliers, customers, and public
Components	Corporate headquarters and divisions
Environment	Government, competitors, labor, suppliers, customers and public
Performance measure	Return on invested capital
Resources	Capital, labor, materials.

Ashby's cybernetic themes and Churchman's system characters/props thus set the stage for the GM-FM analysis. This SDPC scenario unfolds in seven phases:

1. Identifying the goal
2. Formulating the strategy
3. Organizing the structure
4. Training the decision makers
5. Coordinating the components
6. Synchronizing the firm and environment
7. Evaluating the performance.

IDENTIFYING THE GOAL

GM's Design

William C. Durant founded GM in 1908 by trading stock in his successful Buick Motor Company for several additional auto firms. After losing control of the firm to a banking group in 1910, Durant regained the GM presidency in 1916 and maintained direction of GM for almost five years until Sloan and his colleagues succeeded him. Given Durant's haphazard management, GM's new designers faced a monumental task as they assumed command of the foundering GM ship.

The DuPont Company contributed most of the capital resources needed to bail GM out of its immediate financial problems. And with DuPont's involvement came a host of designers, in particular financially oriented planners who would right the GM vessel: Pierre duPont, John Raskob, John Pratt, and—most notably—Donaldson Brown.

In addition, when Durant left, several home-bred designers emerged from GM's wreckage, notably Albert Bradley, Charles Kettering, and C. S. Mott. Most important, however, was Alfred Sloan, whose organizational skill and financial acumen were critical to the DuPont interests.

Since all the members of the Sloan-Brown planning group were either large GM stockholders themselves or closely associated with major stockholders, they quickly identified their principal client as the firm's capital investors. Thus the primary perimeter around GM's "system of decisions" had been well delineated to fit the designers' specialized interests and talents.

The GM customers, of all the nonstockholding groups, received the next highest status, for they exerted the most influence on the stockholders' well-being. However, customers were treated more as a secondary constraint than as a primary client; their well-being was subordinated when it conflicted with stockholder interests.

Given the financial orientation of GM's designers and clients, return on invested capital (ROI) became the logical performance measure, or "financial yardstick," on which to build GM's performance-control system. This return-rate indicator summarized handily how well client resources, or investment dollars, were converted into client benefits, or profits. GM's clients, designers, and decision makers could use ROI to compare the performance of GM with the performance of a competitor, of GM this year with GM last year, of one division with another division, of a big operating unit with a small operating unit. The ROI measure thus provided a unitary, commensurable scale.

From the overall rate-of-return indicator it was a natural step to identifying the particular variables that influenced the client's well-being. Donaldson Brown developed a return-rate model that became the principal conceptual tool used for identifying the system's critical variables and determining their relative "essentialness." This priority ranking, in turn, guided the GM planners in subsequent design stages.

FM's Antidesign

At FM the situation was different. Henry Ford dismissed most of FM's design-oriented executives—for example, James Couzens, Frank Klingensmith, Norval Hawkins, and Ernest Kanzler—and retained only his son Edsel, whom he thwarted with powerful antidesign executives—most notably Charles Sorensen and Harry Bennett. Accordingly, FM was guided by the whims of its dominant antiplanner, Henry Ford.

Similarly, it was difficult to discover a single permanent client profited by the firm. Temporary client groups—such as labor or customers—were discarded abruptly. By the late 1930s isolating any client for FM became impossible. Without designers or a group of clients, there was, of course, no measure of performance for FM.

FORMULATING THE STRATEGY

GM's Design

The GM designers included powerful stockholders (i.e., clients) on the Finance Committee, who seized control over the key decision areas of performance measurement and capital allocation. By stripping headquarters executives and divisional managers of these responsibilities, the planners safeguarded the firm against a repeat of Durant's haphazard accumulation of product offerings.

In retrospect, the Finance Committee matched GM's internal variety to that of its uncontrolled environment. Moreover, the Sloan-Brown planners were careful not to overload the firm's decision makers. Accordingly the Finance Committee avoided movement into basic commodity production: expansion into steel making, rubber production, and glass manufacturing—to name just a few possibilities—offered little improved performance protection yet imposed enormous variety loads.

In more immediate and specialized sectors like input parts production and output credit provision, however, GM's Finance Committee approved more expansive tacks. For example, GM clustered within its operations a wide set of parts units. Even so, the firm did not make all its parts, for capital was better employed in high-return auto production, and outside suppliers were needed to diversify the risk of a production stoppage as well as to generate transfer prices for evaluation purposes. Even before Durant had left, the GM Finance Committee had authorized the creation of a credit source, General Motors Acceptance Corporation (GMAC), for its dealers and customers. Given the firm's extensive experience in such production and financial activities, neither the backward integration toward the parts suppliers nor the forward integration toward the dealers and customers overloaded GM's decision makers.

With respect to product diversification, GM's Finance Committee approved more ambitious plans: they had the firm offer a car for every purse and pocket, a strategy that yielded the protection of both brand and price diversification but added little troublesome variety. As time passed, the Sloan-Brown designers further minimized internal variety by offering a line of cars with many hidden standardized parts; ultimately, the interchangeable body program raised GM's product standardization to amazing heights.

Sloan's annual model innovation not only shielded GM from its environment but also imposed huge variety loads on competitors. Small firms lacked the resources to change body dies yearly, and FM hobbled along with its infrequent model changes and special-purpose machines and plants. Meanwhile, GM managed the added internal variety by using adaptable general-purpose production facilities and by restricting most model alterations to styling superficialities that had high customer impact.

In the middle 1920s, when the domestic market approached its saturation point, Sloan and Brown directed the Finance Committee's attention to foreign markets. Still later, they suggested important though comparatively minor investments in the diesel-locomotive and aviation industries—both potentially damaging auto-industry competitors. No management overloads resulted, as all these activities were based on the mass production and marketing of large durable items employing similar technologies.

Thus GM's planners chose an environmental setting where economic and market fluctuations exerted minimal damage on their clients' well-being. In addition, GM's decision makers held a comparative advantage in setting the variables under the firm's control.

FM's Antidesign

Obsessed with the desire to control all variables related to the production of FM cars, Henry Ford extended his operations into many basic-commodity production processes about which FM's decision makers knew comparatively little. In short, the firm's internal variety exceeded the decision makers' information-handling and computing capacity. Furthermore, firms external to FM could have provided better regulation of the variables in question, just as many such firms did for GM. Ford compounded the problem with an over-aggregated accounting scheme that made no references to competitive market prices. Consequently, the many inefficiencies of FM's operations remained hidden.

Violently opposed to credit, Ford added a consumer-credit plan belatedly and retained it only temporarily. This omission was costly, as such a component could have bolstered his low-cost marketing strategy.

With only the Model T (later the Model A) and the Lincoln, FM lacked the requisite variety of makes that the market demanded. Ford aggravated this deficiency in diversity when he insisted that his low-priced models be offered only in black and without the amenities or options consumers were coming to expect. So where internal variety was needed, Henry Ford eliminated it.

Ford, in sum, concentrated FM's variety along the channel-integration dimension, where it overloaded the firm's decision makers but provided little protection from environmental variations. At the same time he stripped FM's product line of the variety that would have stabilized sales and profit performance but would have imposed minimal managerial loads.

ORGANIZING THE STRUCTURE

GM's Design

The GM designers excelled in structural organization as well. In the vertical direction, Sloan and his Executive Committee determined which decisions most influenced performance and shifted control of these variables from the divisions to the corporate headquarters, where they could be set from the firm-wide perspective with the clients' interests closely in view. In the horizontal direction, a series of subordinate divisions minimized the variety loads thrust upon the firm's decision makers.

Over time, the Executive Committee transferred more and more decisions to the GM headquarters units in New York and Detroit; such critical variables came to include cash, in-process and finished inventory, production schedules, capital investment, personnel selection, product development, styling, engineering, manufacturing, and sales. Since these and other centrally controlled variables generally constrained many other decisions, the central headquarters' power grew far beyond its apparent authority. This hidden influence helped perpetuate the myth that the firm remained decentralized under the Sloan-Brown planners.

At GM, the designers could centralize without headquarters overload because they had judiciously limited the firm's internal variety by integrating and diversifying within narrow bounds. Moreover, this circumscribed strategy allowed GM's planners to configure a highly redundant set of divisions that imposed minimal variety on the central office. The long tenure of the corporate executives also allowed them to take on more divisional decisions as they gained experience.

The Sloan-Brown designers buttressed their centralization efforts by increasing the corporate headquarters' size. They expanded the Executive Committee but eliminated most divisional managers from its membership. To aid the president, the planners employed at various times an executive vice-president plus a series of group vice-presidents. Next, Sloan, Mott, and Pratt built strong Advisory Staff groups which, with the growing status of the corporate office, came to operate more as decision-making bodies than as so-called advisory staffs. Another blow to divisional autonomy came with Pratt and Sloan's Interdivisional Committees, which dealt with the functional interconnections often neglected when a firm employs a pure divisional structure. These committees eroded divisional authority because powerful and prestigious headquarters executives—e.g., Sloan, Brown, Pratt, and Mott—sat alongside second-level divisional executives, bypassing their divisional heads. Later, when the Interdivisional Committees became Policy Groups, divisional executives were excluded completely.

Thus the firm's product divisions all were tightly coupled to GM's corporate headquarters. The vertical couplings were very tight for the crucial automobile divisions, particularly for Buick and Chevrolet.

While the corporate headquarters could control many of the firm's important variables, the planners recognized its limitations. Selectively they left numerous less important variables in the control of the firm's divisions.

GM's designers assigned sets of variables to the subordinate divisions so that decision making was simplified for both the corporate and divisional executives. By diversifying control of the less important decision tasks across more than thirty divisions, the designers kept these components small, thrust relatively little variety on their managements, and thus caused few decision-maker overloads and errors. Configuring the firm's components as a series of unique "profit" and "investment" centers permitted comparisons among the numerous divisions in terms of the unitary ROI scale; this performance commensurability eased divisional evaluations and focused attention on the exceptional units. The divisions' functional similarities, in turn, made it easy to discover the causes underlying particular divisional failures and successes; among the key auto divisions this causal comparability reached startling proportions, for they all mass-produced and marketed a highly similar product. And by division of the firm into a series of independent, i.e. decomposable, components, they all could be managed without concern for the others.

GM's simplified organizational arrangement was achieved via such techniques as standardized cost/benefit accounts, market-referenced transfer pricing and decoupling schemes, a divisional rather than a functional structure, inventory buffers, standardized parts, market segmentation, and geographic separation.

FM's Antidesign

Rather than build the company's headquarters to provide strong firm-wide direction, Henry Ford *dismantled* his central units. He mistakenly believed such subunits offered only increased overhead. Ford, more generally, delighted in announcing that FM had no formal organization and ridiculed the Sloan-Brown designers at GM for their great emphasis on administrative structures and organization charts. Worse yet, Ford permitted his decision makers to compose their own jobs; besides causing major competitive tensions within the firm, this practice yielded an ad hoc organizational structure under constant flux because of Ford's continual executive firings. It seems Ford felt that if FM's internal boundaries remained nebulous, all decisions would have to come to his attention. Unwilling to recognize his limited abilities, he denied the need for a straightforward organizational structure that would have simplified FM's decision making.

TRAINING THE DECISION MAKERS

GM's Design

When it came to amplifying their control, the GM designers again did their work with diligence. The money, time, and effort they expended on training insured that GM's decision makers were properly oriented, thoroughly educated, and carefully selected for the task at hand.

In orienting the decision makers to the clients' value scheme, the designers instituted a variety of lucrative bonus plans; the most notable was the Managers Securities Plan for the firm's top seventy or so decision makers. This program possessed several important design characteristics. The bonus took the form of stock options that made clients of the managers and thus engendered the proper decision-making orientation. The designers based the recipients' stock on their contribution to GM as a whole rather than to more narrow, or parochial, interests. Furthermore, the stock payments were made only after a minimum seven percent return was achieved. Still another client-oriented safeguard prevented the decision makers from selling their GM stock and thus diversifying their personal risk and reducing their motivation to serve GM's stockholders over the long-run. Finally, Sloan's heavy emphasis on bonuses rather than on less variable salaries further increased the designers' motivational leverage.

A breadth of educational schemes provided GM's decision makers with the factual premises for decisions. Initially, the designers met with individual decision makers; Sloan, for example, visited many division managers so he might teach them how to improve their performance. (Once educated properly, these decision makers could be expected to amplify Sloan's control over many future periods of operation.) Later, the Sloan-Brown designers achieved greater amplification by educating groups of decision makers, e.g. via the stockholders' meetings of Managers Securities and the formal sessions of the Interdivisional Committees. Ultimately, GM's own university, General Motors Institute (GMI), was founded to train future decision makers *en masse;* in time GMI graduated many important GM executives.

The designers' careful selection of decision makers went along with their thorough value-orientation and factual-education work. One of the planners' first moves replaced incompetent and/or "uncooperative" Durant managers in the important auto divisions with more able and less independent executives. The selection of these financially able replacements signaled GM's shift of emphasis from the production of cars to profits and helped set the new standard for acceptable manager behavior.

Generally, GM's designers selected managers with strong appreciation for finance or accounting, beginning with the transfer of a number of DuPont Company managers to important GM decision-making positions. Later, GM's parts and accessories divisions served as a basic training ground for positions in the finished-auto divisions and corporate headquarters. The accessory and parts divisions' small size provided both a performance-safe training ground and the opportunity for a manager to understand the entire unit's operations. And since all the divisions were composed along similar lines, understanding a small parts unit yielded familiarity with the biggest auto division. Finally, because GM's decision makers could not restructure their jobs to fit their particular abilities and personalities, Sloan always had a large pool of interchangeable decision makers from which to fill GM's similar executive positions.

With their value orientation, factual education, and selection methods, then, GM's designers amplified their ability to control the firm's performance for the clients' benefit. Many variables were brought under the designers' influence via the control of a limited number of training-related variables. In other words, the planners achieved a significant amplification of control by contracting the vector of variables they had to deal with directly.

FM's Antidesign

Ford created no means of providing value orientation for his managers and fostered rather than hindered a parochial outlook among his executives. Instead of making their decisions for the benefit of FM as a whole, FM executives concentrated on their own concerns and often caused problems for other FM executives. Moreover, no managerial training programs supplied FM's decision makers with the proper factual education. Managers with expertise—other than high-volume production—were distrusted and without exception removed from their decision-making positions. As to selection and deselection, Ford fired competent decision makers while he retained and promoted the incompetent or inexperienced. Apparently Ford wished to set every decision variable personally and felt that inadequate managers would be forced to consult with him directly before reaching any decision. Given that Ford pursued numerous nonindustry preoccupations such as the Ford Museum and folk dancing, FM's managers were left to decide matters they were unqualified to consider. Thus Henry Ford achieved no amplification of control.

COORDINATING THE COMPONENTS

GM's Design

The GM designers also made significant organizational advances in fashioning the communication linkages needed to coordinate the firm's divisional and corporate decision makers. Most notably, Sloan's latticework of committees melded GM's decision makers—with their varying functional views and time perspectives—into a responsive, unified managerial team.

In the horizontal direction, GM's loosely coupled divisional structure minimized the number of high-frequency communication links needed among the divisions. However, where such connections were necessary—as between the auto units and the accessories and parts divisions—the firm's standardized parts specifications gave the participants a *common language* that improved understanding and reduced interactions. Later, the designers installed a centrally operated communication network to handle efficiently the routine coordinating communications among GM's far-flung divisions. After the mid-1930s, three components—Fisher Body, the auto divisions employing the firm's interchangeable bodies, and the corporate Styling Staff that developed the body designs and color schemes for the subordinate divisions—were somewhat more tightly coupled. The Sloan-Brown planners created a corporate-dominated Policy Group to supply the necessary contact points and communication bridges.

Given the divisions' similarity, improvements could be made if the causes behind high or low divisional results were disseminated periodically. At GM, the Interdivisional Committees and Advisory Staffs spread the performance-improvement information gleaned from planned experiments or serendipitous findings.

These committees and staffs also helped link the corporate headquarters and the subordinate divisions. In general, such vertical linkages supplied the divisions with the policy guidelines and directives they needed to regulate their decisions and provided the headquarters with the data, i.e., feedback it needed to correct and coordinate all divisional decisions. Another noteworthy vertical connection was the monthly budget review instituted by Sloan and Brown. The highly summarized nature of this budgetary process permitted the corporate headquarters to monitor divisional operations without overloading the firm's vertical communication lines with trivial details. Furthermore, the simple "threat of review" kept the divisional decision makers adhering to corporate directives. GM's designers further enhanced the headquarters' observational power by having divisional financial staffs report to the corporate office rather than to the division managers. The divisional financial officers thus acted as "reliable local data sources" for the central office. Sloan also used his powerful Advisory Staff to break down "institutions of privacy" protecting divisional autonomy, and he declared that nonfinancial functions should be *audited* just as financial results. To conserve his own audit capacity, Sloan concentrated on the auto divisions, which in turn monitored the parts divisions' price and quality performances.

To supplement these formal communication linkages, Sloan devised a series of annual meetings, where several hundred of GM's most important corporate and divisional decision makers convened to discuss mutual problems. Besides promoting the firm-wide perspective, the Sloan meetings sanctioned and stimulated the informal contacts needed to close any communication gaps the designers overlooked.

By far, Sloan's extensive network of interlocking committees did the most to ensure that GM's decisions and performances were thoroughly explored and reviewed. During the 1920s, for instance, the Interdivisional Committees—in functional areas like engineering, purchasing, sales, advertising, manufacturing, and maintenance—brought together corporate staff officers, their divisional counterparts, and corporate line executives. At the next level up, the Operations Committee linked the divisions' general managers with the Executive Committee members—the principal line and staff decision makers of the corporate headquarters. While GM's Interdivisional Committees focused on functional specifics, the Operations Committee concentrated on overall performances. The Executive Committee, in turn, set corporate-wide policy, reviewed long-run capital allocation needs, and made investment recommendations to the Finance Committee—which controlled the firm's strategy via its capital-budgeting purse strings. When interlocking these committees for coordinated decision making, Sloan himself filled the major linking-pin role. Other notables—in particular Pierre duPont, Raskob, Brown, Pratt, and Mott—also served on multiple committees and thereby helped to craft GM into a well-coordinated decision making body. Here, then, was a thoroughly unified firm, capable of reaching variegated decisions without undue delay.

FM's Antidesign

In contrast to the lavish attention given by the Sloan-Brown designers at GM to recomposition, i.e. coordination and communication, Henry Ford was insulating. Ford cut many of the internal communication lines that had existed when he assumed full control of FM. For example, he disbanded communications units like accounting and authorized a drastic reduction in telephones. Further, he made it difficult for his decision makers to confer by eliminating their offices or desks and, more importantly, by forbidding meetings and conferences, which he abhorred. When coupled with the distrust Ford kindled within his company, such communication prohibitions virtually halted any recomposition contact. Consequently, the firm was left with a set of decision makers who pursued decidedly uncoordinated and suboptimal approaches.

SYNCHRONIZING THE FIRM AND ENVIRONMENT

GM's Design

Just as the GM planners gave serious attention to GM's internal communication networks, they carefully designed the firm's external communication linkages with the immediate economic environment. Contacts were developed with important environmental groups: suppliers, workers, competitors, dealers, and, particularly, customers. Should the customers not be satisfied, there would be little the firm could do to achieve a performance acceptable to the stockholders. Consequently, Sloan and Brown themselves visited GM's dealers to learn firsthand about consumers' preferences. Headquarters units analyzed sales figures to set production levels. GM's new proving ground tested GM's and competitors' cars to correct product deficiencies. An econometrics unit, a research laboratory, and a customer research staff attempted to predict and anticipate future demand levels, product developments, customer tastes, and buying habits. In short, the design team attempted to furnish its decision makers with vital feedback and feedforward information to keep production levels and product offerings synchronized with the surrounding economy.

In addition, the GM designers used their environmental sensing units to supplement and focus the firm's advertising efforts. Thus the customer-research and advertising units worked together developing promotional appeals to induce customers, i.e., environmental decision makers, to purchase GM's product for the ultimate benefit of GM's stockholders.

Performance data on each division plus environmental information were presented to GM's Finance Committee. Intelligence on general economic conditions revealed the effect of uncontrolled variables on divisional performances and guided the Finance Committee in adjusting internal resource allocations to optimize GM's overall performance. To ensure accurate and commensurable performance reports, the planners placed GM's Financial Staff under the Finance

Committee's jurisdiction rather than under the direction of the headquarters executives and divisional managers.

FM's Antidesign

Henry Ford dismantled many of the communication channels needed to gather environmental intelligence. Moreover, when external information such as dealer complaints did manage to filter back into the firm, Ford ordinarily ignored it as long as possible in revising his decisions. Blinded by his monumental success with the Model T, Ford did not seek to anticipate forthcoming auto-market or general economic conditions. Without this feedforward input FM's models soon lagged behind consumers' tastes and competitors' products.

Ford's antagonism toward product advertising further undermined the firm's market position, as did Ford's ill-conceived public-relations efforts. His anti-Semitic and antilabor campaigns, for example, induced many external decision makers to retaliate against Ford and his products.

EVALUATING THE PERFORMANCE

GM's Design

For overall evaluation of GM's performance, the full Board of Directors was involved, aided by the Financial Staff, which supplied the requisite evaluative information. In turn, GM's Board supplied *reliable* evaluative information to the firm's stockholding clients. As early as 1918 Sloan engaged the independent auditing firm of Haskins & Sells to certify that the financial performance data in GM's *Annual Reports* was accurate (and commensurable with other firms'). GM's financial reports thus could be used safely by shareholders to value the planners' achievements against the efforts of other groups competing for capital resources. Because GM fared well in these comparisons, investors continued to supply capital to the firm and permitted it to retain earnings for expansion.

Even customers—though not GM's primary clients—rewarded the designers by doubling GM's market share. Duly proud of their accomplishments in revitalizing GM, the designers described their efforts in numerous papers and speeches so that other businessmen could employ their successful procedures.

GM's planners also attended to the longer-term shifts in the firm's social and political environment. They adjusted GM's policies to meet the protracted problems stemming from the 1930s depression and made an important, though admittedly tardy and somewhat reluctant, effort to establish a long-term accord with labor. Nevertheless, after the bruising labor-relations battle in 1937, Sloan drew GM inward and away from its broader social and political responsibilities. *GM's success thus remained limited strictly to the economic arena.*

FM's Antidesign

Henry Ford did little if anything to evaluate or correct the performance of FM. Not only had he no visible client group nor any performance measure on which to judge his efforts, but he also ignored his Board of Directors. Furthermore, he steadfastly refused to accept the permanent shift in labor conditions that came about in the 1930s. If either sales or profits were his concerns, he failed: FM's market share plummeted and its profits plunged to become monumental losses.

PART II

The General Motors Case

3

Repudiating Antiplanning

William C. Durant founded GM in 1908 by trading stock in his successful Buick Motor Company for several additional auto firms. Durant, late in 1910, lost control of GM temporarily to a group of bankers who criticized Durant's rapidly expanding empire. Given that Durant dismissed the need for a central staff, accounting procedures, and inventory controls, it was virtually impossible to amass "more than twenty companies of varying profitability and uneven management in two years and be entirely certain of what was going on. . . . Durant was operating largely on Buick's success, on his own confidence and salesmanship, and on his infectious optimism" (Gustin, 1973:134).

By 1916 Durant regained the GM presidency by trading Chevrolet stock, another of his many ventures, for GM stock. This time Durant maintained his leadership of GM for almost five years until he was deposed finally for the same causes.

Durant, in spite of his interrupted direction of GM, must be credited for founding GM and for "guiding" the firm through its first decade of operation. Unfortunately, Durant's guidance can be characterized only as strongly antiplanning in nature.

IDENTIFYING THE GOAL

Durant made little effort to control his organization's performance. An overloaded corporate staff, which might have provided the nucleus for a performance-control group, consisted only of Durant himself and a few personal secretaries and special "fire-fighting" assistants. More importantly, Durant had no defini-tive clientele in mind. Hence Durant would have been hard put to identify improved performance since he did not seem to be pursuing a unified measure of system well-being, i.e., an organizational goal.

Durant's activities, however, were not completely unsystematic, for some evidence indicates that he occasionally sought to maximize a performance-related variable: the price of GM stock. For instance, during the 1920 recession Durant "lowered his horns and plunged into Wall Street with all his friends on his broad back, buying. It was hopeless" ("General Motors," *Fortune,* 1938:164). In this ultimately unsuccessful attempt to control the stock-price variable Durant pursued two purposes. First, he felt obligated to the *many* friends he had encouraged to invest in GM (Rae, 1958:266). For his friends' sake, then, Durant wanted to insure that their investment appreciated. The second objec-tive behind Durant's intermittent efforts to control the stock price was to fuel GM's aggressive expansion. Strong stock prices meant high demand for GM shares; in turn, high demand meant that the owners of firms Durant wanted would be eager to trade their businesses for GM stock. In expanding GM's empire Durant would not need to risk encounters with the "dreaded bankers" who always wanted to impose systematic plans on the expansive and specula-tive Will Durant.

Durant's efforts to control the GM stock price reduced significantly the time he had available for improving internal operations. Instead of being concerned with improved performance control, Durant's efforts were devoted to endless dealings with brokerage houses, fellow speculators, and their likes. A bank of ten or more telephones on Durant's desk rang incessantly with calls regarding such manipulations (Weisberger, 1979:249). "At times of unusual price changes, Durant became so involved with his brokers that it was almost impossible for his small cadre of top assistants to get into his office" (Gustin, 1973:200). Often, when division executives were called to the New York or Detroit offices from Flint, Lansing, or elsewhere, they had to wait hours before Durant would confer with them (Sloan, 1941:115–116). Once Durant had asked Walter Chrysler, the very able head of the prosperous Buick division and the executive in charge of improving the other divisions' performances, to discuss so-called important matters in New York. According to Chrysler (1937:156), he waited in Durant's New York office each day for almost a week. Chrysler finally left to return to Buick's Flint headquarters without finding out what Durant wanted;

Chrysler left a message that he was returning to Michigan, but Durant failed to answer Chrysler's note. Critical decision capacity thus was diverted from internal performance control, where Durant had a good long-run hope of strengthening GM's position, to external stock price manipulations, where even Durant's wealth, understanding, and courage produced limited results.

FORMULATING THE STRATEGY

Durant unquestionably devoted considerable attention to GM's expansion, concerning himself with inducing other firms to join the GM assemblage. In the area of channel integration, Durant achieved limited success by buying critical parts-manufacturing firms whose outputs were desperately needed to fulfill his steadily expanding production dreams.

Durant's accessory and parts acquisitions, however, were not without problems. "The biggest blunder was the purchase of the Heany Lamp group of companies. . . . The Heany patent rights were voided, and published estimates of the eventual loss . . . reached as high as $12 million" (Gustin, 1973:135–136). Durant, moreover, paid no attention to the transfer pricing and payment problems of his channel-integrated units, a practice which often hid the poor performances of GM's weaker components.

Durant's product-diversification strategy was even less successful than his channel-integration effort. When he originally created the GM combination, uppermost in his mind was risk diversification (Green, 1933:20; Gustin, 1973:144). "The business of an individual manufacturer was hazardous because the model on which he staked his chances of sales might prove to have some mechanical defect or the body design might fail to strike the fancy of the buying public" (Crow, 1945:74). Durant bought the Cartercar, for example, solely on the basis of its friction drive, which might spell future success for GM (Epstein, 1928:182), and, similarly, he purchased the Elmore because it possessed a two-cycle engine. As Durant expressed his composition strategy, "I was for getting every kind of thing in sight, playing safe all along the line" (pp. 182–183). Durant believed, then, that while a firm that produced a single model might be a risky investment, a merger of a group of car-producing companies would be a sound investment and insure stabilized profits (Crow, 1945:74).

Yet Durant's many acquisitions yielded little safety. First, he lost several competent management teams. "Buick and Cadillac were the only dependable, profit-earning divisions of General Motors. Chrysler and the Lelands had built these divisions and were highly successful in operating them" (Crabb, 1969:382). Durant's methods so disturbed these able managers that they left GM and went on to establish two of its competitors: the Chrysler and the Lincoln enterprises. Second, "apart from Buick and Cadillac, the automobile companies included in General Motors had either already failed (e.g. Cartercar, Elmore, Ewing, Marquette, Welch); were about to fail (e.g. Scripps-Booth, Sheridan); had shut down during the 1920–21 depression (e.g. Oakland, Oldsmobile); or had poor prospects. Chevrolet lost $8.7 million in 1921 and a firm of consulting industrial engineers recommended that it be dropped" (General Motors Corporation, 1968:3).

Yet Durant's most serious strategic mistake was to overlap his products' prices and, worse yet, to concentrate them in the recession-sensitive middle price class. Durant wanted to bring "together a few concerns committed to volume production in the medium priced class, all having a common objective, all heading for a highly competitive field" (Durant, 1965:15). The 1920–1921 product prices, not surprisingly, centered on the mid-priced cars and largely omitted the extremely important low-priced auto. Specifically, in 1921 GM had ten models in seven makes:

Chevrolet "490" (four-cylinder)	$ 775–1375
"FB" (four-cylinder)	$1320–2075
Oakland (six-cylinder)	$1395–2065
Olds (four-cylinder "FB")	$1445–2145
(six-cylinder)	$1450–2145
(eight-cylinder)	$2100–3300
Scripps-Booth (six-cylinder)	$1545–2245
Sheridan (four-cylinder "FB")	$ 1685
Buick (six-cylinder)	$1795–3295
Cadillac (eight-cylinder)	$3790–5690

(Sloan, 1964:59)

While Buick was by far GM's leading unit, Durant had six divisions whose prices overlapped Buick's. Similarly, the Chevrolet "FB," Oakland, and Olds were almost the same in price. Durant's failure to provide price variety yielded two damaging conditions for GM's overall performance: (1) divisions competed against each other (for vital cost-spreading volume), with one unit's performance gains resulting in another's losses; and, more importantly, (2) divisions suffered performance declines simultaneously when there were general economic downturns. Durant created a performance nightmare: during good times performance was below potential, during bad times it was terrible.

Hence Durant's divisions taken as a group failed to possess the variety needed to match that of the economic environment, which was capable of booms and depressions. And when the demand for automobiles dropped in 1920, Durant was struck particularly hard because the demand decrease hit the medium-priced cars more than the low-priced competition. "In his long business career . . . he wrestled many times with the business cycle without apparently becoming convinced of its periodicity" (Pound, 1934:191).

These divisional and corporate weaknesses caused GM's low 1921 market share of 14 percent. In addition, divisional failures and shutdowns greatly reduced GM's rate of return through the forced liquidation and idling of its capital investments.

Durant's product-diversification strategy was not limited to automobiles, however, for he also invested GM's funds in products like refrigerators and tractors. Durant's creation of GM's Frigidaire Corporation was an attempt at sophisticated diversification (testimony of Pratt, *U.S.* v. *DuPont,* 1956:1403). Durant acquired Frigidaire because he feared a war-created shortage of steel would force GM to stop auto production; mostly wooden refrigerators then could be built instead. But in the long run the fact that both automobiles and refrigerators were expensive consumer durables yielded little performance smoothing for the firm. Durant's tractor ventures show why he often acquired worthless firms and products. Charles Kettering, GM's research director, once caught a group of promoters sending false telegrams to Durant regarding their own tractor's still untested performance (testimony of Kettering, *U.S.* v. *DuPont,* 1956:1530–1531). Eventually, GM had to divest itself of its tractor properties because of its inability to meet the competition of manufacturers who specialized exclusively in an array of farm implements; even Henry Ford, with his immense productive capacity and distribution network, could not operate successfully in the tractor market.

ORGANIZING THE STRUCTURE

With respect to GM's internal operations, Durant was always somewhat uninterested. Under Durant's presidency, for instance, the corporate parent, GM, owned the several subsidiaries, e.g., Buick and Chevrolet divisions; but it did not systematically constrain or integrate divisional operations (testimony of Pratt, *U.S.* v. *DuPont,* 1956:1406). Durant (1965) in a 1908 merger conference with other auto manufacturers, notably Benjamin Briscoe of the Briscoe-Maxwell Company, explained his management philosophy regarding the then proposed General Motors Company: "I took the position . . . that there should be no change or interference in the manner of operating, that the different companies should continue exactly as they were. In other words, I had in mind a *holding company*. Briscoe came back jokingly with 'Ho! Ho! Durant is for States' rights; I am for a Union' " (p. 15).

Durant, as is often the case with self-proclaimed proponents of subordinate freedom, occasionally exhibited marked dictatorial tendencies (Fanning, 1964:219; Weisberger, 1979:249). But Durant directed his divisional attention haphazardly and often to trivial matters, such as "passing personal judgment on such details as the wiring diagrams at one of his factories" (Gustin, 1973:127). Durant, then, wasted his capacity on trivialities and overlooked serious problems such as the divisions' bloated inventories and poor products.

Durant established no systematic ranking of priorities to guide the assignment of decision-variable responsibility within GM. Durant's jurisdictional dispute with Chrysler, who was named his assistant and vice-president in charge of all GM operations in 1919 (Gustin, 1973:200), provides the most notorious example of his failure to establish clear-cut hierarchical boundaries. As John L. Pratt, Durant's able assistant, testified: "Mr. Durant . . . brought Mr. Chrysler in . . . to be sort of a general manager, but he and Mr. Chrysler could never agree on the division of the responsibility, because Mr. Durant didn't want to give up the direct dealings with the [division] managers. . . . he wanted to deal with everyone" (*U.S.* v. *DuPont*, 1956:1405–1406).

Durant's holding-company philosophy, haphazard intervention, and unsystematic decomposition, coupled with his astonishing inability to choose strong divisions in composing the firm, created an utter disaster. Lacking any coordination (or even communication) among the numerous divisions and any monitoring of performance of the divisions from Durant's minuscule headquarters, the company found itself in dangerous financial circumstances, literally on the brink of receivership for the second time, with the onset of the 1920 depression. GM as a whole—if this can be imagined—was in even more precarious circumstances as the firm-wide aggregation of production, inventory, and cash problems experienced by each of the individual divisions made corporate management problems almost insurmountable.

The worst of the performance-control problems, according to Sloan (1964:124), was in the massive inventories of raw and semifinished materials. During the postwar inflationary boom, "businessmen bought liberally in expectation of higher prices when the goods were sold" (Soule, 1947:90). The speculative managers running Durant's divisions were no exceptions to the general hysteria (Weisberger, 1979:248). GM's inventories were therefore not only excessive but also grossly overvalued, having been purchased at 1919 inflationary prices. Adding to the speculative inventory glut was an *unwarranted* 36-percent increase in production schedules from 1919 to 1920; the "schedules were made by rule of thumb, or the division manager's ambition" (Sloan, 1964:30), wholly without reference to changing environmental conditions. To meet these overoptimistic production schedules, divisional purchasing agents made heavy inventory commitments (p. 30). During the first nine months of 1920 inventories rose a spectacular $72 million. By October of 1920 the productive inventory reached $209 million, which far exceeded the divisions' current production needs (p. 124). Furthermore, the productive inventory level exceeded the Executive Committee's maximum allotment by $59 million. Eventually, $84 million had to be written off GM's books as an inventory loss.

There were several internal causes for these inventory overages. First, with Durant's small corporate headquarters unit it was extremely difficult to enforce the rulings of the Executive Committee and of an ad hoc Inventory Allotment Committee that futilely had set the upper limit on inventory at $150 million. Second, the Executive Committee members were basically unwilling to limit strictly any divisional expenditures. Under Durant's management the Executive Committee consisted predominantly of the operating division managers themselves, each of whom was oriented toward expanding production of his *own* product with no concern for the firm-wide results of such action (Weisberger, 1979:248). A "horse trading" situation rapidly developed among the division managers (testimony of Pratt, *U.S.* v. *DuPont,* 1956:1402).

The horse-trading approach to critical-variable control was evident even in the area of long-term capital appropriations, where "overruns on capital investment had become the rule" (Sloan, 1964:9). Ironically, Durant had himself argued against several large investment projects (Weisberger, 1979:237, 243). But without rational capital allocation procedures to support his efforts, little could be done to rein in the spending.

With inventories mounting and appropriations overrunning, cash soon *seemed* to be in short supply. In October of 1920, GM found it necessary to borrow $83 million to meet payrolls and payables (Sloan, 1964:31). According to Crabb (1969:389), it was discovered subsequently that several divisions, predictably Buick, actually had excess funds at the time of the borrowing. Sloan (1964) described the cash-control system as "almost unbelievable" (p. 122). Each division held its own cash, to which the corporation had no direct access even though it actually owned the division's assets. When the corporate headquarters needed money, GM treasurer Meyer Prenskey (later Prentis) had to estimate the cash balances of the divisions and then go "hat-in-hand" to such strong cash producers as Buick and plead that they transfer money to the corporate accounts (p. 122).

Durant's failure to install control procedures over his divisions' cash resources demonstrates the depth of his unsystematic character. He had created GM for the expressed purpose of sharing risks and resources, yet when it was necessary to pump cash balances from prospering to foundering divisions, no mechanisms were available for the transfer. No one in GM's New York headquarters even knew excess funds were available within the assemblage of units. Thus, one must question whether Durant really wanted to create a unified group out of the many GM divisions. Far too many critical performance variables were controlled by the division managers who, even when competent and prudent, could not understand the corporation's aggregate problems from their limited vantage points. GM's divisions under Durant remained too uncoordinated and decentralized because of his failure to establish a strong central headquarters, so his holding company could not be expected to generate high performances.

In organizing GM's divisions, Durant made a good beginning by avoiding the functional structure common during those days. "Briscoe took the position that the purchasing and engineering departments" of all the divisions "should be consolidated, that the advertising and sales departments should be combined, and that a central committee should pass on all operating policies. I took the position that this would only lead to confusion" (Durant, 1965:15).

But Durant was inconsistent in dealing with the divisions' horizontal couplings. GM's overlapping prices for its multiple lines of cars tightly coupled the auto divisions. Without a buffering gap between the prices it was difficult to treat each unit separately, yet all of them could not be treated simultaneously without a major corporate headquarters. Durant made essentially no effort to standardize parts moving among the divisions. Such a simplification would have greatly reduced the decision variety flowing across division boundaries, engendered by many needless specification changes. Similarly, buffer inventories and integrated production schedules were not established to decouple the interactions of the finished-auto units from the parts divisions (testimony of Kettering, *U.S.* v. *DuPont,* 1956:1531-1532). Nor were markets interspersed between such divisions to provide buffer production capacity for peak demand periods and uncontrollable failures. And without such markets, transfer prices could only be arrived at through a bargaining process that often masked the true costs of production and contribution of benefits.

TRAINING THE DECISION MAKERS

Durant ineptly oriented, educated, and selected his managers, making it impossible for him to achieve any amplification of control. Without an explicit purpose in mind for GM, Durant was unable to orient his managers to a corporate objective. Left without guidance, the managers—as can be expected—found it easy to supply their own purposes for their units. Durant unwittingly aided this divisional preoccupation by signing contracts that gave individual division managers a large share of their particular units' profits. Accordingly, divisional considerations rather than corporate ends absorbed the decision makers' attention.

As with orientation, Durant also spent little or no effort in the systematic education of his division managers. GM's poor divisional performances, therefore, must ultimately be attributed to Durant's unwillingness and/or inability to teach his managers how to operate their units.

Finally, Durant's means of selecting division managers was probably his most certain control-amplification failure (Sloan, 1941:111). Promoters of new automobile ventures would often come to Durant with a proposition that he back the building and testing of a new entry. Invariably, Durant would agree— unfortunately permitting the entrepreneur to do the testing himself (p. 111). After the "successful" test, the same individual would often be installed as the new division manager. Even if the promoter was honest and a good engineer, that was no reason to establish him as the manager of the production and marketing facilities for which a different personality and intellectual skills were required. Under such a selection procedure the managers in charge of even the best cars were all too often found to be incompetent.

Alone for long periods, except for the most haphazard and random interference, the division managers could not be expected to control the whole range of variables left to them. They simply were not oriented, educated, or selected for such significant decision-making tasks. Hence Durant failed to achieve any amplification of control over GM's performance.

COORDINATING THE COMPONENTS

In dealing with GM's internal information needs Durant did absolutely nothing. No attempt was made to connect divisional offices with each other or with the corporate headquarters. Committees and systematic reporting procedures were practically impossible to unearth.

Neither performance improvements nor even results were transmitted horizontally or vertically. With his minuscule central staff, Durant was unable, and probably not motivated, to provide for the discovery and communication of the best techniques of auto design, production, and marketing among his divisions. Durant consequently failed to develop information for measuring divisional performance; he failed to provide communication channels to transmit such information to the corporate level; and he failed to create a group of corporate-level executives to monitor these performance data from the divisions and GM as a whole so adjustments could be made to improve performance. In the area of new plant construction, for instance, "valuable experience and techniques developing in one plant were not transmitted to the other factories" (Chandler and Salsbury, 1971:469).

The creator of GM made "cost accounting, forecasting, scheduling, inventory control, and budgeting appear as unnecessary refinements in the booming automobile business of the day" (Chandler, 1966:367–368). Furthermore, "Durant did not have a sound concept of accounting . . . and did not realize its great significance in administration." GM's corporate headquarters accordingly "did not have adequate knowledge or control of the individual operating divisions. It was a management by crony." Without procedures for generating information and disseminating it, for establishing the quality of performances as well as their causes, it was impossible "to control the various parts of the organization" (Sloan, 1964:26).

SYNCHRONIZING THE FIRM AND ENVIRONMENT

Durant's lack of interest in internal information for coordination was matched by a similar disdain for providing external information for environmental synchronization. Durant's telephones were connected to the stock market, not the automobile markets.

He attempted to synchronize GM with the automobile public's tastes, as mentioned, by accumulating an array of firms. A firm might be purchased because of its patent on a novelty item, such as a friction drive (Gustin, 1973:144): Who knew what might catch on with the fickle automobile public? Durant certainly did not, as he made no attempt to find out what consumers desired.

Similarly, Durant failed to synchronize production and pricing with current economic conditions. Car sales (along with prices in general) declined substantially after the summer of 1920; yet the GM corporate or division headquarters had little or no information on the demand decline (Crabb, 1969:385). Therefore, divisions purchased supplies fully expecting to be able to use the inventories shortly. The inventory glut was aggravated by Durant's and the division managers' refusal to adjust prices downward in a clearing effort. Though Durant prophesied doom (John, 1924:250; Brown, 1957:41), he failed to design an information network to control the production, pricing, and inventory variables adequately. Consequently, GM accumulated massive inventories and huge loans.

EVALUATING THE PERFORMANCE

Besides making little or no effort to evaluate the divisions' performances, Durant also did not think to provide the firm's stockholders with evaluative information. The fact that GM even issued annual reports was more Sloan's than Durant's doing, as Sloan arranged for the verification and publication of this information. Nor did Durant make any effort to gauge long-term shifts in GM's social and political environment.

In sum, both Henry Ford and William C. Durant scorned systematic performance control, but Ford had initially synchronized his firm with prevailing market conditions: cheap, rugged transportation. FM also performed better during the 1918–1921 period with respect to shorter-run inventory, production, and pricing synchronization, as Ford had yet to dismantle his firm's performance-control system and had to manage only one make (Crabb, 1969:385). Consequently, Henry Ford was able not only to weather the postwar recession but also to secure complete ownership of his company by buying out FM's minority stockholders. Durant, in contrast, was forced to resign from his presidency forever when GM's major stockholders required him to sell his substantial holdings.

The exact details of Durant's resignation are clouded except that the time was November 30, 1920 (Sloan, 1964:38), and that Pierre duPont was induced temporarily to replace Durant as GM's president (Sloan, 1941:131). Pierre duPont was elected president because the duPont family and firm held the controlling interest in GM.

4

Identifying the Goal

GM's new designers faced a monumental design task as they assumed command of the foundering GM ship at the close of 1920. "As things stood, the corporation faced simultaneously an economic slump on the outside and a management crisis on the inside" (Sloan, 1964:42).

The lack of performance-control mechanisms was exposed by the crises in inventory, capital appropriations, and cash. "To meet these specific emergency problems," GM's planners developed "new methods of financial coordination and control" (Sloan, 1964:118). The control of nonfinancial operations accompanied, with minimal lag, the establishment of controls over finances.

The first task undertaken in improving GM's performance was to select an executive group to design the firm's control system. Second, the designers had to identify the system's clientele. During this design phase there were two other concomitant subtasks: defining an operational goal, or performance measure, that summarized the system's overall state, and identifying the critical performance variables that influenced the system's goal.

SELECTING THE DESIGNERS

With the DuPont Company's involvement in GM affairs came a whole host of design and decision-making talent, most notably Donaldson Brown. Other designers coming to GM via the DuPont connection were Pierre S. duPont, John Raskob, and John Pratt. When Durant left, several of GM's own designers emerged from the ruins. This team included Albert Bradley, Charles Kettering, C. S. Mott, and the most important designer anywhere on the scene, Alfred Sloan. Sloan, along with Brown and their codesigners (the Sloan-Brown team), became the primary source of GM's recovery and its ultimate ascendancy over FM. Eventually, all of these individuals were selected to serve on GM's Board of Directors, the policy-setting body ultimately responsible for designing the firm's performance-control system.

The Principal Designers

Pierre S. duPont, coaxed out of retirement and continuing as GM's president only under duress (Sloan, 1941:133), was quite eager to relinquish his position to Alfred P. Sloan. In announcing Sloan's selection as his successor, duPont explicitly praised Sloan's design work, "I greatly admire Mr. Sloan and his business methods and look upon him as one of the most able partners in the management of General Motors Corporation" (Forbes, 1924:760).

Sloan, indeed, was unusually well-qualified for organizational design work. The precision he had developed during his undergraduate days as a student electrical engineer at the Massachusetts Institute of Technology helped him build and operate his own extremely successful firm, Hyatt Roller Bearing Company. In his early days at Hyatt, Sloan worked nine to ten hours a day, six days a week, making the business his whole life and feeling no need for hobbies or relaxation (Douglass, 1954:131).

Hyatt was Sloan's training school where he learned the "draftsmanship, designing, engineering, production, sales, advertising and executive direction that fitted him for the career that followed and eventually put him at the head of General Motors" (Sinsabaugh, 1940:180). Since two of Hyatt's customers were Ford Motor and Cadillac (run by the Lelands), Sloan had long been exposed to the mass production of high-precision, interchangeable parts for the automobile industry (Sloan, 1927b, 1941).

To protect his growing investment, Sloan wanted to affiliate his concern with a larger firm. Consequently, he sold Hyatt to Durant's United Motors Corporation (UMC) in 1916, simultaneously agreeing to become the president of this GM subsidiary.

As UMC's president, Sloan developed, installed, and operated a performance-control system for "a multiple-unit organization with different products made by separate divisions" (Sloan, 1964:47). As with GM, "all that held United Motors together in its beginnings was the concept of automotive parts and accessories" (p. 47). Sloan strengthened both the affiliation among UMC's components and the units themselves. When he discovered, for instance, that Dayton Engineering Laboratories' (Delco's) books did not provide a clear cost picture, he devised a completely new accounting system ("Alfred P. Sloan, Jr.: Chairman," *Fortune*, 1938:110).

Sloan's position as UMC president, more importantly, gave him a good vantage point for viewing the larger operations of GM. So during the Durant days, Sloan (1941) "observed the daily moving picture of General Motors operations ... to wonder why it was done this way or that way. To consider whether it would not be better ... done some other way, and to build castles in the air" (p. 132). While Sloan never became a Durant "crony," he was included within the group of top GM executives who met with Durant during his visits to Detroit (Fanning, 1964:215). These discussions taught Sloan much about GM's far-flung operations and resulted in several valuable organizational planning papers, the most important being Sloan's "General Motors Corporation: Study of Organization" (General Motors Exhibit [GMX] 1, *U.S.* v. *DuPont*, 1956:6532-6560). This document became the firm's fundamental design outline for the next eighteen years' work. Sloan, then, "stood out as an 'organization man' in contrast to Durant, who gave much attention to the stock market" ("Alfred P. Sloan, Jr.: Chairman," *Fortune*, 1938:110).

Sloan had come of age "at almost exactly the time when the automobile business in the United States came into being" (Sloan, 1964:17) and had been intimately associated with the industry for many years by the time he came to GM. And because of his additional experience with GM under Durant, Sloan's "concept of General Motors was clearly focused: a richly efficient mass-production enterprise, founded on the teamwork of scientists, engineers and manufacturers, in a promising industry" (Fanning, 1964:220). Moreover, GM became his occupation, avocation, and recreation (Nielsen, 1972:192).

Surely to the duPont interests, the most important dimension of Sloan's experience was on the financial side. As W. S. Carpenter, a president of the DuPont Company, remarked: "He came up through the small corporation, and was, therefore, always impressed with all aspects of business management, particularly the financial aspects. ... During his regime ... we had from the operating side ... a sympathetic appreciation of the financial aspects of business" (Government Trial Exhibit [GTX] 1238, *U.S.* v. *DuPont*, 1956:5186-5187).

In passing, it should be noted that Sloan's astute interpersonal skills and his sophisticated organizational structures masked for many years (at least to the academic community) his strong and autocratic personality. Accounts of his later work at the Sloan Foundation unearth this aspect (Nielson, 1972:194–195). Sloan, for instance, used his small foundation staff simply to carry out his instructions. To clinch his control, he appointed only his closest friends to an executive committee that screened proposals for the entire board, which rarely if ever overruled or even modified these recommendations (p. 194). To his death Sloan "found it impossible to break his deeply ingrained habit of absolute rule, and he rarely surrendered his personal control over final decisions. He remained an autocrat to the end" (p. 195).

Sloan's close associate at GM was Donaldson Brown, vice-president in charge of the Financial Staff. Of Brown, Sloan (1964) said: "He and I shared similar views on the value of detailed, disciplined controls in the operation of a business. From the time of his arrival in the corporation we recognized this affinity and began a long and congenial relationship" (p. 118). As Brown (1957) characterized their relationship, "Mr. Sloan and I worked in close harmony over the years, we both recognized that while in theory the designated functions were separate and distinct, the fact was that in the management of a business as complex as GM there are, inescapably, areas in which the line between them cannot be clearly drawn" (p. 67).

Brown, by most people thought to be a very cold (Hickerson, 1968:71) and analytical person, joined GM from the DuPont Company on January 1, 1921. Because of the duPonts' immense GM investment, Pierre duPont and John Raskob believed it was "essential that . . . Brown be transferred from the DuPont Company to the General Motors [Company] to act as their Vice President in charge of finance" (GTX 180, *U.S.* v. *DuPont*, 1956:3406). At DuPont, Brown had been the treasurer, a member of the DuPont board, and a member of the important Executive and Finance Committees (Brown, 1957:33–34).

While at DuPont, Brown created a method of analysis that could be used to control a firm's financial performance (Brown, 1957:26). Coleman duPont, the company's president, was impressed greatly, and Brown's brilliant financial-performance-control model soon culminated in the DuPont Company's *chart room*, where "statistical data pertaining to each segment of the company's operations were displayed. Meetings were held regularly with department heads, and extended discussions were held regarding the possibility of improving specific cost and expense items, in relation to the end-result of return on invested capital" (p. 27). Just before joining GM Brown had finished his work on the DuPont performance-control system.

Brown, in addition, had been concerned with other important design tasks. One of the most important was his work on the DuPont Company's subcommittee on organization, which was considering DuPont's structural layout. Brown and another organizational subcommittee member, Harry Haskell, wisely argued against a functional structure and for a divisional decomposition schema (Brown, 1957:37).

Other Designers

While Sloan and Brown, working as a closely integrated team, carried the bulk of the GM design load, they received important help from other GM executives.

Pierre S. duPont was the interim president, between Durant and Sloan. As such, duPont played a somewhat lesser role as a GM designer. His primary efforts were to arrange the GM rescue operation. His respected position with the financial community facilitated GM's refinancing; his experience in success-fully restructuring DuPont years before, "his ability to choose men and his knowledge of organization" were also of "great help" to GM (Dale, 1960:241). Like Sloan, he had received an MIT education that introduced him to "modern technology" and gave him "a rational and analytic approach to management and finance" (Chandler and Salsbury, 1971:600).

John J. Raskob had long worked closely with Pierre duPont, starting as his personal secretary. An ambitious man, Raskob moved up rapidly in the DuPont Company hierarchy; he held the positions of assistant treasurer and treasurer as well as a Finance Committee membership (White, 1931:254–255). Raskob had promoted the duPonts' interest in the GM investment as a use for their excess capital after the war. At GM, Raskob's primary contributions concerned external financial relationships, acquisitions, and financially oriented motiva-tion schemes.

John L. Pratt, an engineering graduate from the University of Virginia (Weisberger, 1979:246), was another GM designer who had begun his career with DuPont rather than with GM. Pratt eventually joined the DuPont Company's Motor Development Section, which was supposed to aid Durant in systematiz-ing GM's operations. Shortly thereafter, Pratt moved to GM and worked directly with Durant (p. 248). Thus Pratt provided an extremely valuable addition to the new design team, for he had worked closely with Durant, vainly attempting to bring some order to GM, especially in inventory control and central-staff development. Brown (1957) considered Pratt "as perhaps" his "closest friend and business associate over the years" (p. 36).

Albert Bradley joined Brown's Financial Staff shortly after Brown himself moved to GM. Since Durant had created few financial and statistical controls, GM's Financial Staff required additional talent. Bradley became an ideal assist-ant for Brown, as he had received his Ph.D. in 1917 from the University of Michigan, specializing in finance and economics (Bradley, 1927:412), and had obtained practical experience in accounting with the army. In 1922, Brown had Bradley transferred from GM's Detroit offices to the New York headquarters where he worked on the firm's financial control policies (GTX 199, *U.S.* v. *DuPont*, 1956:3443). Together with Brown, Bradley "developed cost-accounting techniques by which G.M.'s prices, profits, working capital, costs, and return on investment could be kept in calculated balance with market demand" (Stryker, 1952:129).

C. S. Mott had been with GM from its earliest days but was not a "Durant man" (Young and Quinn, 1963:83). Mott was a stickler for performance control. He instructed: "You must exercise eternal vigilance in watching overhead. . . . Get facts. Never guess. . . . Know every month exactly what your business has done in all its departments. Don't merely have these statistics compiled: study them, analyze them, use them as a basis . . . for your vision of the future and your planning. . . . Devote careful attention to training other men to shoulder and properly discharge responsibilities" (Forbes and Foster, 1926:206-207). Similarly, in running his Mott Foundation years later, he "applied the techniques of modern industrial management—production planning, product promotion, statistical control, and follow up—to its philanthropies" (Nielsen, 1972:203). Mott, like Sloan, was an *autocrat*. His foundation was "a disciplined instrument of the donor's will" whose small board and staff he dominated, tolerating no interference (Nielsen, 1972:204). As can be expected: "Sloan and Mott liked and understood one another; each was an engineer, . . . each was capable of large vision, while not losing sight of the very down-to-earth practical detail work necessary to hammer vision into reality" (Young and Quinn, 1963:91). About Mott, Sloan (1941) wrote: "I liked to work with Mott. . . . Neither of us ever took any pride in hunches. We left all the glory of that kind of thinking to such men as like to be labeled 'genius.' We much preferred the slow process of getting all the available facts, analyzing them . . . and then deciding our course" (pp. 49-50).

IDENTIFYING THE CLIENTS

The Sloan-Brown planning group's primary client is clearly identified as the GM stockholder. Evidence for the stockholder-as-client thesis is found in: (a) the value orientation of the Sloan-Brown team; (b) the extensive stockholdings of the designers and principal decision makers as well as their intimate relationship with the duPonts, who were the major stockholders; and (c) the financially oriented performance-control system ultimately developed. (Since the financial control system is discussed in the next sections of this chapter as well as throughout subsequent chapters, attention here is concentrated only on the orientation of the Sloan-Brown group and their stockholdings.)

The Value Orientation of the Sloan-Brown Designers

The attitude of the Sloan-Brown designers toward nonstockholding groups involved in GM operations was continually one of tolerant antagonism. While GM's laborers and customers were all necessary to its operation, they were certainly not to be served as the firm's primary clientele.

Though GM paid high hourly wages and aided workers in building homes, the underlying motivation could not be considered altruistic. High wages were needed to make up for the cyclical and seasonal nature of automobile work. Similarly, since housing had always been a problem in the rapidly expanding auto industry, housing was a necessary constraint to be satisfied in obtaining workers.

The GM designers also strongly opposed unionization. During the 1937 General Motors sit-down strike Sloan, at first, refused to deal with the workers and was rebuked by President Roosevelt (Sloan, 1964:393). Similarly, Mott felt that the workers "had no right to sit-down there. They were illegally occupying" GM's property. "The owners had the right to demand from the Governor to get those people out. It wasn't done" (Terkel, 1970:135). After organized labor, with the support of federal and state governments, demonstrated its capability to halt completely the firm's service to its stockholders, the GM designers agreed to enter into collective bargaining with union representatives.

The GM customers, of all the nonstockholding groups, probably received the status closest to that of system client. This noncaptive group, after all, exerted most influence on the stockholders' well-being. During the early days of multifarious automobile makers, the customers' demands had to be heard if performance was to be adequate.

But even here, the customers were more a constraint than a client. "The primary object of the corporation . . . was to make money, not just to make motor cars" (Sloan, 1964:64). A subsequent interpretation of Sloan's extreme pecuniary philosophy would be that "the General Motors men were all technicians and applied their knowledge of technology strictly to making money, not to engineering the best possible cars" (Lundberg, 1968:95). Making money and making good cars were often completely compatible. Under such happy circumstances the customer could expect to get improved transportation values; Sloan, for example, introduced four-wheel brakes to draw Ford's customers. In the late 1920s he created GM's Art and Color Section to satisfy the public's rapidly emerging taste for style, form, and color. Needless to say, when customer and stockholder interests were incompatible, the customers' well-being took the back seat. Sloan eventually used GM's yearly model changes, by then only cosmetic in nature, to keep customers transferring their money to stockholder accounts: " 'Each year we build the best car we possibly can to satisfy the customer,' said Alfred P. Sloan, Jr., when he was chairman of General

Motors, 'and then the next year we build another to make him dissatisfied' " (Livingston, 1958:28).

Although GM concentrated on financially beneficial cosmetic changes, it often ignored more important safety items (Ayres, 1970:49-50). In 1929 Sloan, for instance, delayed the introduction of shatterproof safety glass. Letters between Lammot duPont—who was promoting the use of the new glass as his DuPont Company made its plastic inner layer—and Sloan provide ample evidence of the customers' secondary position after the stockholders'. Sloan's response to Lammot's first suggestion countered: "Two or three years ago I would have felt that perhaps it was the desirable thing for General Motors to take an advanced position similar to what it did on front wheel brakes, but . . . with our volume increasing at a decelerated rate, I feel that such a position can materially offset our profits" (Hearings, 1968:965). Sloan strengthened his response to Lammot's second query: "Accidents or no accidents, my concern in this problem is a matter of profit and loss. . . . I will go further . . . and say that what I fear . . . is that the advent of safety glass will result in both ourselves and our competitors absorbing a very considerable part of the extra cost out of our profits. You, of course, are familiar with the comparatively large return that the automobile industry enjoys and General Motors in particular" (p. 966). So GM's use of safety glass, in Sloan's words again, "would have reduced the return on . . . capital and the public would have obtained still more value per dollar expended" (p. 967).

Three years later (i.e., 1932), Sloan was still fighting Lammot regarding GM's refusal to use safety glass:

> I feel very strongly that if we adopt safety glass it will be very materially at the expense of the stockholders. . . . I would very much rather spend the same amount of money in improving our car[s] in other ways because I think, from the standpoint of selfish business, it would be a very much better investment. You can say perhaps that I am selfish, but business is selfish. We are not a charitable institution—we are trying to make a profit for our stockholders. (p. 968)

Finally, after Ford adopted and the public demanded safety glass, Sloan began to capitulate. But even in retreat he protected his stockholders against the public by promoting a "manifestly inferior" (Sloan's words, p. 968) safety glass which could be installed at minimal cost if all GM divisions standardized on it.

Beyond the decidedly negative attitude expressed toward laborers and customers, further evidence emphasizing the stockholders' primacy at GM is found in the designers' positive attitudes. As Raskob (1927) summarized it, "The central motive in the business is the permanent welfare of the owners" (p. 134).

Sloan in his GM organization-design study similarly placed the stockholders at the pinnacle of the organization (GMX 1, *U.S.* v. *DuPont*, 1956:6536):

1. Stockholders
2. Directors
3. (a) Finance Committee
   (b) Executive Committee.

Moreover, he stated, "It is in their interests first of all that the corporation is supposed to be run in the private-enterprise scheme of things" (Sloan, 1964:213). Accordingly, in coordinating one part of the corporation with another, reflection on "what the interests of the stockholders require enables us to obtain the correct answer" (Sloan, 1929:96).

Brown too, as a finance-oriented executive, envisioned GM's client as its stockholders. "Since business owes its existence to its owners," he wrote in 1927, "it is expected to operate for their benefit." "In industrial management" there was but "one central motive," namely "the permanent welfare of the owners of the business" (p. 5). Thirty years later Brown's (1957) views remained unchanged: "The basic purpose of management is to serve the interests of the owners of business, whom management directly represents" (p. 3). Relating the stockholders' interests to performance control, Brown stated: "The responsibility of a Board to the stockholder-owners of a business . . . is to see to it that the business is operated in the true and long-range interests of the owners. Therefore, the proper organization of control is forced by absolute necessity" (p. 59).

The designers' value orientation toward *all* stockholders, and not just the special interests of a few, can be documented partially by considering the duPont ownership of GM stock. In promoting the GM investment to the duPonts, Raskob emphasized that GM would be a certain outlet for DuPont Company products (GTX 124, *U.S.* v. *DuPont,* 1956:3221). And, indeed, DuPont sold such items as paints, antifreeze, fabrics, brake fluids, adhesives, and heat-treating and copper-plating chemicals to GM in rather large quantities. The other GM designers, in contrast, seem to have been surprisingly adamant and quite adept in opposing the duPont influence. First, John L. Pratt, in spite of his one-time employment at DuPont Company, fought a running battle with the duPont interests as chairman of the GM Purchasing Committee. Pratt, at one point, even mockingly directed DuPont's "desk warmers" to leave their desks, get on the road, and sell the divisions rather than attempting to exert pressure on them through him (GTX 373, *U.S.* v. *DuPont,* 1956:3915).

In addition to such interpersonal struggles, the GM designers also used formal procedures to limit DuPont influence on GM purchasing: a multiple source of supply policy was implemented that not only insured a constant supply but also provided a market test for the price and quality. Thus DuPont Company had to meet the price and specifications of competitors. Should DuPont and a competitor offer identical prices and specifications—as was often the case after the multiple-source policy forced DuPont into line—DuPont would certainly receive the larger share of the business. But this favored treatment would not be at the expense of smaller stockholders.

More generally, Green (1933), after extensive study on the DuPont-GM financial relationship, concluded that "the organization has been directed in a course which was best for it as a company. Such action could only help the minority interests pro-rata with the controlling faction or factions" (p. 295).

The Stockholdings of the Principal Designers and Decision Makers

Since GM's principal designers and decision makers were all substantial holders of GM stock themselves, in addition to being intimately involved with the major stockholding group, the duPonts, they found it extremely easy to accept and identify with the stockholders' values, desires, and concerns. After Durant left GM, the duPonts were by far the major stockholders. Their original investment dated back to 1914, when Raskob interested Irénée and Pierre duPont to make modest personal investments in GM stock. During those early days "Pierre duPont owned 2,200 shares of General Motors and John J. Raskob about 1,200 shares" (U.S. Senate, 1956:26). This ownership resulted in board seats for both. It was not until 1917 that the DuPont Company became intertwined with GM when Durant invited the duPonts to invest $25 million in the firm (p. 27). With this investment the duPonts secured a 23.83 percent interest in GM's common stock. Other interests were purchased subsequently so that at the end of 1919 DuPont owned 28.74 percent of the outstanding General Motors stock (p. 28). Through further purchases and the DuPont Company's efforts to provide GM with additional capital during Durant's 1920 crisis, DuPont ownership of GM jumped to 35.8 percent common-stock interest (p. 29). DuPont's direct control continued at about this level for the next several decades. It meant that a rational, professional management would use "its energies to produce earnings for the stockholders, thereby benefiting "all branches of the duPont family" (Chandler and Salsbury, 1971:565).

According to *Fortune,* Sloan "with some 750,000 shares" was "the largest individual G.M. shareholder" ("General Motors," *Fortune,* 1938:178). Sloan acquired many of his GM shares through the sale of his prosperous Hyatt Roller Bearing Company to Durant. The original payment was largely United Motors Corporation stock, then a subsidiary of Durant's GM (Sloan, 1964:23-24). Because his fellow Hyatt stockholders were wary of the automobile industry in general and GM in particular, Sloan accepted a disproportionate share of stock, leaving the far safer cash to his partners. Sloan took the risk willingly: "My position as an individual toward the future of the Corporation is fully evidenced by the fact that I have practically everything that I have in the world in General Motors, and not only that, . . . I have recently made a very large commitment at the present market increasing my interest still further and have incurred a very heavy obligation" (*U.S.* v. *DuPont,* 1956:1224). Over the years, Sloan kept his faith in the GM stock, often expressing to his fellow Sloan Foundation trustees that selling it would not be in the Foundation's best interest (Zurcher, 1972: 91). Through avid accumulation and dogged retention, the Sloan holdings reached an estimated $200 to $400 million (Lundberg, 1968:42). The Sloan Foundation assets, valued at 1964 market prices, were estimated at $297.7 million (Weaver, 1967:48).

In 1939, C. S. Mott was listed by the United States Federal Trade Commission (USFTC) *Report on the Motor Vehicle Industry* as owning 523,087 shares of GM stock or 1.22 percent of the total (p. 429). Mott began to accumulate his GM stock in 1908 when General Motors acquired 49 percent of the stock of Weston-Mott Company, a firm that manufactured axles Durant needed for his Buick cars. Mott, the firm's sole owner, exchanged his remaining 51 percent for GM stock in 1913. Over time he saw his original number of shares multiply fifty times (Forbes and Foster, 1926:202), become worth millions of dollars, and make possible the Mott Foundation (Young and Quinn, 1963:61). Mott's personal fortune was estimated (by Lundberg, 1968:44) to be between $75 and $100 million; the Mott Foundation assets were valued in 1964 at $418 million (Weaver, 1967:48).

In 1939 Charles Kettering owned 447,198 shares of GM stock or 1.04 percent of the total (USFTC, 1939:429). Kettering occupied an important position in the GM corporate Advisory Staff, vice-president in charge of the corporation's vital Research Section. Kettering obtained much of his substantial GM holdings through the sale of his company, Delco, to Durant in 1916 (Leslie, 1983:57). Lundberg (1968:42) placed the Kettering holdings between $100 and $200 million. Like Mott, he created his own foundation (the Kettering Foundation) and also provided funds with Sloan for the Sloan-Kettering Foundation. In running his foundation Kettering "was no less the autocrat than Sloan, his lifelong friend. The small board was made up of close friends he could count on to be pliant to his wishes" (Nielsen, 1972:198).

Kettering, as GM's research chief, participated little in the organization design efforts; however, he was an important GM decision maker. Kettering made such contributions to improved performance as the development of quick-drying lacquers, which reduced in-process inventories immensely and introduced more colors to automobile styling. In his pursuit to improve GM's financial performance, Kettering (1928) always kept in mind that "the man responsible for the functioning of a research laboratory must never lose sight of the fact that research work, done with corporate funds, has to justify itself economically" (p. 737).

Another of GM's large stockholding groups was the Fisher brothers. The brothers had long been in the automobile-body business and had accumulated extensive properties. They transferred the first 60 percent of their holdings to GM during Durant's 1919 expansion; GM acquired the remaining 40 percent in 1926 (Sloan, 1964:161). In return, the Fishers received 664,720 shares of GM common stock, about 11.46 percent of GM's entire common-stock issue, with a market value of $130,000,000 (Seltzer, 1928:218–219). This price reflected Sloan's desire for complete control of the critically important Fisher Body plants and more intimate Fisher involvement in GM operations. While several of the brothers did accept Sloan's invitation, even then they remained more decision makers in the system than its designers. Fred Fisher alone became a GM designer.

John Raskob, John Pratt, and Donaldson Brown were all key executives who joined the GM family without having their own firms to sell. Raskob's start was particularly humble, for he was the son of an immigrant cigar maker (White, 1931:250). But "as Raskob was a pecuniary man to his fingertips with no other apparent interest in his life, his fortune before he started redeploying it may well have exceeded $75 or $100 million" (Lundberg, 1968:71). Much of this money was tied up—see Chapter 8—in GM stock. Similarly, Pratt, starting with little capital and a relatively low salary, amassed a fortune somewhere between $100 and $200 million (p. 42). Donaldson Brown, too, joined the GM team without extensive tangible assets yet gained a fortune of $75 to $100 million (p. 43). In *Some Reminiscences of an Industrialist* Brown outlined how he acquired such wealth: accumulating GM stock via the Bonus Plan, investing his savings chiefly in the company which employed him, borrowing against future income to buy GM stock, and "buying a large block of GM stock from the duPont Company on a deferred installment plan" (p. 5).

As previous owners of firms or as managers closely acquainted with owners, the Sloan-Brown group, as well as GM's more prominent decision makers, easily agreed about whom the GM performance control system was to serve. Too, they had extensive personal holdings in GM: Sloan, Kettering, Mott, Pratt, and Brown alone owned at least 3 percent of the GM stock (Finn, 1969:110; Sheenan, 1965:49). Moreover, the Sloan-Brown designers placed great faith in the future value of their holdings. Not only did they readily accept the stockholders as the firm's primary client, but in addition, as stockholders, they understood the value structure intimately.

DEFINING THE PERFORMANCE MEASURE

The Need for a Concrete Performance Measure

Having accepted the primary client to be served by the GM performance control system, the planners next had to move from the abstract concept of the client to a concrete performance indicator capable of measuring the firm's success in serving the client.

To understand the importance of identifying such an indicator, one need only observe Sloan's frustration many years later about lacking an objective criterion for measuring his Foundation's success: "Mr. Sloan pointed out that there was not even the most rudimentary yardstick for measuring foundation accomplishment or failure. He was not averse to adding that foundation staffs knew how to spend money but had no way of knowing whether the money they spent yielded a return" (Zurcher, 1972:76). Similarly, "Sloan was convinced that 'professors' lacked incentive, and that their profession suffered from having no tangible way in which performance could be measured—no profit-and-loss statement at the end of the year" (Weaver, 1970:123). But Sloan's desire for a concrete performance measure went far beyond the yearly profit-and-loss figure.

Specifically, Sloan wanted a performance measure that told him exactly how well his management used the stockholders' capital resources to generate returns in their behalf. He stressed, "the strategic aim of business [is] to earn a return on capital and if . . . the return in the long run is not satisfactory, the deficiency should be corrected or the activity abandoned" (Sloan, 1964:49). As quoted by Pound (1934), "When we invest our stockholders' money as trustees, we must do it in the firm belief that the capital is safely invested and that the return to the stockholders, as a result of the investment, will be fair and equitable; otherwise we have no right to make the investment at all" (p. 339). In scolding Lammot duPont regarding his expensive nonshatterable glass, Sloan was even more emphatic: "I think you will agree with me that it is a very easy thing to reduce . . . return. I am fighting it not only every day but practically every hour of every day, the fact that our relation between cost and selling price is more unfavorable this year than it was last is . . . my prime duty to correct" (Hearings, 1968:966).

Bradley (1927), Brown's principal assistant, expressed similar feelings: "The true basis for measuring the commercial success of any enterprise . . . is the return on the capital employed." Stockholders, for instance, would not be satisfied with sales of "a million cars a year merely for the edification and amusement of the manufacturing and engineering departments. The stockholders themselves must also get a run for their money" (p. 421).

Rate of Return on Invested Capital as GM's Performance Measure

In moving from the abstract idea of earning a return for the GM shareholder to a more explicit and quantifiable performance indicator, Sloan and Brown employed the rate of return on invested capital. "The general test of efficiency of management of any business," Brown (1957) instructed, "is the rate of return on capital employed" (p. 60). This measure, still used at DuPont and GM as the primary performance indicator, spread to many other firms, although it now has attracted countless academic critics.

Commensurability and the Rate-of-Return Measure

Exactly what is the *rate* of return on invested capital? In simplest terms it refers to the percent figure that results from dividing dollar profits by the total dollar equivalent of working capital, plant, and equipment used to generate those profits. It is called "return on investment" or ROI for convenience; but one must always keep in mind that it is *rate* of return.

The Sloan-Brown designers, of course, realized that "there are other measures for the running of a business" (Sloan, 1964:140). Sloan dismissed Durant's revered stock price quickly: "Naturally, I like to see General Motors stock register a good price on the market, but that is just a matter of pride. Personally, I consider its price fluctuations inconsequential. What has counted with me is the true value of the property as a business" (Sloan, 1941:103). Still other measures were considered, "for example, profit on sales, and penetration of the market, but they do not supersede return on investment" (Sloan, 1964:140). Actually, as described below, Donaldson Brown demonstrated how many such indicators as profit on sales could be incorporated, or aggregated, into the overall-return-rate measure.

Rate of return is not a dollar figure; it is a rate or a ratio of dollar numbers to other dollar numbers. It can measure the commercial success of enterprises widely different in size, in different times, or in different countries dealing in different monies. When rate of return is used as the measure, the commercial success of compared enterprises is *commensurable.* The comparison of company with company, division with division, or company in 1985 with company in 1975 is meaningful and valid (assuming, of course, such factors as inflation hold constant). Gross capital investment or net profit does not permit meaningful or valid comparison and might mislead an investor or analyst. Sloan (1964) labored this point:

> An operation making $100,000.00 per year may be a very profitable business. . . .
> On the other hand, a business making $10,000,000 a year may be a very

unprofitable. . . . It is . . . a matter of . . . the relation of . . . profit to the real
worth of invested capital within the business. (p. 49)

So performance comparisons over time and across units became possible
because the ROI percent measure permitted referencing client benefits, i.e.
profits, to a common resource denominator, the amount of the clients' capital
employed. In short, ROI measured how well the business employed its clients'
resources to generate client benefits.

Long-Run Return on Invested Capital

With his usual foresight, Sloan directed his decision makers to assume a
long-run stance toward the rate-of-return measure: "The question is not simply
one of maximizing the rate of return for a specific short period of time. Mr.
Brown's thought on this was that the fundamental consideration was an average
return over a long period of time" (Sloan, 1964:140–141).

IDENTIFYING GM'S CRITICAL PERFORMANCE VARIABLES

Brown's Financially Oriented Critical-Variable Model

When Donaldson Brown joined GM he brought with him the "financial yardstick"
(Sloan, 1964:141) and analytical model that became the basis for performance control at GM. As Sloan saw it, Brown had developed "a method of crystallizing facts"
(p. 141). Specifically, Brown's financial model identified the critical financial
variables that affected, or produced, the system's ROI performance and, in turn,
the level of stockholder well-being. Viewed in the extreme, Brown's method
allowed GM "to determine at any instant the current profitability of any nook
or cranny in its vast maze" because Brown's analytical interest went beyond
"the rate of return on *total* invested capital, a fuzzy figure that conceals many
financially inefficient operations" (Lundberg, 1968:833). Sloan (1964) put it more
moderately: Brown brought "about a recognition of the structure of profit and
loss in operations. Essentially it was a matter of making things visible" (p. 142).

The basis of Brown's model for evaluating the performance of GM and its
various divisions was a relatively straightforward, but highly innovative, rate-of-
return formula (Brown, 1957:6):

$$R = T \times P$$

where R = rate of return on invested capital,
T = rate of turnover of invested capital, and
P = percent of profit on sales.

Brown, like Sloan and the duPonts, thought the R variable was the final and fundamental measure of industrial efficiency in terms of management's primary responsibility (p. 26). "In gauging the effectiveness of management the first approach always is to examine the overall result, which is the rate of return on capital employed. If this be subnormal, having due regard to the character of business and the competitive situation, it is self-evident that something is wrong with the management. The second step is to identify the cause—and correct it" (p. 61). The aggregate variables T and P in Brown's simple rate-of-return equation helped the GM planners and decision makers identify the financial causes of good and bad performances reflected in R.

Both the turnover of invested capital and the percent of profit on sales were broken down in what Sloan (1964:142) described as a "deaggregating" process. Figure 4-1 depicts graphically Brown's disaggregation schema for R. Brown separated the T variable into various fixed-capital variables such as plant and equipment, and into working-capital items such as cash balances, in-process and finished inventory, and accounts receivable. These various investment variables were reported in terms of a ratio to sales, the reciprocal of which represented the rate of turnover T. Finally, Brown also factored the cost variables into significant categories. When these costs were deducted from and divided by sales, the percent of profit on sales P was obtained.

So Brown's analytical model yielded "a specific disclosure of causes and effects for the return on investment. . . . Effective control, or lack of it, for any item on either side of the equation [T or P] could be identified, thus making possible efforts to improve conditions" (Brown, 1957:27). Or, if the targeted return rate rose, there were two approaches to achieving the goal. First, P could be increased either by reducing expenses or by boosting sales revenues, holding other things constant. Brown's method of analysis, then, "lent itself to measured calculations of the effect, in terms of stockholder benefits, of price reductions which might result in expanded sales volume" (p. 27). Second, T could be boosted either by raising sales without increasing investment or by holding sales constant while reducing investment, for example through inventory reductions. According to Sloan (1926), a "number of different angles" could be used to solve "this problem of speeding up the rate of turnover, of making each dollar do more work every year," including "purchasing in right

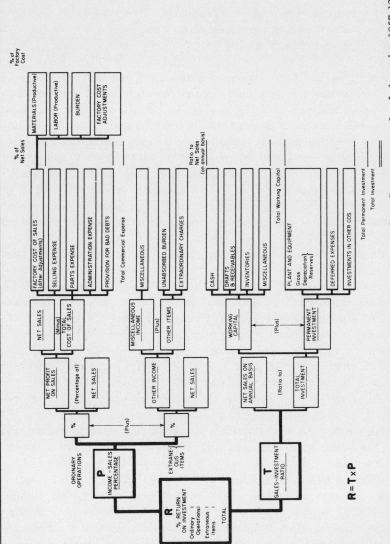

Figure 4-1. Brown's return-on-investment form. *Source:* Reproduced from Brown, *Some Reminiscences of an Industrialist,* 1957:129.

quantities" as well as "thrifty arrangement in time" and "manufacturing by processes which perform the same operations in materially lessened time" (p. 993). The analytical method by which Brown disaggregated the ROI variable made it possible to understand exactly how much the relevant variables influenced the stockholders' well-being. Consequently, it provided GM's designers and decision makers with a "penetrating look into the business" (Sloan, 1964:141) as well as its priorities.

The Nonfinancial Variables and the Brown Model

While Brown's financial model provided the first step in identifying the relevant variables that had to be controlled for satisfactory performance, it did not elucidate all the variables that required decision-making attention. Brown naturally realized that the success of GM often hinged on its ability to control nonfinancial variables. He mentioned the importance of manufacturing, advertising, and marketing. He added that producing an article "exclusive in design, possessing superior engineering qualities and carrying with it a peculiar appeal to fashion" afforded the opportunity for a favorable rate of return. Hence risk could be minimized through "skill in progressive engineering improvements" and "ingenuity in anticipating the changing tastes of the buying public" (Brown, 1957:60–61). So to control performance the firm often had to adjust in ways that were nonfinancial, though the results of the adaptation were ultimately reflected in the financial rate-of-return measure and in the underlying critical financial variables. That is, nonfinancial variables could be mapped onto the more critical financial variables so as to inform the designers and decision makers what performance improvements for the stockholders might be expected. The value of a reduction in in-process inventories from a new quick-drying paint, for instance, was easily assessed with Brown's model.

Performance Control at DuPont and GM

GM planners and decision makers ultimately used Brown's model to generate specific standards for each of the working- and fixed-capital investment variables and for the various cost and revenue categories. They based their reference points on past performance experience tempered with future expectations. In addition, Brown and his Financial Staff developed tables illustrating how the levels of inventory, working capital, and the like influenced capital turnover in the division, and, in turn, the divisions' return rates. The critical-variable standards established by Brown's staff when compared to actual performances yielded feedback information vital to the designers' effort to control GM's performance. Actual performance data were submitted by each component's comptroller in the form of monthly reports following the schema shown in Figure 4-1. "The heart of the financial-control principle lies in such comparisons" between expected and actual performances (Sloan, 1964:142).

In summary, Stryker (1952), in an article entitled "Planning and Control for Profit" published in *Fortune,* concluded that "among corporate giants, only two—General Motors and duPont—have an extended record of systematic and consistent adherence to the concept of coordinating controls toward a planned profit objective" (p. 129).

5

Formulating the Strategy

Another boundary-definition problem confronting GM's designers dealt with partitioning the uncontrolled or environmental variables away from the controlled variables or firm decisions. When composing, or selecting, the activities to include within the controlled domain, GM's designers wanted, to the extent possible, to encompass sufficient internal decision variety to counter external disturbances without overloading its management capacity. In formulating the firm's strategy via *channel* integration and *product* diversification, the planners also had to establish procedures for deciding: (a) how much capital would be allocated to the firm in the aggregate, and (b) how the performance of its various investments would be measured.

SETTING THE RESOURCE–ALLOCATION AND PERFORMANCE–MEASUREMENT POLICIES

The stockholders as a body, of course, had neither the time nor the skills to control GM's performance. Similarly, the GM Board of Directors, though skilled, lacked the channel capacity to control the corporation's rate of return and filled more of a final-oversight role than a management function. Still, at GM not all policy decisions were delegated to active, full-time corporate executives. Even Sloan, by his own design, had limited jurisdiction in those areas most vital to the clients' interests. For instance, "in 1920 Sloan perceived the need to separate the power to finance projects, whether originated at the corporate or division level, from those people who advocated them" (Schnapp and Cassettari, 1979:147). Under this policy, proposed investments approved by GM's Executive Committee, if they required an allocation of corporate capital, had to be reviewed and approved by the Finance Committee before being submitted to the Board of Directors (p. 147).

While Sloan dominated the operations side of the corporation's activities via the Executive Committee and his presidency, the duPonts along with their associates and representatives supervised the entire financial policy of the corporation through the Board of Directors' more powerful Finance Committee (Brown, 1957:66-67). Consequently, the most critical of variables, those closely related to the stockholders' financial investment and benefits, were not included within the set of variables controlled by GM's full-time managers, but were retained by the Finance Committee. "The central responsibility of the Finance Committee," as Sloan (1964) explained, "is the corporation's purse." The by-laws granted this committee jurisdiction over GM's financial policies and affairs, "all capital appropriations and . . . entrance into any new line of business." Pricing policies and procedures formulated by the Executive Committee were subject to its approval, and it was responsible for assuring that GM received "adequate capital for its needs," and earned "a satisfactory return on its investment." This committee also made "the dividend recommendation to the board" (p. 187).

Allocation of Capital to GM

Thus, the "outsiders" who dominated the Finance Committee controlled several extremely important sets of resource allocation variables. First, they decided the firm's dividend policy. In disbursing payments to the stockholders, they also had to deal with a different, but related, variable: namely, how much of the clients' resources would be retained for GM's expansion. Current benefits had to be balanced against future benefits that would accrue from the reinvestment of retained earnings. Brown (1924a) put it this way: "Capital must be derived from undivided profits or from the sale of securities, and the amount available from these sources must be measured in the light of its economic cost, with due regard to dividend income to which the stockholders are entitled" (p. 196). In dividing profits between dividends and reinvestment, Pierre duPont "wanted expansion to be financed from within, dividends to be regularly maintained, with extra dividends declared only when the surplus account was ample" (Chandler and Salsbury, 1971:573).

Measurement of Performance

The second set of decision variables controlled by the Finance Committee included those associated with corporate and divisional performance measurement. Unless such performances were both accurate and, more importantly, commensurable, the Finance Committee could not hope to allocate resources to or within the firm rationally.

The work of measuring both corporate and divisional performances was delegated by the Finance Committee to its Financial Staff, but the Committee—which the partial proprietors dominated—supervised the financial and accounting departments instead of giving that jurisdiction to Sloan (Dale, 1960:255). As early as 1918, "the Secretary, Assistant Secretaries, Treasurer, Assistant Treasurers, Comptroller, Assistant Comptrollers, the Accounting Department and their respective offices [were] under the direct control and supervision of the Finance Committee" (GTX 130, *U.S.* v. *DuPont*, 1956:3249). Much later it was still being reported that "the various financial officers such as the vice president in charge of finance, the treasurer, comptroller, divisional comptrollers, and other financial officers, all report to the chairman of the financial policy committee" (U.S. Senate, 1956:33).

From the duPont perspective, the fundamental concern was to protect the DuPont Company and the duPont family investments in GM (Chandler and Salsbury, 1971:434). After the Durant debacle, Pierre duPont and his associates had learned that exercising financial control without adequate administrative and statistical techniques meant "hardly any control at all" (p. 434). Raskob, accordingly, started to improve GM's financial-measurement procedures in 1919 by appointing as GM comptroller an executive who had been with the DuPont Company for thirty-two years, Frank Turner. Though they were not realized until a full three years later, Raskob also outlined plans at the time of Turner's appointment to strengthen the financial and accounting organization, to establish uniform accounting methods, and eventually to create a centrally controlled financial and accounting organization. Turner was charged with organizing the Accounting Department, installing proper accounting procedures for the main office and all divisions and branches, and devising effective reports for the central and divisional offices as well as for the Finance Committee.

Raskob undermined the divisions' autonomy by insisting that their financial officers "while directly responsible to the general manager of their particular division or company, . . . render their fullest support and cooperation to the Treasurer and Comptroller of General Motors respectively" (p. 473). Furthermore, the treasurer's and comptroller's approval would be needed for all future financial and accounting appointments in the divisions.

Even though Raskob had jurisdiction over the Financial Staff, the ultimate design of adequate administrative and statistical procedures for financial control waited for Brown's craftmanship, since "Raskob's strength lay in external financing rather than in development of internal financial procedures" (Chandler and Salsbury, 1971:497). Brown, in contrast, excelled in internal-performance measurement, having devised DuPont's basic return-on-investment formula in 1912 and developed uniform accounting and other statistical procedures for DuPont's new diversification ventures.

Since Durant had thwarted most of Raskob's efforts to establish GM's financial controls between 1919 and 1921, the Financial Staff required the addition of more offices and the appointment of more executives to GM's corporate headquarters (Chandler, 1966:169). To staff his new Financial Analysis and Statistics Section of the Financial Staff, Brown secured the transfer of several key statisticians and economists from the DuPont Company's financial control group (Brown, 1957:48). Accountants were also added to Turner's Accounting Department.

Brown's major task, as outlined by Raskob, was to develop and enforce uniform accounting procedures throughout GM. In order that Brown's financial-control model could be used, the designers had to ensure that similar classification accounts and valuation procedures were used by all divisions in generating the requisite performance data. When Sloan headed United Motors Corporation for Durant, for instance, he instituted uniform accounting methods for his various divisions; he wanted to ensure that each division's return-rate performance was comparable to all the others' performances as well as to UMC's overall performance (Sloan, 1964:48). Similarly, throughout the entire family of GM divisions, uniformity was "essential to financial control, since without it comparisons" were "difficult if not impossible." An "immediate" task, "therefore, was . . . to institute standard accounting practice throughout the organization." By the beginning of 1921 the classification of all GM accounts had been standardized and "a standard accounting manual, specifying a uniform set of procedures, became effective . . . on January 1, 1923" (p. 143).

Since performance commensurability was essential, the designers made sure that these accounts and procedures were applied consistently among divisions and over time. "To co-ordinate financial organizations of the divisions and the central Financial Staff," Sloan and his colleagues "reaffirmed in 1921 the principle of dual responsibility for the divisional comptrollers," originally "introduced in 1919 to make those comptrollers responsible not only to their divisional general managers, but to the corporation comptroller as well" (Sloan, 1964:143). Divisional comptrollers now followed the guidance of the Financial Staff's Cost Accounting Department on account classification and valuation procedures (Chandler, 1966:177). So the data developed by the new corporate accounting methods provided the Finance Committee with consistently accurate and comparable figures for costs, sales, profits, and returns (pp. 177–178).

Allocation of Capital within GM

The development of commensurability among GM's many investments was extremely critical to another set of variables remaining in the hands of the Finance Committee: the internal allocation of GM's resources. The Finance Committee's power to approve major appropriations was critically important, for the internal allocation of capital (and executive talent) generated much of the firm's overall performance and influenced the profits available for both current and future benefits.

The Finance Committee, however, did not become involved with small capital appropriations. The Executive Committee handled smaller appropriations; still smaller sums of capital could be allocated by the division managers themselves. What the Finance Committee concentrated on were major capital investments that defined the general "character" (testimony of Pierre duPont, *U.S.* v. *DuPont,* 1956:665) of GM as a business. Along these lines W. S. Carpenter said that the Finance Committee "had the responsibility of approving certain large appropriations which may have been approved in advance by the Executive Committee for the construction of new plants or going into new business, or anything of that character" (*U.S.* v. *DuPont,* 1956:2730).

These strategic investment decisions determined which variables the firm's management ultimately decided, i.e., which activities were circumscribed within the decision makers' controllable set. Had the Finance Committee deferred these investment choices to the corporate executives, the result might well have been deviation from the business, or composition, strategy deemed best. Hence, management was not left to make its own strategy selections. Rather, the outsider-dominated Finance Committee, on recommendation from the insider-oriented Executive Committee, chose the business ventures that the corporate and divisional executives operated.

The advice of Sloan and the other Executive Committee members became vital to the Finance Committee's internal resource allocation, for these executives ultimately had to provide the decision-making capacity to handle GM's controlled set of variables (Brown, 1957:66-67). If performance was to be adequate, Sloan and his fellow operating executives must not be swamped with excessive internal variety.

In advising the Finance Committee about the firm's composition, the designers at GM did not naively think they had a free hand to dial in any desired rate of return. They were well aware, to the contrary, that GM, though growing increasingly strong, was still somewhat at the mercy of certain uncontrollable variables. Sloan (1964:140), for instance, was cognizant that competition from other auto manufacturers, particularly Ford, influenced prices in those early days. Since competitive prices could or could not result in a desired return, results had to be planned and evaluated in the light of competition. Furthermore, other uncontrollable factors, such as the general level of business activity and supplies of raw materials and parts, had to be planned for in attempting to achieve a desired return for the stockholder.

The GM planners attempted to protect the stockholders against such exogenous factors by selecting the firm's controllable variables so that the remaining uncontrollable variables could not damage performance and, in turn, the clients' well-being. Two underlying themes shaped their work. First, scarce capital resources had to be allocated to gain control of those variables that were most critical to overall ROI performance. Second, equally scarce decision-making capacity had to be assigned to set the variables included within the controlled boundary; should the requisite choice capacity be unavailable, performance deteriorated.

More specifically, GM's designers needed to decide what products and services would be purchased from external sources and what products and services would be generated internally for subsequent exchange with the larger economy. These strategic boundary-location decisions fell into two categories: channel integration and product diversification.

SETTING THE CHANNEL-INTEGRATION POLICY

GM's allocation of capital and management resources within the automobile production-distribution channel will be discussed in terms of: (a) the production of basic raw materials, such as steel, glass, and rubber; (b) the fabrication of auto parts, such as bodies, starters, and wheels; (c) the assembly and marketing of finished automobiles out of raw materials and parts; (d) the provision of credit for the sale of autos; and (e) the distribution of the products to retail customers.

Production of Basic Raw Materials

The Sloan-Brown designers, unlike Ford, decided that little performance protection could be bought by GM's entry into the production of basic auto materials such as steel, glass, and rubber. There were several reasons for not investing in these primary production processes. First, since there were usually no shortages with respect to such basic commodities, long-term contracts could be written quite readily to obtain the requisite quantities at low prices. As

Crowther (1920), a student of the auto industry and an observer of Henry Ford, wrote about GM's designers: "What they need most of is steel, and, with the world to choose from, one is in normal times, given the money, always able to buy steel" (pp. 616–617). With the raw material of glass, the GM planners faced a slightly different situation. In acquiring Fisher Body Corportion in 1926 GM became involved in glass manufacture since Fisher owned National Plate Glass Company, but by 1931 GM was negotiating its sale with both Pittsburgh Plate Glass and Libbey-Owens-Ford Glass, contingent upon an agreement "which would assure a supply of glass at a favorable price." In July of that year "such an agreement was consummated with Libbey-Owens-Ford" (U.S. Senate, 1956:31). With abundant supplies and tight contracts, precious little environmental variety resulting from variations in price or quantity of raw materials could filter through to injure stockholders.

If environmental shocks did hit GM, the planners attempted to soften the blow. When the price of natural rubber skyrocketed in 1925, for instance, Sloan authorized some preliminary research into the potential of artificial rubber (testimony of Sloan, *U.S.* v. *DuPont*, 1956:1323–1324). But when the price dropped, Sloan eventually reduced the funds for this basic research. While the investment of capital could be used to stabilize prices and flows, it was always a strategy to be avoided wherever possible.

Further reasons for avoiding capital investments in basic commodity production stemmed from the *standardized nature* of the items and from the *large number of producers* for each good. Additional protection against supply fluctuations or high prices could be had by using several suppliers; should one fluctuate because of fire, strike, price gouging, or the like, other sources could be tapped immediately to stabilize the flow or price. To formalize this policy, Sloan had his General Purchasing Committee establish a two-source-of-supply policy in the early 1920s. The policy required multiple sources for both parts and basic commodities. GM's purchases of carbon-steel bars in 1923, for example, were divided as follows:

PERCENT OF TOTAL REQUIREMENTS	TONS	SUPPLIER
70	105,000	Carnegie Steel Co.
15	22,500	Cambria Steel Co.
15	22,500	Open to Divisions

(GMX 159, *U.S.* v. *DuPont*, 1956:7108)

The final and by far the most important reason for not investing capital in the primary commodity products was the concomitant investment of executive talent. To control the vast number of variables added to GM's decision set even by an investment in only one industry, e.g. steel, the GM designers would have had to divert great amounts of decision capacity from the vital finished-auto divisions. The diversion would be particularly massive because GM, while it had automobile men, did not have steel men. Moreover, it did not want them, since they would do little to improve performance.

Accordingly, GM's planners "quickly voted against a policy of backwards integration beyond parts and accessories to the control of basic supplies" (Chandler and Salsbury, 1971:514). And GM became firmly entrenched in lines that related directly to the construction of the car, truck, and tractor, but did not invest in primary industries (*GM Annual Report,* 1920:8). Many years later Sloan (1964) still could write that GM "does not engage in the production of raw materials, as do some of its competitors, and we purchase a large portion of the items that go into our end products, because there is no reason to believe that by producing them we could obtain better products or service, or a lower price" (p. 433).

Fabrication of Automobile Parts

In definite contrast to the Finance Committee's reluctance to invest in basic industries, GM included a sizable number of specialized-parts units within the organization. First, GM's designers decided to retain nearly all the parts and accessories divisions collected by Durant because Pierre duPont wanted the supply of all *unique* items essential for assembling an automobile to be assured (Chandler and Salsbury, 1971:514). Second, GM added to its holdings (along with other less important units) a manufacturer of highly specialized major components: the Fisher Body Corporation.

GM planners were willing to include parts and accessories divisions within the firm's controlled domain, for (a) they were needed to safeguard the production of the finished-auto divisions, (b) they added little troublesome variety, (c) they helped smooth GM's overall performance, and (d) they could be evaluated easily. The Finance Committee willingly invested in the specialized-parts-producing units because the corporation had made a massive commitment to finished-auto production and distribution. Every finished-auto division, regardless of price, was committed to the mass production of automobiles. Furthermore, to maximize return on capital employed, in-process inventories were pared to the bone. Yet, "every piece of the motorcar is essential in the sense that the automobile is not complete unless every part is available. Delay in delivery of any part stops the work." To miss delivery was thus tantamount to a sin for the supplier and a calamity for the buyer because production stopped and capital idled (Sloan, 1941:48–49). As Sloan mentioned, when describing his days as the supplier of Hyatt Roller Bearings, "Eventually we kept two men in Buffalo, just to make absolutely sure no cars of Hyatt freight could go astray" (p. 58).

Similarly, uninterrupted supply became so important at Buick that its traffic men went "to the junction yards and transfer points several hundred miles away to trace through cars of material needed by one of the factories" (Buick, 1925:41). Thus the desire for channel integration into specialized-parts production was not merely to guarantee price, which long-term contracts could arrange, but to insure an item's future delivery (White, 1971:78). In short, "the parts manufacturer that you operate yourself is more of an assured supply than an outside supply" (testimony of Sloan, Hearings, Part 7, 1955:3526). Pierre duPont felt that this "insurance"—which also yielded cost information and helped control quality—"should provide enough capacity to produce at least $1/3$ of the parts and accessories required for the expanded automobile-making capacity" (Chandler and Salsbury, 1971:551).

The most important venture extending GM's controlled boundary was the Finance Committee's decision to secure full control of the Fisher Body Corporation in 1926. Durant had originally purchased 60 percent of the Fisher stock, but he failed to gain total control because of a voting-trust agreement that permitted the Fishers complete managerial prerogatives.

The least of the Fisher problems for the GM planners was an immensely lucrative "cost plus 17.6 per cent" (Seltzer, 1928:218) contract granted to the Fishers by Durant. That contract, Sloan testified, "became burdensome." GM's increased sales and turnover "resulted in cost and selling prices that were no longer competitive" but could not be adjusted "because we always had to respect the forty percent outstanding interests" (*U.S.* v. *DuPont*, 1956:1233).

The Fishers' other managerial freedoms caused the GM designers the most severe worries. The closed body for automobiles increased rapidly in popularity from 10 percent in 1919 to 85 percent in 1927 since it made the automobile an "all-year-round vehicle" (Sloan, 1964:160). Unfortunately for GM's planners, the independent Fishers retained contractual freedom to sell their closed bodies to other manufacturers—most notably, Chrysler's young but growing firm (testimony of Lawrence P. Fisher, *U.S.* v. *DuPont*, 1956:579).

Worse yet, the Fishers remained reluctant to increase production capacity to meet GM's mounting demands. In particular, with Chevrolet's growing popularity during the 1920s, GM wanted to construct "throughout the country assembly plants, because it was an economical method of production" (testimony of Sloan, *U.S.* v. *DuPont*, 1956:1234). "Where we had a chassis assembly plant," Sloan continued, "we had to have a Fisher Body assembly plant, but the Fisher Body Corporation was unwilling to put in an investment in these assembly plants. That handicapped us considerably."

Sloan was intensely concerned, for Chevrolet's use of the closed body was potentially the decisive weapon in the fight against the outmoded open-body FM cars. Specifically, "with its light chassis," the Model T "was unsuited to the heavier closed body, and so in less than two years the closed body made the already obsolescing design of the Model T noncompetitive as an engineering design" (Sloan, 1964:162).

By 1926 Sloan was also pushing hard for greater integration within the GM family. Coordination was particularly important between the Fisher factories and the auto-division assembly facilities since the Fishers fabricated such a significant component. Moreover, with the advent of the yearly styling change, coordination was required on a continuing basis. Ongoing control and coordination on an in-house basis would be easier than the arm's-length transactions that took place between more-independent companies (White, 1971:78). GM "had to have an integrated operation" (testimony of Sloan, *U.S.* v. *DuPont,* 1956:1233).

Sloan's participation in the Fisher acquisition "was to impress upon Mr. Raskob and Mr. Brown, who were not on the operating side, the importance of getting the job done as soon as possible" (p. 1234). "It was of vital importance to General Motors that the consolidation take place" (p. 1233). Pierre duPont and his Finance Committee, through a long series of negotiations (detailed by Chandler and Salsbury, 1971:575–578), finally succeeded in purchasing the last 40 percent of Fisher Body stock. GM now had complete control over the critical closed-body production. To cement this relationship and to integrate Fisher operations with GM's overall operations, Sloan moved several of the brothers into other corporate and divisional positions.

Sloan (1941) aptly summarized such GM channel-integration acquisitions: "Bit by bit, we were able to see a constant evolution bringing the manufacture of the motorcar itself and the manufacture of its component parts into a closer corporate relationship. All were to cohere as if drawn together by some magnetic force" (p. 44).

A further reason for retaining the parts suppliers was that they added little variety for which there was not already adequate decision-making capacity. The technology of mass producing an automobile's component parts was exactly the same as that of producing the finished automobile itself. Moreover, many of the designers, men like Sloan and Mott, had originally run these firms very profitably, and for the most part they were extremely well-designed units. Sloan's old Hyatt Roller Bearing and United Motors Corporation were models of efficiency and, furthermore, sources of well-rounded executive talent (i.e. managerial variety) for the larger and more important auto divisions. Often "companies manufacturing parts," as Sloan (1941) himself explained, "grew out of enterprises engaged in business long before there were automobiles." Their "years of experience and successful operation" gave them managements "superior to that of their big customers." Having faced sharp competition, for instance, their purchasing agents "were keen and aggressive," exacting "the last ounce of flesh." But whereas there was "urgent pressure for efficiency in parts manufacture," the automobile manufacturers' "main problem was how to produce enough. It is not surprising, therefore, that as integration continued, the experience and ability of the former parts executives brought them to the top" (pp. 101-102).

Still another—and perhaps an initially unplanned—advantage of GM's possessing its own parts-manufacturing units was the moderate performance smoothing the (replacement) parts units contributed to the firm. Along these lines, Colston (1939) reported: "In 1932, when auto sales were down 75 percent from 1929 and resulted in a thumping net loss, sales of parts and accessories were off only 47 percent and yielded a profit equal to 15.7 percent of gross. Indeed, it was the company's accessory business that was largely responsible for keeping overall operations out of the red that year" (p. 609). The inverse relationship between automobile and parts profits stems from the fact that especially in depressed times replacement parts become a substitute for new cars.

The final reason the GM designers retained practically all of the parts divisions accumulated by Durant was that their benefits to the stockholders could now be evaluated economically via the designers' newly instituted managerial methods. Pierre duPont, Sloan, and several of the other designers "argued that the new organizational structure and financial controls would make . . . a separation unnecessary and the bankers" on the Finance Committee "quickly agreed" (Chandler and Salsbury, 1971:514). Inefficient operations could not now be hidden as they were under Durant. The contributions these internal suppliers made to the other units could be priced and the returns evaluated without undue complication.

Along with the new uniform accounting procedures, GM's newly designed transfer-pricing scheme became the key to accurate evaluation of the parts suppliers. Sloan, at the time United Motors Corporation was integrated into the GM family of divisions, was particularly disturbed by Durant's failure to provide a method of transfer pricing between divisions. His concern culminated in a report written in 1919 (Chandler and Salsbury, 1971:473). Sloan found that if he followed GM's prevailing practice of transfer pricing, he would be unable "to determine the rate of return on investment for those accessory divisions individually or as a group" (Sloan, 1964:48); for in late 1918, material "was passing from one operating division to another at cost, or at cost plus some predetermined percentage." In Sloan's United Motors Corporation, his divisions had sold "at the market price" to both outside customers and other divisions. And as head of one of GM's few truly profitable units, Sloan did not want his "operating results on interdivisional business swallowed up in the extra bookkeeping profits of some other division. It was a case of keeping the information clear." That is, when Sloan sold to GM as a "relative" outsider, he could judge the return performance of his yet unintegrated UMC and its various subdivisions. In contrast, after UMC was fully integrated into Durant's

GM, Sloan would lose his ability to evaluate and control his area of operations (p. 48).

At the corporate level too, "no one knew how much was being contributed—plus or minus—by each division to the common good of the corporation." Because no one "could prove where the efficiencies and inefficiencies lay, there was no objective basis for the allocation of new investment" (p. 48).

Like Sloan, Brown (1927) felt that pricing products between divisions was important and that *markets* should be used to arrive at the interdivisional prices. "Unless a true competitive situation is preserved, as to prices, there is no basis upon which the performance of the divisions can be measured. . . . Prices at which its [the division's] products are sold must be based upon actual competitive values" (pp. 8–10). Otherwise, the return-on-investment figures "would not reflect accurately the use of the resources of either division" (Chandler, 1966:174).

Not all the planners were convinced of the value of the Sloan-Brown approach to interdivisional billing. Pierre duPont, for one, initially wanted to treat the firm "as an enlarged factory. The producing division, therefore, should bill the receiving one at cost" (Chandler and Salsbury, 1971:493). Pierre—as the GM president—asked several important Financial Staff executives, notably Turner and Meyer Prenskey, the treasurer, together with Pratt to investigate the matter (p. 500). They agreed with Sloan and Brown that the market-referenced transfer prices were vital in evaluating divisional performances (p. 500) and for allocating investment dollars (Sloan, 1964:48).

The Turner, Prenskey, and Pratt committee added "careful definitions of the terms cost, profit, and capital employed," in part to aid Turner in "developing a uniform accounting system for all the divisions within the corporation" (Chandler and Salsbury, 1971:501). And the new transfer-pricing scheme was intimately intertwined with the previously mentioned uniform accounting methods, which also were directed toward improving performance commensurability.

The use of market-referenced transfer prices helped to promote performance commensurability among divisions because, instead of just being cost centers, the parts divisions became benefit/cost centers, as were the auto divisions that sold their products into evaluative markets. In addition, the prices charged to the auto divisions now reflected the costs of capital employed by the GM parts units; no longer would auto-division performances be inflated by inaccurately cheap parts prices.

Providing market-referenced transfer prices between divisions also permitted divisional and corporate performance to be evaluated on the same scale. The interconnected markets surrounding both the divisions and the corporation provided the common value scale. Thus resources could be allocated as efficiently within the firm as they were to it.

In sum, GM used the parts markets to value intraorganizational transfers in a simple and straightforward manner. Had the evaluative markets not existed the resultant negotiation process would have siphoned managerial time, not just from the interacting divisions but also from the corporate headquarters. Furthermore, this easy evaluation demanded less direct supervision of the parts divisions by the corporate headquarters. "The stimulating effect of competition," Sloan (1944) stressed, "more than offset . . . a tendency toward complacency leading to decay" (p. 8).

To make interdivisional transfers economical, the designers had to end the wasteful use of cash in interdivisional transfer payments, which had prevailed under Durant. "Formerly payments were made by check, forwarded through the mails. . . . Clearance took additional time, and, as a result, four or five days were consumed on each check transaction. More accurately four or five days were wasted, to say nothing of the added expense of accounting and other services in handling the transactions" (Swayne, 1924:22). As a consequence, divisional cash balances had to be maintained to cover such payments; and while the payments were in transit, the covering balances could not be used for more lucrative capital investments. To correct this obvious waste of cash resources, the GM designers established a new *internal* procedure in which "the settlement of interplant balances by check through the mails was discontinued in favor of a credit slip system" (p. 22). "We set up an intracorporation settlement procedure," as Sloan (1964) put it, "under which the Financial Staff at headquarters acted as a clearing house for the settlement of interdivisional claims and payments" (p. 123). In this way intercompany remittances were minimized, clerical labor saved, and funds in transit reduced (Swayne, 1924:22).

The vital importance of the designers' new transfer-pricing and transfer-payment programs is indicated by the immensity of GM's interdivisional sales. In 1928, for instance, such sales amounted to more than $310 million or 21.0 percent of total sales. By 1937, figures had jumped to more than $387 million or 24.1 percent (USFTC, 1939:527-528).

By no means did GM attempt to supply internally *every* part needed for its cars. First of all, such a policy would have absorbed massive capital investments, diverted at least some decision capacity from the more important auto-assembly and marketing activities, and eliminated the markets needed for transfer-price setting. Consequently, the finished auto divisions were required to purchase at least some internally available material outside.

Besides requiring outside purchases of the divisions, GM also promoted the development of the parts markets via its parts *standardization* and *publication* work. The GM General Purchasing Committee provided much of this effort. Prior to Sloan's creation of the General Purchasing Committee, the Factory Division of the Advisory Staff published a *Book of Standard Parts* "containing 196 pages descriptive of standard parts, 100 pages on materials, and about 50 pages of miscellaneous information" (Baird, 1923:336). (For an example of these specifications, see Baird, 1923:335.) Probably to encourage the development of competitive parts markets, GM was, as one executive wrote, "very generous in giving out copies of the volume to all who have requested it, so that there [were] some 2,000 copies in existence" (p. 336). Further proof of GM's willingness to disseminate such information comes from the National Industrial Conference Board's (1929:276-277) list of manufacturers doing organized work on standards. Of the sixty firms listed by the Board, fifty-eight used nonpublished and sometimes confidential standards; only the Baldwin Locomotive Works and GM published their standards. The overt actions of the GM designers fostered the creation of the very markets they needed to evaluate the firm's components as well as to supply its remaining needs.

Could GM's designers expect other firms to accept their specifications? It certainly seems so. For with hundreds of firms involved, the wisdom of uniformity could not be denied, neither by manufacturers, nor parts suppliers, nor automobile owners; to all it was desirable that "any spark plug would fit any engine, and that such elementary components as nuts and bolts should conform to generally accepted specifications" (Rae, 1965:39). Accordingly, "many manufacturers of commercial parts asked for and received permission to adopt the [GM] system for numbering their own stocks and published new catalogues in which they identified their goods by this numbering plan. The Maxwell Motor Company . . . copied the system, verbatim in part, while other automobile manufacturers [used] some of its features" (Baird, 1923:336).

As is implied above, GM's parts-standardization work was a cooperative undertaking, carried on by GM, other parts suppliers and automobile manufacturers, and the Society of Automotive Engineers (Baird, 1923:336). Consequently, the Society of Automotive Engineers (SAE) became an extremely important *interfirm* decision-making body. SAE activities "eliminated many of the evils of production at that time, resulting in the standardization of spark-plug sizes, tire sizes, wheel sizes, and the like" (Sinsabaugh, 1940:164). Eventually, the SAE *Handbook* made stringent engineering specifications, i.e., variety-reducing standards, available to all car designers and purchasing agents (Epstein, 1927a:170). So a second important contribution—discussed more fully in Chapter 7—of the GM standardization effort was that it quite literally reduced the environmental variety GM faced on the parts and accessories border of its controlled domain. GM had erected what Ashby calls "walls of constancy."

In GM's case the environmental variety reduction obtained via parts standardization was particularly significant. The SAE standardization program conformed closely to the pattern GM preferred; for as early as 1915 Karl W. Zimmerschied, then a GM vice president, was chairman of the SAE Standards Committee on interchangeable parts (Barnes, 1921:208). In January, 1918, Charles Kettering, GM's famous researcher, had been elected president of the SAE. By 1923, more importantly, GM's staff held seats on sixteen of the twenty-one SAE committees dealing with motor vehicles, and five of the Standards Committee chairmen or vice chairmen had come from GM (Thompson, 1954:13). Thus the GM planners could expect a close correspondence between SAE and GM standards. And "because the automotive industry was the major consumer of industrial products, the work of the SAE automatically extended the drive for standards into many other industries" (Noble, 1977:79). In this way, much of the environmental variety on GM's input interface came to match its internal variety.

One could say simply that GM extended its boundaries to control, if only partially, the auto industry's standardization practices and parts specifications. That many GM managers occupied important SAE decision-making posts corroborates this view, for GM executive talent needed to make sure these so-called industrywide decision variables were set to the firm's advantage.

This effort was worthwhile since the existence of precise parts and material specifications greatly eased GM's decision-making problem. Because of product homogeneity, suppliers to GM had only one dimension on which to compete: price. Consequently, GM had a very simple managerial problem. There were no complex calculations, nor large number of variables to negotiate; it had only to choose the lowest bidders and obtain the necessary quantities.

GM's parts-standardization work also gave the firm multiple input sources to decouple it from its external supply environment and to separate GM's parts and accessories divisions from the finished-auto divisions. Without industrywide and interdivisional standardization "of carburetors and ignition apparatus," for example, "major alterations in engine construction were required to change sources of supply in these accessories" (Bachman, 1921:356). With the multiple-source rule operating here too, the standardized parts markets could be treated in much the same way as the low-variety basic industries.

The last advantage of GM's market-oriented standardization effort was that it improved GM's bargaining position in the supply markets. Prior to standardization, each division bought "specialized parts in small lots at a long price," sometimes even insisting "on certain dimensions for such items as bolt heads which necessitated the rolling of special bars at the mills" and, of course, a premium price. Moreover, since "each division listed and numbered its parts according to its own system," GM bought "as many as fifteen or twenty identical articles in small quantities, under different parts numbers," when it could have bought "them all in one large order, under one . . . part number" (Baird, 1923:334). It was important, then, to standardize parts and materials because specialized components commanded premium prices whereas standardized items sold very close to cost when purchased in GM's quantities. Thus with standardization GM could exert a degree of monopsony control over the many price variables on its input boundary.

The existence of GM's parts and accessories divisions as well as its increasing volume further increased the firm's control over its input markets. "Should it be necessary to press matters, General Motors [had] the possibility of supplying a material or component to itself. The option of eliminating a market is an important source of power for controlling it" (Galbraith, 1967:29). GM's extreme market power allowed its corporate Purchase Section to write "requirements contracts" in which the supplier agreed to furnish whatever amount the firm needed of a certain material or component. Accordingly, GM did not need to absorb unwanted inventories during downturns nor chase additional supplies during upturns. In short, its variety load was minimized.

Assembling and Marketing of Finished Automobiles

With many external supply sources available, GM's parts and accessories division took a back seat to the auto-production divisions when the clients' funds were allocated internally. Priority was "given to the car-making divisions, where as high as 25% return on investment could be expected" (Chandler and Salsbury, 1971:551). "Expansion must be limited within the resources of capital," as Brown (1924a) explained early on, "and in a corporation such as General Motors, the available capital must be directed into channels that hold the greatest promise" (p. 196). Finished automobiles garnered their high returns simply because they did not need to be treated as standardized goods but could be given a differential appeal. "Capital employed in the production and sale of a product that is bought on the basis of exact specifications, and in which no highly specialized knowledge is required for manufacture and distribution, contents itself with a relatively low rate of return" (Brown, 1957:61). To avoid this fate, GM would invest considerable capital in differentiating its finished automobiles—as will be shown shortly in the product-diversification section of this chapter.

The extreme importance that both Sloan and Pierre duPont attached to the auto divisions meant that these received priority not only in capital allocation but also in executive attention, which influenced how well capital was converted to profit. Sloan as GM's president was particularly lavish in devoting his talents to the auto divisions. Since the auto divisions were the focal point of attention for the corporate headquarters, investments in other GM activities were made to insure that these central auto divisions functioned unencumbered by such environmental loads as parts shortages, excessive prices, or poor quality, which would damage their conversion of resources into benefits for the stockholders.

Complete protection of the auto divisions by the designers was impossible, of course, for some variables strayed from their grasp. Organized labor demonstrated this to the designers by initiating on April 23, 1935, "the most significant work stoppage in the history of GM to that date" (Fine, 1969:48). Not surprisingly, the workers seized "one of the most important links in the GM chain of shops at that time since it was the corporation's sole producer of Chevrolet transmissions" (p. 48). The GM designers strongly opposed union activity "because it threatened to circumscribe the customary prerogatives of management" (p. 29). (See Chapter 11 for further details.) In the end, having been dealt a defeat in this arena, GM's planners increased the firm's investments in buffer inventories and dispersed its production facilities ("Auto Industry Decentralizing . . . ," *Business Week*, April 11, 1936:28-29).

Provision of Credit

Environmental hazards existed for the central auto divisions on both sides of their position in the marketing channel. Not only could parts suppliers or organized labor paralyze production, but in addition banks could withhold the credit needed to purchase automobiles. Because of their basic conservatism and distrust of the automobile and its industry, many bankers refused to grant credit for the purchase of automobiles (Sloan, 1964:302). Credit was withheld from automobile dealers, who required it to carry their very expensive inventories; similarly, credit was not granted to the dealers' customers who needed it to spread the cost of this relatively expensive consumer durable. The consumer-loan problem was particularly galling to GM's duPont-dominated Finance Committee, inasmuch as readily available credit could be used to reduce the initial cost of GM cars in the battle against the cheap FM Model T. "It was reasonable to assume that consumers," as Sloan put it in 1925, "would lift their sights to higher levels of quality. Installment selling, we thought, would stimulate this trend" (Rothschild, 1973:83). The solution was General Motors Acceptance Corporation (or GMAC).

John J. Raskob, then chairman of GM's Finance Committee, played an instrumental part in starting GMAC (Sloan, 1964:303). He concluded that sales could be increased greatly "by the development of the installment method, and especially by the creation of a finance company . . . in close touch with the parent manufacturing company" and limited "to financing the paper of those who dealt in any of [its] cars" (Seligman, 1927a:48). Accordingly, GMAC was formed in early 1919; its first president was a DuPont executive, J. Amory Haskell, and its next head, C. C. Cooper, had been a DuPont legal advisor. The primary activity of GMAC became the extending of credit to dealers for carrying their inventories and to consumers for covering their car purchases, simultaneously earning GM a return on its investment.

In financing sales to dealers the GMAC investment was safeguarded by the corporation's retention of title to the financed cars; if, as Sloan (1964) put it, the dealer failed "to pay his obligation upon demand, or to comply with other agreed-upon terms and conditions," GMAC had "the right to take back the product" (p. 305). In 1921 the financing charge to dealers was one percent a month, a good return even after costs were deducted, for there was very little chance for loss. Similarly, in financing sales to consumers GMAC was protected by the ability to reject any customer contract a dealer offered to it and by the requirement that the dealer guarantee the customer's payment to GMAC (Getz, 1924:871; Grimes, 1926:36; Sloan, 1964:308). In short, the control burden was shifted to the dealers (Pound, 1934:385-386). Under such a default-proof procedure Sloan (1964) could boast: "In the case of GMAC, the retail loss on installment paper from 1919 to 1929 was approximately one third of 1 per cent of the retail volume purchased." Speaking of the GMAC ratio, not of dealer losses after a car had been repossessed, he added that "in the worst of the depression the rate of loss never reached 1 percent of volume—a remarkable indication of the safety of the system" (p. 304). Through the careful selection of risks and through the safety factors built into the operation, GMAC was able to ensure a low-loss ratio and a high rate of return on money invested in financing sales.

GMAC's importance in increasing auto sales is shown in the fact that installment sales rose from 29.2 percent in 1920 to 55.9 in 1926 (Filipetti, 1927:425). These increased credit purchases were concentrated in the important low-price class, for "the higher the price of the car, the smaller [was] the percentage of cars bought on installment" (p. 426).

After its investment in GMAC, GM could reset the critical demand-stimulating variables that, when under environmental (specifically, the bankers') control, had been set at values damaging to the mass-production-oriented auto divisions. Thus the planners widened GM's umbrella, i.e. the controllable set, to include more variables, in particular variables that exerted a strong force on how well the firm's resources generated benefits. Including these new variables within the controlled domain did not adversely affect the internal variety balance, as the duPonts had provided GM with an already strong Financial Staff easily capable of managing GMAC and insuring a substantial investment return. In addition, earnings from its financing activities—like the income from its parts and accessories units—helped GM show an overall profit during the 1930s when even Chevrolet's comparatively strong showing could not compensate for the other auto divisions' losses (Cray, 1980:268-269).

Distribution of Automobiles to Retail Customers

The last link in the GM marketing chain was, of course, the distribution of automobiles to the retail buyers. While the firm willingly extended inventory credit to its dealers, the Finance Committee avoided investing significant amounts of capital in the direct ownership of retail outlets. Limited capital-equipment loans were made, however, to carefully selected dealers via General Motors Holding Corporation (Pound, 1934:376); in 1930 this amounted to "$1 million invested in . . . 50 dealers" ("G.M. Will Find Capital," *Business Week*, July 2, 1930:12). This unit was starting to fill a few critical, uncovered territories "at a time when local bankers, awed by the saturation of the market, were advising ambitious garage helpers to stay out of the dealer business" ("G.M. III," *Fortune*, 1939:110). So even though GM's planners wished to avoid direct marketing, "the several divisions had not been able to attract to these open points individual dealers willing to risk the necessary capital" (Brown, 1957:56). Naturally, to protect its investment GM's Financial Staff exercised "a considerable measure of control" (Pound, 1934:377) over the dealers in the program.

Three important reasons existed for limiting the investment of GM capital at this juncture in the production-distribution chain. First, the *amount of capital needed* for the twenty thousand retail outlets would have been astronomical, for each dealer made "a substantial investment in facilities and organization" (Sloan, 1964:281). Accordingly, "such a course was unthinkable. . . . The capital requirements would be of unmanageable proportions" (Brown, 1957:55).

Even if GM's influential Finance Committee might have raised the capital, a second and more compelling reason prevented them from extending GM's investment into the retail environment: The *high variety* encountered in the dealers' worlds could not be handled by the firm's executives. GM's many dealers were "scattered over the entire world" and employed "over 1,000,000 individuals" (Sloan, 1929:92). Moreover, at first thousands—and eventually millions—of trading relationships would have overloaded the firm's information-handling and decision-making capacity, thereby damaging the return on all capital. Sloan (1964), for one, felt that it would have been very difficult for automobile manufacturers "to merchandise their own product," especially after the early 1920s when the many used car trade-ins made merchandising automobiles "more a trading . . . than an ordinary selling proposition." To organize and supervise "the necessary thousands of complex trading institutions would have been difficult," for "trading" did not "fit into the conventional type of a managerially controlled scheme of organization" (p. 282). Fine-tuning the price and product to the individual customer, accordingly, was better left to the local retailer in the community.

A third and final reason for GM's avoiding direct dealership investments was that, where needed, *control could be exerted without direct ownership* via the franchise agreement. While the GM designers and decision makers could not handle the full variety of the dealers' transactions, important dimensions of the channel-emptying retail business did concern them. The one-sided economic advantage GM enjoyed over its dealers permitted the corporation to dictate critical variable settings through obligations specified in the franchise contract. Thus, GM capital did not have to be risked to specify the dealers' accounting procedures, the information they would supply to GM, the repair equipment they would purchase, and the like. "In no retail trade" was "the dealer so closely controlled" (Pound, 1934:370). Consequently, GM gained control over those variables which were vital to its own performance without investing capital—capital that would be endangered because its control was beyond the firm's managerial capacity.

SETTING THE PRODUCT-DIVERSIFICATION POLICY

Added shock-absorbing capacity was built into GM's performance-control system through finished-product diversification. Since attempts to diversify add great amounts of internal variety quickly, designers soon encounter their organization's control limits. In trade-off terms, performance protection resulting from diversification must be balanced against the performance safeguards lost because of poorer internal control.

The Sloan-Brown group, as conservative planners and decision makers, tipped the scales away from excessive finished-product diversification. Yet they had GM ingest adequate variety to protect its stockholders against the remaining uncontrolled variables. Sloan termed this controlled-domain boundary-extension activity "broadening the profit base" ("General Motors," *Fortune*, 1938:167). The GM designers' diversification strategy addressed (a) the domestic markets, (b) the foreign markets, and (c) the allied industries.

Diversification in the Domestic Markets

Sloan and his fellow designers realized only too well that the early automobile business was dangerous and uncertain. Financial resources were always a problem for the capital-starved industry, and even a firm with a good product could falter because of a single year's economic downturn. Makes came and went rapidly. Consumer tastes were uncertain and no one knew exactly what new innovation might net their flighty sales dollar. Indeed, Epstein's exhaustive study of the automobile industry painted a depressing picture of a manufacturer's long-term chances for success: "*Mediocrity, once surmounted, is soon reverted to;*

incompetence surmounted is also soon reverted to; but leadership once lost, is hardly ever seized again" (1927b:285). More specifically, of the 181 automobile manufacturers operating between 1895 and 1926, only 36 percent "lasted 10 years or more" and only 19 percent "lasted 16 years or more" (1928:168). While automobile-manufacturing firms failed at an average rate of more than 8 percent per year from 1903 to 1927 (from a low of zero in 1907 to a high of 26 percent in 1910), general business failures during the same period ranged only from 0.04 percent to 1.1 percent annually (p. 177).

To cautious planners like GM's, such statistics obviously called for a diversified stable of entries that varied with the times and the economy. During the early 1920s GM's planners ended the price overlap among the firm's cars and spread them out to cover the full market. In the mid to late 1920s they added considerable variety to the product line as "the sale of higher-priced cars rose with national prosperity" (Sloan, 1964:438). In the early 1930s when "the demand reversed itself and became concentrated in the low-price area," GM's planners eliminated some of the middle-price makes added during the late 1920s and shifted those remaining toward the low-price market segment. As the economy revived during the late 1930s, GM began to shift its offering back to the middle- and high-price ranges.

The early phase of GM's automobile-diversification strategy clarified the firm's basic product policy. In short, GM needed "a product line that would make money" (Sloan, 1964:64). Sloan's solution resulted in a broad spectrum of differently priced makes. In 1921, from the least to the most expensive, these offerings were "Chevrolet, Oakland, a new Buick 4, Buick 6, Olds, and Cadillac" (p. 69). After 1921, "only the price-class positions of Chevrolet and Cadillac . . . were to be permanent." The inexpensive Chevrolet would offset its low profit margin with high volume, the expensive Cadillac would offset its low volume with high profit margins, and the makes in between would combine moderate margins and volumes. Sloan summarized the firm's new marketing policy in his now famous dictum: "a car for every purse and purpose" (p. 438).

Unlike Durant's naively diversified and haphazardly assembled group of divisions, Sloan's offerings had to complement each other to provide strong firm-wide performance protection. GM under Durant, as Sloan (1964) criticized, had "no established policy for the car lines as a whole" (p. 59). In particular, Chevrolet's price and quality were not "competitive with Ford" in the low-price area "where the big volume and substantial future growth lay." And "in the middle," GM's makes "were concentrated with duplication," taking "volume from each other" (p. 60).

To develop a more "rational policy," Sloan appointed an ad hoc pricing committee—described more fully in the next chapter. Within a month Sloan's special committee developed three recommendations that were adopted immediately: (1) GM should produce a car in each price class from the lowest to the highest that would still permit quantity production; (2) there should be no overlap in the price steps among divisions; yet (3) the price steps should not leave wide gaps between the offerings (Sloan, 1964:65). A bracketed price structure resulted that opened up a new classification at the low end and reduced GM's middle offerings from eight to four. Accordingly, in 1921 GM sold Sheridan, dissolved Scripps-Booth, and dropped the Chevrolet "FB." Sloan (1964:67) listed GM's remaining six passenger-car price ranges as follows: $450-$600, $600-$900, $900-$1200, $1200-$1700, $1700-$2500, and $2500-$3500.

The special pricing committee also urged that GM's cars be positioned at the top of the price ranges to attract sales from below and above that class (Sloan, 1964:67). By attracting customers from both market segments volume could be increased substantially, thereby reducing unit costs. Spreading fixed costs over additional units, in turn, justified the selling of higher-quality cars at prices lower than high-price competitors offered. Volume for each division was increased by reducing GM's internal competition and by attracting sales from GM's external competition.

GM's variety marketing policy was especially effective since it introduced a GM car into the extremely high-volume low-price field previously dominated by Ford's Model T. With the new GM price brackets, Chevrolet's price fell from $820 to $525 while the Model T sold for $355. On a comparable-equipment basis, Chevrolet and the Model T were separated in price by only $90. With the rising expectations and the mounting affluence of the 1920s, the $90 difference became insignificant to many consumers, especially now that most of them had a used-car trade-in and installment financing to help cover initial cost (Sloan, 1964:163). Furthermore, as Sloan outlined GM's strategy, "It was our intention to continue adding improvements and over a period of time to move down in price on the Model T" when the volume "justified it" (p. 154). The Sloan-Brown designers had launched Chevrolet on its way toward unseating Ford from his long-held position of industry dominance. Placing Chevrolet in the low-price class was key to GM's long-run success. Chevrolet's position was to prove essential during the 1930s depression environment when many mid-price cars suffered significant sales declines and profit losses.

Spreading its entrants over all price classes also aided GM's adaptation to environmental booms. During the expansionary 1920s GM was protected since it offered a price stairway which buyers could climb as their incomes and expectations rose, moving from a Chevrolet to an Oakland and so on. The firm was *almost* ready for the "*mass-class* market," with its "increasing diversity" (Sloan, 1964:150).

Work remained to be done, however, in fine-tuning GM's product strategy. In particular, the firm's cars often lacked "class," as it was then coming to be seen by consumers. These shortcomings became unmistakable during the 1924 recession when GM, especially Buick and Chevrolet, suffered a severe sales decline. Fortunately for Sloan and his colleagues, the 1924 recession was milder than the previous 1920–1921 downturn that toppled Durant. Not only was there time to recoup, but the disparate results of 1924 pointed the way to a successful future. The market was starting to turn.

Indeed, sizable gains by several manufacturers in 1924 so "astonished the industry" that they "were examined with care by rival producers" (Thomas, 1973:128). "The striking, countercyclical successes of Chrysler, Dodge, and Hudson had two elements in common: the introduction of new models to a declining market, and the concentration upon the production of closed automobiles." As early as 1919, Hudson (in conjunction with GM's own Fisher Body) had developed "a low-cost closed automobile body" (p. 129). Subsequent advances in large hydraulic presses and electric welding machines helped reduce costs further. By 1924 Hudson could offer the closed and open models of its Essex make at the same price (p. 129). Though Dodge espoused Ford's static-model policy, the firm—somewhat fortuitously—relented in 1924; furthermore, it foresaw the emerging trend to the closed body so "Dodge did not even offer an open model when it introduced its new car" (p. 130). The brand new Chrysler make combined a well-engineered power plant and four-wheel hydraulic brakes with a Fisher-built body design that was particularly pleasing.

The "aesthetic appeal" (p. 132) even appeared to work for cars lacking Chrysler's sophisticated engineering. "Auburn," for instance, "introduced new closed models featuring advanced styling and coloring. A sweeping belt line . . . often painted in contrasting colors . . . gave a striking impression." When "the red ink on Auburn's books" changed to "black for the 1924 season," there followed an "increased industry emphasis on the aesthetic appeal of its products."

The 1924 model year provided still another important lesson for the attentive automobile manufacturer intent upon "getting the facts." With *Saturday Evening Post* ads, like the famous "Somewhere West of Laramie," the Jordan Motor Car Company pushed its otherwise ordinary Playboy model. "Promoted as more than a utilitarian machine," the car became "an escape mechanism, a psychological crutch, and a status symbol." Jordan's advertisements "hit a responsive note": sales rose 61 percent in 1924 (pp. 133–134).

At GM, meanwhile, only Oldsmobile achieved a sales increase for 1924. Besides offering GM's newest and sportiest model, Oldsmobile undoubtedly benefited from its own Jordan-type ads. Of GM's other divisions, Oakland suffered the smallest decline. Though it offered a mechanically inferior car with dated styling, Oakland managed to hold most of its sales by pioneering the use of the new Duco colored lacquers. The Cadillac Division came next in the sales-decline ranking. It built an excellent car that had received technical improvements such as a balanced crankshaft for 1924; but its styling, though revamped, remained conservative and dated. Buick faced a similar situation. Although this excellent automobile featured 1924 refinements such as four-wheel (albeit mechanical) brakes and a substantial power increase, its styling "improvements" seemed copied from the Packard, especially the radiator and hoodline. By far, Chevrolet did the worst in 1924. Besides being poorly made, it had an obsolete engine and a drab, boxy body unchanged in years.

Only the Model T was more out-of-date. But because it was well made, reliable, and cheap, it remained a good buy for a recessionary year, and FM's 1924 sales slipped only slightly. So Henry Ford dismissed the early warning signals and persisted with the same old, open-bodied black Model Ts. Ford hobbled his firm further by not providing it with a price stairway of makes like GM's.

The most striking facet of the 1924 success stories was their concentration in the middle price range: even during a recession these larger cars, generally powered by six- and eight-cylinder engines, made advances; and "the medium price class remained the hub of the industry" (Thomas, 1973:127).

Sloan took the 1924 lessons seriously. GM's Finance Committee was displeased about the firm's dated products, lost volume, inventory excesses, and higher unit costs (see Chapter 10). So Sloan pushed GM's divisions relentlessly to improve the sales appeal of their products with new models, closed bodies, colored paints, styling refinements, as well as technical improvements.

During the 1924–1929 period, he also backed the development and introduction of attractive new makes to increase the variety of lines in the middle-price range where more and more buyers clustered with the growing affluence of the times. "The [price] list for the still-dominant [open-bodied] touring cars in 1924 was: Chevrolet, $510; Olds, $750; Oakland, $945; Buick 4, $965; Buick 6, $1295; and Cadillac, $2985" (Sloan, 1964:155). To fill the gaps, GM's planners added four cars to the product line: the Pontiac produced by the Oakland Division, the Viking by Oldsmobile, the Marquette by Buick, and the La Salle by Cadillac. In addition, the firm's planners extended the price range upward by backing Cadillac's development of its classic V-16 (see Hendry, 1983; Schneider, 1974). So at the end of 1929 GM offered no less than "ten distinct makes of passenger cars," when counting the Cadillac V-8 and V-16 separately, "with a total of 137 styles and body types, ranging in price from $495 to $9,700 at the factories" (GM Annual Report, 1929:8).

Of the four cars added to GM's line, the Pontiac and the La Salle filled the most important openings. "From the strategic standpoint . . . the most dangerous gap in the list was between the Chevrolet and the Olds." It could accommodate a "competitor against whom we then had no counter" (Sloan, 1964:155). Since it appeared to Sloan that the Essex might soon invade this potentially high-volume territory, he proposed that the new Pontiac car cover the field for

GM (p. 159). GM's planners thus would increase the firm's product variety and plug one gap in its line of offerings.

The other big opening existed between the Buick 6 and the Cadillac. GM's planners became especially concerned about this sector of the environmental front when Packard introduced a less expensive car in 1926 (Mandel, 1982:186). To defend this slot, Sloan "proposed that Cadillac study the possibility of making a family-type car to sell at about $2000, which eventually resulted in the famous La Salle car introduced in 1927" (Sloan, 1964:155).

Not only did the Pontiac and the La Salle help keep GM's product variety matched to the growing strength of the middle-price market segment, but they also offered an opportunity to experiment with newly emerging product concepts. "We designed our first Pontiac car," Sloan (1964) observed, "exclusively with closed bodies, a coupe and a coach" (p. 159).

Of even more importance to GM's strategic evolution was the fact that the proposed six-cylinder Pontiac was "designed in physical co-ordination with [the four-cylinder] Chevrolet so as to share Chevrolet's economies, and vice versa" (p. 155). Chevrolet thus would enjoy a lower unit cost to fight Ford's Model T as would the new car in meeting its own competition. Without this coordination the new car could have drawn "some volume away from Chevrolet, reducing its economies," and resulted in "a loss ... for both cars." In other words, the Pontiac and Chevrolet were to be so close in price that they might have hurt each other in the market place.

Since GM's planners realized "that the future favored sixes and eights," they developed the Pontiac design on "a Chevrolet chassis with a six-cylinder engine" (p. 156). Specifically, by copying much of the Chevrolet chassis, the Pontiac could be built using many "Chevrolet components, [manufacturing] plants, and assembly plants" (p. 157). The end product was to be "the lowest priced 6-cylinder car that [was] possible, constructed with Chevrolet parts" (p. 159). To facilitate coordination, "the proposed Pontiac was assembled and road tested at Chevrolet and then assigned back to Oakland with full responsibility to that division for its final development, production, and ultimate sale as a companion car to the Oakland" (p. 158). As early as 1925, then, "the Pontiac, co-ordinated in part with a car in another price class [the Chevrolet], was to demonstrate that mass production of automobiles could be reconciled with variety in product" (p. 158).

The Oakland experience also proved valuable when Chevrolet introduced its own six-cylinder engine in 1929. In stark contrast to FM's Model A changeover, which did not even involve a larger engine, Chevrolet's switch went smoothly.

The La Salle, for its part, further substantiated the value of GM's emerging coordinated-, or common-chassis policy. "Sharing the Cadillac chassis, as Pontiac did the Chevrolet, any La Salle sold was, in effect, a boost to Cadillac volume and profits" (Cray, 1980:252). (For full details on the extensiveness of the La Salle/Cadillac integration program, see Hendry, 1983:165-166.)

The La Salle made another important contribution to GM's rapidly developing product strategy: *styling.* "The car," according to Sloan (1964), "made a sensational debut in March 1927." Its introduction marked a turning point in "American automotive history" as "the first stylist's car to achieve success in mass production" (p. 269). The La Salle's creator was Harley Earl, a Los Angeles custom-body designer recruited by Lawrence Fisher to improve Cadillac styling. When the stylish Packards began eroding Cadillac's market in the mid-1920s, Fisher himself had been appointed division manager to upgrade Cadillac styling—long downplayed because of Henry Leland's early emphasis on engineering rather than appearance (Hendry, 1983:131). In developing the La Salle, Earl, with Fisher's backing, drew heavily from prominent European manufacturers and coachbuilders (Schneider, 1974:13). Deeply drawn "Flying Wing" fenders, rounded corners, reproportioned side windows, a new beltline molding, and other design details gave the La Salle a "unified appearance" (Sloan, 1964:289). Compared to the 1926 Buick sedan, "the La Salle looked longer and lower."

From then on Harley Earl's "primary purpose" was "to lengthen and lower the American automobile, at times in reality and always at least in appearance" (p. 274). "GM's stylists reasoned that the automobile was a machine of motion and should reveal that purpose with long, gentle and curving horizontal lines to suggest speed and power" (*General Motors: The First 75 Years,* 1983:54). As a secondary styling objective, Earl extended the body "in all directions in an attempt to cover some of the ugly projections and exposed parts of the chassis" (Sloan, 1964:274). GM's stylists, for example, covered gas tanks with "beaver tails," hid radiators behind grilles, and removed exterior sun visors.

GM's new Art and Color Section, under Earl's direction, confirmed earlier discoveries for the firm's planners. Sloan (1941) boasted: "In 1927 Cadillac caused a tremendous commotion with an announcement to the effect that customers might have a choice of 500 color and upholstery options. Today there are about a million possible options" (p. 183). (See also Cadillac Motor Car Company, 1927.) So besides price, make, and model variety, GM would present consumers with a multifarious choice of styles, colors, upholstery, and equipment. "The market made it clear that appearance was selling cars," Sloan (1964) put it. "Chrysler was getting good results with color, and so were we" (p. 272).

The growing popularity of Sloan's variety marketing policy devastated Ford, who liked to think in terms of his black Model T. Shortly after the Model T's demise, in fact, Sloan (1964) wrote to William A. Fisher, then president of Fisher Body Corporation, that GM's future rested on "the attractiveness" of its bodies with respect to their "luxury of appointment," their eye-pleasing qualities, and their distinctiveness from the "competition" (p. 272). Product differentiation, as Sloan undoubtedly knew, had to be achieved since a product which *appeared* standardized would be bought and sold as just another basic commodity. Consequently, its manufacturer could not demand a premium price. Better yet for GM, consumers increasingly shunned standardized products like Ford's all-black Model Ts. Instead, they wanted vehicles that differentiated them from their neighbors in terms of taste, status, and income. They willingly paid for these differentiated products, for ability to pay attested to one's position in the social hierarchy. And while "it is easy enough now to laugh at the business of gussying-up divisional body architecture into dreams measurable in dollars," it "meant the odd Joe on the street had his life as a consumer enhanced" (Bayley, 1983:120).

Beyond offering a wide variety of makes, models, colors, interiors, and equipment, the GM designers increased variety further by presenting the consumer with *restyled* models each year. "It should be emphasized," at the outset, "that the establishment of annual model changes represented an increase in the rate of model obsolescence, not the establishment of a new practice" (Thomas, 1973:134). Furthermore, whereas "it is often argued that GM's annual model changes revealed cold-blooded manipulation of the public," the annually restyled model arguably evolved from "a symbiotic relationship between business and consumers" (Meikle, 1979:14). A suddenly taste-conscious and affluent public set the pace in the middle and late 1920s (see Chapter 15). They preferred the new models over the old. It is not surprising, then, that "the 1925 season witnessed the introduction of a wave of new closed models as rival firms attempted to imitate" the 1924 successes of "Chrysler, Dodge, and Hudson" (Thomas, 1973:130–131). Moreover, "the fact that new models generally did better during their first year than during their second was not likely to be lost on the industry as a whole" (p. 136). Engineering-oriented businessmen like Sloan did their best to keep abreast of the consumers' rapidly changing desires. *GM's evolution of the annual model policy allowed the firm to regularize or routinize change so that it became manageable within a corporate administrative structure as a simple repetitive process.*

Throughout the late 1920s Sloan's biggest managerial problems were to convince the Finance Committee to reinvest capital for GM's perpetual retooling and to cajole the divisional managements to change their products annually. Few of these designers and decision makers were styling-oriented.

Even Sloan's top salesman, R. H. Grant of Chevrolet, opposed the annual model concept (Sloan, 1964:165–167). He much preferred the continuous and unannounced introduction of changes. The divisional engineers, in particular, resisted change because it disrupted production. By no means was their resistance to innovation limited to styling and appearance matters, for they often opposed more significant product improvements. After all, any innovation meant increased work, and more importantly, lowered productivity.

To promote orderly change, GM's planners adopted the use of general-purposed machinery wherever possible. Hence they avoided Ford's stagnation-inducing policy of using highly specialized tools and facilities. Kettering, in particular, opposed special-purpose machinery since he felt that the divisional production people used its existence as an excuse for avoiding or delaying his innovations. "We now avoid gigantic, costly special machines," as he put it. "We prefer to get the same result by a combination of mechanisms which can, at any time, be redistributed to other jobs without loss" (Kettering, 1928:447).

Another internal managerial problem confronting Sloan stemmed from his multiplicity of models. GM's increased offerings had to take volume from each other. To keep costs manageable, it became necessary to standardize additional parts across the divisions. Standardization across divisions, however, restricted the individual engineer's freedom "to inject his own personality and ideas into the picture" (Sloan, 1964:156–157). As Sloan had to admit, "The more you co-ordinate, the more questions you draw up into the [corporate] policy area" (p. 181). Centralization of engineering and styling eventually provided a solution.

Once Sloan had solved these internal administrative problems, GM's planners realized they had unearthed a vast variety reserve from which to draw an endless stream of styling changes to attract consumers (Sloan, 1964:274). With no intention either to condone or excuse GM's subsequent actions, one can argue: "If the business sector later grew proficient in manipulating style trends, consumers in part brought the curse of planned obsolescence down on themselves" (Meikle, 1979:14). Not surprisingly, GM soon mastered the strategy appropriate for the emerging environmental situation. As Sloan (1941) himself eventually acknowledged, "We want to make available to you, as rapidly as we can, the most advanced knowledge and practice in the building of motorcars; we want to make you dissatisfied with your current car so you will buy a new one" (p. 177). "General Motors in fact had annual models in the twenties, every year after 1923" (Sloan, 1964:167). "With this concept went the need for salesmanship" (p. 163).

Creating the impression, or image, of bigger and better cars required "salesmanship," for the annual model changes, more and more, would become only as skin deep as the interdivisional differences would be. Underneath, the cars came to be essentially the same year after year. Throughout the industry, mechanical improvements were slower in coming as time passed and, at GM, were introduced only when they promised a sales advantage and increased profits.

Sloan, however, did not shun incorporating mechanical changes into GM's restyled models when he knew consumers would pay for them. As a formally trained engineer, he held abiding faith in scientific research and engineering development. More importantly, he backed his belief with GM's investment dollars, and he granted the very able Kettering almost complete freedom in conducting GM's research activities. As a result, the firm made numerous contributions to the automobile's evolution in the 1920s. For instance, GM developed a crankcase ventilation system which, by removing corrosive bypass gases, greatly increased engine life (Sloan, 1964:237). Kettering's staff developed ethyl gasoline and the high-compression engine, thereby improving power and efficiency (p. 221). Kettering also discovered the ingredient for the new Duco lacquer, a product that extended finish durability (p. 236). Sloan himself pushed the development of four-wheel brakes to improve stopping and stability. The firm's engineers worked on perfecting the balloon tire (p. 231), and GM introduced the synchromesh transmission in 1929 to smooth gear shifting.

With the pace for the introduction of major product developments slackening, Harley Earl's styling efforts became increasingly important. It was also fortunate for Earl that his first GM car came "at a time when the corporation was revising its model line so that it could offer a finely graduated spectrum of cars to meet every nuance of the mass class market" (Bayley, 1983:36). Sloan wanted to pack as much variety as possible into *all* GM's cars. Accordingly, Earl's Art and Color Section, later to be called the Styling Staff, would become a corporate-level staff activity. Thus styling "took its place behind financial controls, centralized buying and the other procedural techniques" (p. 120). As such, it became "a business tool which interpreted with minute degrees of differentiation each level of the product line for each level of consumer." Eventually, however, diffentiating the product line from year to year would gain the styling staff's primary attention.

Probably more than anything else, GM's planners evolved the annual styling change to handle what Sloan (1964) assumed to be "a rising curve of used-car trade-ins" (p. 152). By 1927 the automobile industry had matured to the extent that most new cars were being sold to replace still functioning used cars (Ayres, 1927:372). Viability in a replacement-demand market required that the new-car manufacturer offer something different (Sloan, 1964:270). GM "met the used-car trade-in early" and "tried to make [its] models more attractive each year" (p. 152). In essence, GM's annual model changes kept its cars differentiated from the growing mass of inexpensive but reliable used vehicles reentering the market. Yearly alterations in GM's offerings thus assured the firm a steady demand from customers desiring the latest variations in styling, comfort, and convenience. The variety of GM's automobiles thus would match the variety demanded by an increasingly taste-conscious motoring public.

Sloan's annual model change introduced into the environment of GM's competitors an awesome variety load which was most difficult for them to counter. Henry Ford with his "static-model utility car" (Sloan, 1964:167) and his specialized factories was ill-equipped for annual changes. Even after Ford permitted the Model T to be replaced by the Model A, Sloan expected that Ford would repeat his past pattern and keep the Model A unchanged for too long a period. Ford held that "changing models every year [was] the curse of the industry" ("Mr. Ford Doesn't Care," *Fortune*, 1933:67). Since his standardized, low-variety policies were at odds with prevailing industry practice, FM's market share started to dwindle. As FM lost volume, only the vast cash reserves that had been accumulated before 1926 sustained its operations.

Smaller manufacturers than FM faced even greater hardships, given their limited resources. The widespread adoption of the closed body and acceptance of the annual model increased the industry's yearly expenditures for model changes from $12 million in 1921 to $36 million in both 1928 and 1929, years in which Ford made few if any changes (Thomas, 1973:122). By the late 1930s GM alone was spending $35 million annually on the new machinery and dies needed to create its restyled models (Sloan, 1941:177). On confronting this variety, GM's small competitors who could not or would not restyle their products each year were eliminated. Those manufacturers who did attempt to keep pace because they were afraid "that skipping a yearly model would cost too many sales" ("Mr. Ford Doesn't Care," *Fortune*, 1933:67) often encountered a similar fate; a one-year production run did not generate enough volume to cover their retooling investments. "An independent like Packard or Studebaker, for example," had to "sell something like 100,000 cars to absorb tools and dies at a practical figure per car" ("General Motors," *Fortune*, 1938:156). Otherwise they constantly ended up scrapping tools and dies with too much useful life remaining. In turn, when the high-cost producers began to fail, the stronger of the survivors could use their volume gains to accelerate "the rate of style obsolescence" (Thomas, 1973:126).

In the early days, as Sloan had already realized from his Pontiac and La Salle experiences, only one solution existed: pooled volume, obtained from the use of interchangeable parts, chassis, bodies, and plants among a variety of makes and models. Accordingly, "Oldsmobile had introduced its higher-priced Viking to share parts with Buick, while Buick had a low-cost car, the Marquette, ready for 1929" (Cray, 1980:263). But, other than GM, only Walter Chrysler had adopted this diversified-integrated strategy. GM's annual model changes thus resulted in much more than planned obsolescence for many cars (Snell, 1970a:83–86), as it made their manufacturers obsolete too.

With the conclusion of the prosperous 1924–1929 period and with the dawn in 1930 of the depression era, Ford and GM's smaller competitors hit their roughest times. Sales volume dropped precipitously. "Six of the fifteen best-selling automobiles of 1923 no longer were in existence a decade later" (Cray, 1980:267). GM also would not survive unscathed; but through further refinements in Sloan's high variety/high interchangeability marketing strategy, General Motors would emerge in 1938 as the world's foremost manufacturing enterprise.

In the 1933 automobile market the low-price group encompassed "73 per cent of the industry's unit car sales, as compared with 52 per cent in 1926" (Sloan, 1964:179). For GM this sizable environmental shift "meant, for the old car line, that [it] would have four lines in 27 per cent of the market and one line in 73 per cent of the market." To correct this imbalance, Sloan and his colleagues on GM's Finance Committee reallocated resources to the low-priced range, with all models moving down in price simultaneously. Oldsmobile's four-door sedan prices, for instance, dropped from $1046 in 1928 to $750 in 1934; and as economic conditions improved, Olds' prices climbed to $944 in 1938 (USFTC, 1939:898).

During the first years of the depression, several of GM's mid-price offerings were discontinued so as not to siphon volume from the tightly packed car lines remaining. The makes dropped during the depression era were Buick's Marquette, Oldsmobile's Viking, and the Oakland. "Marquette and Viking died after only one year, victims of anemic sales and a corporate decision to concentrate efforts on Buick and Oldsmobile. Oakland, overwhelmed by its offspring Pontiac, succumbed in 1931" (Cray, 1980:268). La Salle hung on until 1940, largely because it was so much cheaper than the Cadillac. In fact, Sloan almost dropped the Cadillac instead of the La Salle (p. 268). To improve its chances for survival in 1934, "La Salle ceased being a Cadillac sibling and became an upmarket brother to Oldsmobile, one segment down on the sheet metal social scale" (Mandel, 1982:186). More specifically, "the La Salle for the years 1934, 1935, and 1936 was virtually a high-class Oldsmobile straight-eight, with special styling [done] under Harley Earl's direction" (Hendry, 1983:239). The La Salle's eventual demise can be traced to the depression's long-term halting of the "volume" demand for really expensive cars. With this demand gone, the Cadillac dropped down in price to cover what had previously been La Salle's territory. The goodwill long associated with the Cadillac reputation thus was saved for GM's overall benefit.

By shifting its offerings toward the low-price segment and by reducing superfluous makes, GM's planners kept the firm's product line in proper relation to the depressed demand of the market place. But reduced volume necessitated further adaptations. Accordingly, Sloan increased the interchangeability among the divisions' products so that the pooled volume would cover fixed costs. (See, for example, "General Motors," *Fortune*, 1938.) After the middle 1930s, then, GM's product variety was literally only skin deep. And even the exteriors of its cars were not much different once the corporate headquarters had rigidly standardized the firm's automobile production via a restrictive common-body program imposed on its once autonomous divisions. By 1938, the small Cadillac, the La Salle, the Buick, plus the larger Oldsmobile and Pontiac, shared the same body shell. Moreover, the chassis and engines of these cars followed "the same factory routing without the lay eye's being able to tell one from another" (p. 152). The smaller Oldsmobile and Pontiac sixes used a body identical to the Chevrolet (p. 156). Divisional "originality" was maintained with grilles, hoods, front fenders, cowls, finishes, and trim.

But even as GM's cars were forced to occupy a very narrow price band and to adopt common chassis, bodies, parts, and plants, Sloan still strove aggressively to keep the firm's product diversity as high as possible. To the Finance Committee Sloan often warned against placing all GM's "eggs in one basket" and advised instead that "different things appeal to different people" and "that not all the sound engineering ideas can be incorporated in one unit" (Sloan, 1964:180). Consequently, he emphatically recommended that "the utmost consideration must be given to the importance of the greatest possible diversity . . . so as to build the strongest foundation of acceptability to the consumer" (p. 181).

Despite the proliferation of interchangeable parts, chassis, and bodies among GM's cars in the late 1930s, Harley Earl's "designers achieved remarkable levels of differentiation" (Bayley, 1983:70). Hence, the firm's products still gave the technically unsophisticated buyer the impression of wondrous variety. Sloan (1941) himself attested that unknowledgeable visitors to GM assembly plants were "bewildered as they observed the variety of the production to discover eleven kinds of chassis taking final shape as Oldsmobiles, Pontiacs, and Buicks. And all this happens on the same assembly line" (p. 183).

Sloan—by combining his financial acumen, marketing knowledge, production talents, and organizational skills—solved the problem that had long baffled automobile manufacturers: "In the automobile industry a delicate balance had always (and still has) to be struck between a policy of making too few models or 'chassis types' to satisfy the demands of the market and one of simultaneously making too many types to permit a sufficiently economical production and a satisfactory control over manufacturing processes" (Epstein, 1927a:161).

Managing GM's diversity was not all that difficult, for more and more its cars came to share basic components. Not only did these common features reduce the variety of items that needed to be designed and produced, but they also reduced costs and thereby raised profit margins. In sum, GM's planners minimized the firm's internal variety to aid its decision makers, but they did so in a disguised manner that insured a match with the uncontrolled environment's variety demands.

That Sloan's price and model diversification worked successfully for GM can be seen in Buick's reversal of positions between 1923 and 1933. Buick in 1923 produced more than 200,000 cars. More importantly, to the finance-oriented GM designers, "the Buick organization won first place at the New York Automobile Show in each of nine successive years 1918–26, inclusive, by reason of leadership in financial volume of sales over all other members of the National Automobile Chamber of Commerce; in 1927, it was displaced by the Chevrolet organization" (Seltzer, 1928:137). In short, Buick was the early backbone of GM. But "in 1929 the factory presented a new model whose slightly bulbous contours caused it to be dubbed at once the Pregnant Buick" ("General Motors," *Fortune*, 1938:148). Supposedly, the problem resulted from the factory's making some unauthorized changes in Harley Earl's design for the car (Sloan, 1964:272-273). Regardless of the cause, sales declined substantially. To make matters worse, the depression, which hit the whole automotive industry, "hit Buick viciously. The production . . . for 1933 was only a little more than forty thousand, less than it had been in twenty years and only one-sixth of the peak production of 1927" (Crow, 1945:79-80). Not surprisingly, Buick lost money in both 1932 and 1933 (USFTC, 1939:531).

Yet GM survived the depression with little serious difficulty. Other GM car makes, the once lowly Chevrolet for example, were now pulling more strongly than ten years earlier. Indeed, Chevrolet earned a profit throughout the depression. And by the end of the 1930s Buick had once again regained much of its old strength, primarily due to its GM membership "which could absorb the losses of the Flint property without passing a dividend, and could advance money [$30,000,000] for the improvement of the plant and the development of new models" (Crow, 1945:82). To update and brighten Buick's image, Harlow Curtice—the new division manager in 1934—worked closely with Harley Earl.

Cadillac's experience closely paralleled Buick's. Though it always was a strong earner in the 1920s, the severely depressed demand for expensive cars in the early 1930s caused Cadillac to lose money in 1931, 1932, 1933, and 1934 (USFTC, 1939:531). The first move was to reduce drastically Cadillac's proliferation of models. And then with considerable corporate support, a new division manager, Nicholas Dreystadt, worked to cut costs that during the prosperous 1920s had never received much serious attention (Hendry, 1983:242). Simultaneously, Dreystadt "redesigned Cadillac's system of production, renewed a stress on quality that would have done the late Henry Leland proud, and transformed a failing automobile into the premier symbol of the upwardly mobile" (Cray, 1980:279).

Revitalizing GM's middle- and high-price makes would soon pay dividends. With the economy's recovery, consumers again bought a larger "proportion of higher-priced lines, and in the years 1939-41," immediately before the United States entered World War II, "the low-price group was accounting for only 57 per cent of the over-all market, or about the same proportion as in the year 1929" (Sloan, 1964:438). Sloan's high-variety marketing plan again enabled GM to respond appropriately.

Diversification into Foreign Markets

As the domestic automobile market's growth began to slacken during the 1920s, GM's Finance Committee shifted some of the diversification activity outside the United States. New markets were sought in the countries of Europe and Asia, then far behind the United States in auto use, to counter the possibility of a stabilizing domestic replacement market.

At first GM exported fully assembled cars while its "strategic plans were directed principally toward the development of a distribution network abroad" (Donner, 1967:12). After the 1920s GM's Finance Committee contemplated a deeper foreign involvement. Overseas assembly of cars out of parts produced in the United States became necessary because of high import duties and shipping costs on fully assembled automobiles (p. 13). GM's designers still avoided parts manufacturing abroad, husbanding the firm's precious capital resources and executive talent for the domestic automobile market (p. 14).

By the late 1920s, however, fully integrated foreign production facilities were being studied by Sloan and his colleagues. The additional involvement was necessitated by the growing differentiation between GM's foreign and domestic marketing environments. The disparity was most pronounced in countries like England and Germany, which possessed strong manufacturing establishments. As demand grew, local producers began to *tailor* their cars, to increase their output, to raise their efficiency, and to become more competitive with United States producers (Donner, 1967:17). While smaller foreign cars forced GM to include more variety within its controlled domain, "seasoned automotive engineers and management personnel were very scarce. At the time, it was not practical for General Motors to consider the large investment of management skill and engineering talent which completely new ventures overseas would require" (p. 18).

Rather than create fully integrated production facilities abroad, then, Sloan recommended and the Finance Committee approved foreign investments into several already completed foreign firms. First, Vauxhall Motors in England was purchased in 1925. Since GM's designers were interested in foreign "manufacture on a mass production basis" (Donner, 1967:18), Vauxhall was a less than ideal acquisition, for its production volume in 1925 fell short of 1,500 passenger cars. Yet the Vauxhall acquisition did "provide the opportunity to acquire experience in manufacturing automobiles overseas" (p. 19).

With the Vauxhall experiment providing useful information (Chandler and Salsbury, 1971:580), GM's planners were ready in 1929 to extend the firm's boundaries further by purchasing Adam Opel, "the largest motor vehicle plant in Germany at the time. In 1928, it had produced nearly 43,000 vehicles, about 30 per cent of the cars and trucks produced in Germany." Moreover, with its "modern, efficient plant" and "its effective dealer organization," Opel "fitted the General Motors concept better than the smaller Vauxhall operation" and "provided a real basis for growth through the use of the General Motors network of assembly facilities which had been established overseas" (Donner, 1967:19).

Neither Vauxhall nor Opel—nor any of GM's other foreign facilities, for that matter—ever really assumed major significance within the firm. *The GM designers allocated little attention to the relatively small foreign investments.* Instead they chose to concentrate on domestic automobile production.

GM's planners retreated from foreign involvement for several reasons. First, while the foreign market accounted for 18 percent of United States automobile sales in 1927, it declined thereafter (Soule, 1947:287). Second, in the 1930s "the trend in automobile designs in the volume markets overseas and in the United States became more divergent" (Donner, 1967:21–22). Third, foreign capital investments became quite precarious during the depression years and those before World War II. Foreign governments did not always provide GM's clientele the same safeguards found in the United States. Precious capital investments might be expropriated or stockholders' profits might be frozen inside foreign countries. "As the depression spread," worse yet, "more severe international trade and foreign exchange restrictions were imposed" (p. 22). Many years later, Drucker (1973) chided GM for not going further with its foreign-investment scheme, but true to form, the firm would not take such risks.

Diversification in Allied Industries

Besides staying close to the domestic markets, the Sloan-Brown designers never permitted GM to stray from the automotive expertise of its decision makers. In 1938 *Fortune*'s editors could write, "It appears . . . that G.M., for all its varie-gated sidelines, is overwhelmingly an automotive concern" ("General Motors," *Fortune*, 1938:47); "first and foremost a producer of automobiles" (Colston, 1939:608). USFTC (1939) breakdowns of the firm's sales, profit, and invest-ment show without a doubt that GM was primarily concerned with automobile production. From 1927 to 1937, for example, 61.4 percent of GM's net profits came from the motor-vehicle group and 22.4 percent from parts and accessories, compared with only 16.2 percent from other products (pp. 509, 527, 528, and 530).

GM's nonautomotive ventures, in fact, received such a low priority that in some instances they were pursued with an uncharacteristic lack of formal planning. As Sloan (1964) himself admitted: "We had, of course, some natural interest in diversification which might afford us a hedge against any decline in automobile sales. But we never had a master plan for nonautomotive ventures;

we got into them for different reasons, and we were very lucky at some crucial points" (pp. 340–341).

Nonetheless, a managerially conservative pattern emerges, for GM rarely ventured into products that could not be mass produced, did not have an engine or a motor, or were not sizable. In a 1930 article, "General Motors Diversifies Only Along 'Natural' Lines," Sloan, for example, stressed that "it must be fully recognized that General Motors is engaged primarily in the automotive business and its affiliated industries" (*Business Week*, October 1, 1930:20). An affiliated industry was one "in which parts, products, or processes of automobile manufacture may be used. Delco-Light illustrates. A Delco-Light plant on a farm is just a gas engine in another place. Manufacture of that engine is no different" (p. 20).

When GM researchers discovered or developed any products outside this highly constricted class, they were quickly sold to or managed by external units. Kettering's ethyl additive for increasing the knock resistance of gasoline, for instance, resulted in the creation of a partially owned firm, the Ethyl Corporation, which paid royalties and returned dividends to GM, yet required minimal managerial effort on GM's part. Sloan felt the ethyl-additive business was foreign to GM's basic managerial expertise. "In the first place," Sloan testified, "I recognized that General Motors organization had no competence whatsoever in chemical manufacture. We were mechanical people dealing with metal processing. . . . I have always taken the position against our stepping too far out of those things in which I felt we had competence" (testimony of Sloan, *U.S.* v. *DuPont,* 1956:1248). Thus, the Ethyl Corporation with "its own management of chemical engineers and chemical marketers" became the supplier worldwide. "GM, in effect, made money on almost every gallon of gasoline sold anyplace by anyone" with only a minimal capital investment and no threat to its "essential unity and manageability" (Drucker, 1973:712).

In a similar fashion, GM formed the Kinetic Chemical Company in 1930 along with DuPont to produce and market freon refrigerants. Though Kettering's researchers had developed these compounds for GM's own Frigidaire Division, the firm's planners did not want to become involved in this nonautomotive venture either (testimony of Pratt, *U.S.* v. *DuPont,* 1956:1491). Hence, they created another separate entity beyond GM's immediate corporate confines. And, here again, GM "received a substantial royalty for its initial development and patent position" (Leslie, 1983:225).

In contrast, the several products which evolved from Kettering's pioneering development of diesel engines were welcomed within the GM controlled domain. The acquisition of the Winton Engine Company and the Electro-Motive Engineering Company, besides being the natural outgrowths of Kettering's interest (Sloan, 1964:341), provided GM a hedge, for the GM planners remained uncertain "about the future of the U.S. automobile market, which had not been expanding during the late 1920s" (p. 346). Consequently, they "had a natural interest in any enterprise within" their "scope that offered . . . a reasonable opportunity to diversify" (p. 346). Pratt thus advised the Finance Committee: "The Winton Company has a capable management and would not require any additional personnel immediately"; purchasing it "will give us a vehicle for capitalizing the developments of our research organization, . . . and if expansion continues, as most of our engineers believe it will, we should ultimately make a good return on the investment" (p. 347). Five months after the Winton acquisition GM acquired the Electro-Motive Company and again retained "the old management . . . to run its affairs" (p. 347).

GM's entry into the new field was extremely cautious. First, the firm entered the field only after Kettering's staff exhibited considerable strength in this area of research. Second, both of the acquired firms were the industry's best, not only technically excellent but also extremely well-managed. Thus, little managerial skill would have to be withdrawn from the auto divisions. Third, GM handled the industrial diesel just as it designed, produced, and marketed its automobiles. It was but a very minor variation on a common theme. Hence, Kettering's "two-stroke diesel, modified for endless applications but still the same basic design, became the backbone of a profitable line of General Motors products" (Leslie, 1983:273). Along with Kettering's research and engineering prowess, "much of the credit for such commercial success belonged to John Pratt, who applied automotive marketing techniques to the diesel—long-term financing, standard designs using common parts and technology, factory-trained maintenance men" (p. 273).

By holding fast to the firm's basic strategy, GM's Electro-Motive Division completely revolutionized American railroads. Beyond replacing the old steam locomotives with diesels, GM single-handedly introduced a line of rigidly standardized locomotives. Until the diesel's introduction, most locomotive construction was a high-variety job-shop enterprise with each unit or small batch built to the various customers' individualized specifications. If at all possible, GM did not want to leave its mass-production technology behind when it ventured forth into this new field. But "mass production of course require[d] a standardized product" (General Motors Corporation, 1973:16). Hence, Pratt recommended to Sloan "one fundamental policy," that "the Electro-Motive Corporation will build a standardized product" and avoid "the many different standards and specifications on which each railroad demands to purchase" (Sloan, 1964:351). Because of significant cost reductions and a thorough demonstration program, GM did not have to yield to specialized construction (Reck, 1948:126; *General Motors: The First 75 Years*, 1983:76). All GM products, even its massive locomotives, would be fabricated by the mass production technology it understood so well. Eventually, visitors to the Electro-Motive facilities were surprised to find how important a role mass production principles played even with "something as large as a locomotive power plant." Ten subassemblies were involved and "cylinder heads, cylinder liners, pistons, and connecting rods" were all "turned out on production lines." For assembly "a moving conveyor line" was used, like that for automobile engines, although some of these powerplants weighed fifteen tons and were "higher than a tall man" (Reck, 1948:120).

In reality, the railroads' needs were sufficiently varied that they could not be met with a single locomotive model. A comparatively small switching engine, for instance, could not handle high-speed passenger service or heavy-haulage freight work. To meet this environmental variety, Electro-Motive offered a line of locomotives from low-horsepower yard switchers to high-horsepower main-line engines. Even so, GM mass-produced the varied line by using many standardized components across models as well as over time. It was just a matter of applying and extending "its know-how from the automobile industry" (General Motors Corporation, 1973:15). "Variations in horsepower from engine to engine" were achieved "simply by varying the number of standard cylinders (and standard cylinder components). Other major components, such as electric traction motors, . . . had exactly the same dimensions regardless of the size or type of locomotive" (pp. 16–17). Standardization was achieved over the years simply "by designing product improvements into components having the same physical configuration and exterior dimensions" as those replaced (p. 17).

So by using standardized parts in various combinations across models and over time, GM's planners minimized the firm's variety loads yet matched the market's product demands. The railroads also benefited from less varied parts inventories and more routine repair procedures. Not surprisingly, then, the Electro-Motive market share soared from 4.0 percent in 1934 to 54.1 percent in 1938 (p. 75).

Just as they were doing with the firm's increasingly similar automobiles, GM's corporate stylists, working in conjunction with Electro-Motive's engineers, enveloped the standard machinery of the larger diesels with a streamlined body, thereby "sweeping away every wind resisting obstruction" (*General Motors: The First 75 Years*, 1983:76). And though GM avoided building its diesel locomotives to the railroads' widely varying specifications, its Electro-Motive Division willingly finished their streamlined form in the bright colors or metals specified by the railroads. After a five-hour session during which GM's representatives refused to relax the strict standardization policy, for instance, a top Electro-Motive executive told the head of one railroad: "We'll build you a locomotive. You tell us what color you want it painted and we'll be responsible for everything else" (Reck, 1948:126). For the third passenger diesel delivered, GM's corporate stylists "unveiled a stunning color scheme . . . of black, cobalt, sarasota blue, golden olive, and pimpernel scarlet" (Leslie, 1983:272). It would not be too long before GM had delivered its passenger and freight locomotives in 40 varied liveries that attracted considerable publicity. *The nation's mostly black steam engines with their clutter of protuberances would go the way of the black, unsightly Model T. GM simply replayed the earlier auto industry scenario with nary a change, and Sloan's automotive marketing strategy would extend even to its industrial products.*

Minor investments were also made in the aviation industry. In 1929 GM purchased 24 percent of the newly formed Bendix Aviation Corporation, 40 percent of the Fokker Aircraft Corporation of America, and "the entire capital stock of the Allison Engineering Company" (Sloan, 1964:362). "During the 1920s," Sloan explained, "it became steadily clearer that aviation was to be one of the great American growth industries." Of particular concern was "talk about developing a 'flivver' plane—that is, a small plane for everyday family use." Because "such a plane would have large, unforeseeable consequences for the automobile industry, . . . we had to gain some protection by 'declaring ourselves in' the aviation industry" (pp. 362–363). As one would expect, GM "was not entirely a stranger to the aviation industry at that time," for Buick and Cadillac "had combined to manufacture the famous Liberty aircraft engine" of World War I (p. 362). Still more experience had been gained from GM's (originally Kettering's) Dayton Wright Airplane Company, "which had produced a total of 3300 airplanes during the war period. Fisher Body also . . . was an important manufacturer of military airplanes" (p. 362). And most importantly, "throughout the 1920s, Kettering supported aviation in every way he could" (Leslie, 1983:88). But with little direct corporate involvement during the 1920s, GM had to reenter the field with Bendix, Fokker, and Allison.

The manner of reentry dramatizes GM's unwillingness to move far from its basic automotive technology. First, even with considerable prior experience in the field, only limited investments were ever considered. In 1939, for instance, while GM's aviation business accounted for about 25 percent of the aircraft industry's, the firm had invested less than two percent of its total assets. "The company could either get out of aviation entirely or gobble up the whole industry without material effect to its overall sales and earnings" (Colston, 1939:609). Second, there was no intention "to operate either Bendix or Fokker as a division of General Motors"; instead, the "investments were made as a means of maintaining a direct and continuing contact with developments in aviation" (Sloan, 1964:363). Furthermore, since "the engineering techniques of the automobile and aircraft industries were still quite similar," the aviation ventures would yield "valuable technical information" for GM's "automobile operations." In particular, Bendix controlled "important patents for devices applicable to the automobile industry" and included in its accessory lines automobile components such as "brakes, carburetors, and starting devices for engines." Bendix's "superb technical staff" made the "investment all the more attractive" (pp. 363-364). In time, GM "encouraged" Fokker's successor, North American Aviation, to mass produce "the ubiquitous AT-6 Texan trainer." "As automobile men," explained Sloan, "we naturally thought in terms of 'standardized' production models which could realize the inherent economies of volume production" (p. 366).

As was mentioned, GM also acquired Allison Engineering in 1929, and this unit eventually became GM's principal link to the aircraft industry (p. 369). Initially, this firm had operated as a supporting machine shop for racers at the Indianapolis Motor Speedway. Later, its highly skilled engineers, machinists, and mechanics built marine engines and reduction gears for high-speed boats and aircraft. "The organization," as Sloan put it, "possessed valuable mechanical skills that we could use" (p. 370). During the middle and late 1930s Allison developed a V-12 liquid-cooled engine more similar to Cadillac's larger V-powerplants than to the radial air-cooled airplane engines then prevalent. "Until the V-1710 engine was developed the Air Corps had taken for granted the superiority of the air-cooled [radial] engine" (p. 370). During World War II seventy thousand of these Allison engines powered such famous fighter planes as the Curtiss P-40 Warhawk, the Bell P-39 Aircobra, the Bell P-63 King Cobra, the North American P-51 Mustang, and the Lockheed P-38 Lightning. The latter plane even supplied Harley Earl with the inspiration for the automobile tailfin.

GM's Frigidaire Division again illustrates the designers' preoccupation with mass production and automobiles. Frigidaire Division had been added by Durant during World War I when he feared a steel shortage; home refrigerators, then made of wood and growing in popularity, would offer GM profit protection. Unfortunately, refrigerators, like cars, were consumer durables and therefore were postponable purchases. So once the war was over and the initial demand filled, Frigidaire did little to stabilize the cyclical and even seasonal fluctuations in GM earnings. Not surprisingly, then, this nonautomotive division did not receive the Sloan-Brown groups' rapt attention. "By 1940 . . . Frigidaire's share of the refrigerator market—which had been above 50 per cent in the 1920s—was down to 20 to 25 per cent" (Sloan, 1964:360). In retrospect the problem was simple: "During the years 1926-36 a number of Frigidaire's competitors gained an advantage" by expanding their lines to include "radios, electric ranges, washers, ironers, and dishwashers, while Frigidaire concentrated on refrigerators" (p. 360). Of all the designers only Pratt recognized the problem, but he was overruled. Sloan, for instance, admitted that "Mr. Pratt had suggested that Frigidaire get more actively into air-conditioning; but his suggestion did not register on us, and the proposal was not then adopted" (p. 361). As usual, the designers had their eyes focused on automobiles, and the only reason they did not drop the line was because it could be mass produced and marketed.

In sum, the Sloan-Brown group shunned disparate conglomerate-type hold-ings from which GM's decision makers could not advantageously extract client benefits and husbanded the firm's decision capacity for the automotive activi-ties its capital supported. *GM's decision-making structure dictated its market strategy.*

6

Centralizing the Corporation

Once chosen by GM's designers on the Finance Committee, the firm's controllable variables had to be assigned to the various decision units within the organization. Variables that exerted the greatest influence on performance needed to be set by decision-making units over which the designers had the most control, i.e., those situated in or run by the corporate headquarters.

The central theme of this chapter is that the Sloan-Brown designers incessantly centralized GM's decision making after Durant's removal. As they themselves readily admit and as is well known, they started with numerous financial controls. But, in reality, their centralization efforts extended much further into the decision-making domains of the once decentralized divisions. Over time the variables controlled by the corporate headquarters came to include, among others, prices, purchasing schedules, in-process and finished inventory levels, weekly production rates, cash balances, capital expenditures, research development, styling, engineering, manufacturing, and marketing.

With many of these decisions, corporate control *greatly enhanced the internal variety,* or flexibility, needed to counter environmental fluctuations. In inventory control, for example, with headquarters setting purchasing and production schedules, GM no longer became saddled with nonliquid stocks: inventory levels could be altered immediately to match environmental demand shifts. The centralization policy entailed significant risks, however, since it diminished the control diversity within GM. The corporate headquarters had to be right; otherwise the results would be disastrous. While Sloan and his experienced colleagues were in charge no problems developed, but difficulties surfaced after their retirement.

There are two reasons GM could centralize so vigorously. The first stems from GM's conservative diversification strategy. Faced with little variety and much similarity among its divisions, GM's corporate headquarters had the capacity to delve deeply into divisional affairs.

Second, Sloan and his corporate colleagues enjoyed extremely long tenures in their corporate positions. With each passing year they became more and more proficient in their jobs. Formerly novel situations soon became repetitive, routine, and rapidly resolved; past experience pointed to quick solutions. Hence corporate capacity became increasingly available for further incursions into divisional affairs. There was little the executives in the supposedly decentralized divisions could do about this erosion of their power. They remained comparatively unfamiliar with their jobs because they rarely held them very long. "The term of office for a general manager of Buick, Oldsmobile, Cadillac, or Pontiac, for example, . . . since 1922 averaged a little over four years, that of their sales managers even less" ("General Motors II," *Fortune*, 1939:40). While Chevrolet executives lasted a little longer, their tenure did not begin to approach that of men like Sloan, Brown, Bradley, Mott, Pratt, and Kettering.

GM would become extensively centralized through several techniques: some masked, some manifest. The corporate headquarters, for example, *leveraged* its power by controlling a comparatively few but critically important variables. These variables, in turn, determined or constrained many of the decisions left within the still theoretically decentralized divisional domains. So if Brown's Financial Staff controlled each division's cash access, many other divisional decisions were severely limited. If Sloan set the prices to be charged for the divisions' automobiles, then the divisional engineers were tightly restricted in terms of design and production variables. And when the corporate headquarters eventually saw fit to style the divisions' cars, the divisional freedoms were even more circumscribed.

While several of the corporate control procedures discussed in this chapter dealt with all divisions' critical variables, many were designed specifically for the auto divisions, singled out because they received the bulk of the stockholders' capital investment. Another important reason for increasing the tightness of the vertical relationship between the corporate headquarters and the auto divisions stemmed from the divisional interconnections. The production schedules for the finished-auto units, for example, dictated the production schedules of the many parts units.

Most obviously, GM's designers centralized the firm by building a very large corporate headquarters—housed in both New York and Detroit. Almost immediately after their takeover, Pierre duPont and Sloan started replacing Durant's "tiny personal headquarters" with a corporate headquarters "consisting of a number of powerful general executives and large advisory and financial staffs" (Chandler, 1977:460). Numerous corporate-dominated Interdivisional

Committees, later called Policy Groups, were also instituted. The future of the executives filling the corporate positions, manning the staff units, and sitting on the committees would only be enhanced by an extension of their influence. Simply, they had little or no incentive to limit the centralizing activities. If the supposedly decentralized divisions continuously lost autonomy in the process, that was just too bad.

So GM became centralized when the variables most tightly coupled to the firm's return rate were specified by the (a) Executive Committee, (b) Executive Officers, (c) Financial Staff, (d) Advisory Staff, (e) Interdivisional Committees and Policy Groups. Most of these units are shown in the organizational chart presented in Figure 6-1.

THE EXECUTIVE COMMITTEE

In Chapter 5 the establishment, under the Board of Directors, of a Finance Committee and an Executive Committee was discussed. The Finance Committee, for the most part, included "men of large affairs" (Brown, 1927:7), identified with banking and big businesses other than GM: men concerned with representing the stockholding clients, but unable to control the firm's internal operations. With their extensive outside responsibilities, Finance Committee members could control only a few of the most important performance variables: dividends, retained earnings, performance measurement, and allocation of resources to GM's strategic activities.

The Executive Committee was responsible for internal operating policy. To it fell the early problem of controlling the performances of Durant's unwieldy aggregation of divisions. Sloan in his "Study of Organization" outlined the Executive Committee's role: it had "entire supervision over the operations side of the Corporation's activities, constituting as a whole practically the entire operating staff . . . and through its own deliberations or through Committees appointed for special study" it determined "the broad policies of the Corporation so far as they affect one or more operations" (GMX 1, *U.S.* v. *DuPont,* 1956:6538). Pierre duPont, along with Sloan, wanted the Executive Committee to be small and all-powerful (Chandler and Salsbury, 1971:496). Though the committee grew in size, it remained all-powerful with respect to internal matters until its demise in 1937.

The Executive Committee's charge, in contrast to that of the Finance Committee, required that most of its members devote all their time exclusively to GM's affairs (Brown, 1927:7). The work load in the early days of GM's design was extremely heavy. Throughout 1921 the committee had 101 formal sessions, and the members "individually and together . . . were absorbed in the innumerable problems of the emergency and of the future, and were constantly on the go visiting divisions and their plants in Detroit, Flint, Dayton, and elsewhere" (Sloan, 1964:56).

The many meetings and visits mentioned by Sloan included biweekly conferences with the auto divisions' executives, who now were barred from Executive Committee membership. When first taking over control of GM, Pierre duPont and Sloan did not trust Durant's division managers enough to allow them to occupy such critical policy-setting positions (Cray, 1980:192). But even later, when Durant's cohorts had been replaced by his own appointees, Sloan still excluded most division managers from the Executive Committee because they did not possess a corporate outlook and instead "used their membership on the Executive Committee mainly to advance the interests of their respective divisions" (Sloan, 1964:49).

Durant's old Executive Committee, composed almost entirely of division general managers, was downgraded to the Operations Committee, which reported to the Executive Committee (p. 45). The Operations Committee was inactive during the management crisis of 1921.

After Durant's departure, the Executive Committee was reduced to Pierre duPont, Sloan, Raskob, and Haskell. Haskell, who died shortly after the reorganization of the committee, and Raskob, who was primarily an externally oriented financial expert, played relatively minor roles on the Executive Committee.

Improving the divisions' operations then fell on the shoulders of Pierre duPont and Sloan. Pierre duPont became the corporation's new president as well as board chairman. In addition, he became division manager of the foundering Chevrolet, having taken over Chevrolet temporarily because of problems with its manager, Zimmerschied. DuPont and Sloan worked and traveled together. Sloan rose in stature rapidly, and after six months came to be vice president in charge of all operations, reporting directly to duPont (Sloan, 1964:56). Given that duPont was an unwilling president, Sloan assumed more and more authority. In 1923, as GM president, Sloan chaired the Executive Committee.

Since duPont and Sloan possessed limited time and also since duPont (and Raskob) had comparatively little experience in the production, assembly, and marketing of automobiles, the Executive Committee's channel capacity needed to be increased. DuPont was particularly concerned that it had "not been possible for Alfred Sloan to give personal attention to all the ramifications of the Corporation" (GMX 32, *U.S.* v. *DuPont*, 1956:6658). Accordingly, Sloan

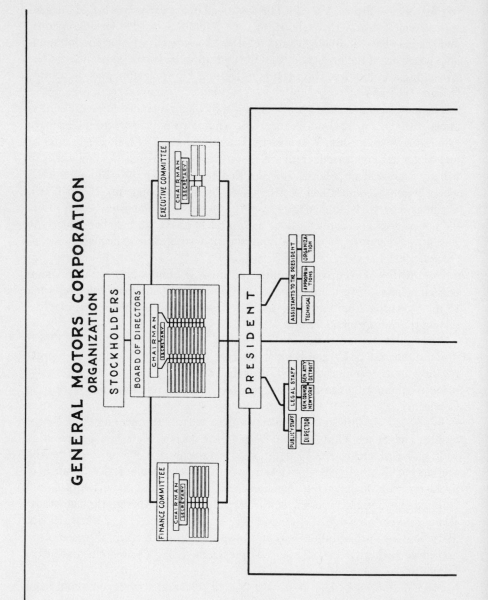

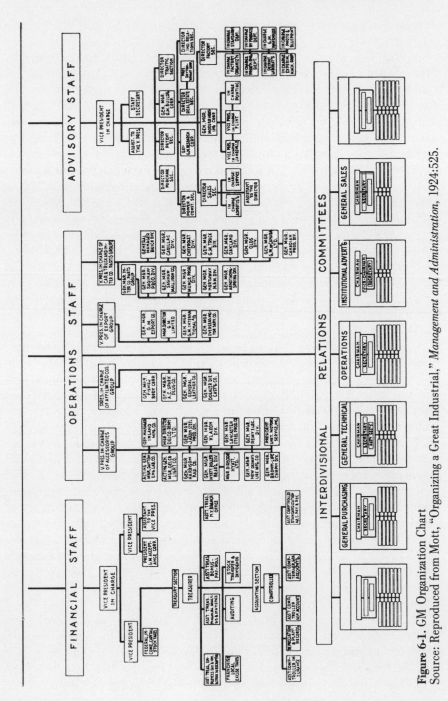

Figure 6-1. GM Organization Chart
Source: Reproduced from Mott, "Organizing a Great Industrial," *Management and Administration*, 1924:525.

with duPont's approval carefully engineered the required Executive Committee additions.

In 1922, C. S. Mott and Fred J. Fisher were elected to Committee membership. Fisher was the head of the critically important Fisher Body Corporation, a unit that dealt with all the GM auto divisions, needed to be integrated into the GM family more intimately, and accounted for a large portion of GM's capital. Furthermore, Fred Fisher's long experience in the industry provided Sloan with much-needed technical ability. Fisher's membership in the Executive Committee also added weight for the corporate view, since the Fisher brothers were major GM stockholders. A corporate viewpoint was enhanced further when Fisher was elected to GM's proprietor-dominated Finance Committee in 1924. Two years later, at the time of complete integration of Fisher Body into GM, Sloan complimented Fred Fisher's GM loyalty: "Fred's whole attitude is so enthusiastically for General Motors, its present position and its future. . . . He has, in every sense of the word, become one of us" (GMX 34, *U.S.* v. *DuPont,* 1956:6661).

Mott, like Fisher, also had long experience in the automobile industry; as early as 1905 he manufactured automobile wheels and axles (Chandler, 1966:142). In 1922 Mott headed the Detroit units of the Advisory Staff; consequently, he too was more concerned with corporate than divisional concerns. Mott's substantial GM stockholdings heightened his corporate perspective.

While Pierre duPont and Sloan believed that "the top operating committee" should include "only general executives" and "be a policy group detached from the interests of specific divisions" (Sloan, 1964:113), two exceptions were made in 1924: Harry Bassett, division manager at Buick, and Lawrence Fisher, shortly to become head of the Cadillac division, were added to the committee. There were probably three reasons for these atypical additions. First, in 1924 integrating Fisher Body into the firm was problematic; another Fisher on the Executive Committee would help the exchange of views. Second, Buick's and Cadillac's excellence needed to be extended via policy decisions to the other auto divisions. Third, these units, especially Buick, were of critical importance, which required that their executives be brought into close corporate contact so that Sloan could influence their decisions. (Not too surprisingly, years later when Knudsen had replaced Sloan and Marvin Coyle had taken over for Knudsen at the then crucial Chevrolet division, it could be said: "Mr. Knudsen is still very close to Chevrolet and Mr. Coyle gets perhaps more corporate attention than the smaller division heads" ["General Motors II," *Fortune,* 1939:46].)

The firm-wide perspective of the Executive Committee was furthered when Sloan added several important corporate executives to its membership. In 1924, Donaldson Brown joined the group. Brown, like Sloan, opposed permitting the Executive Committee to be ruled by the principal divisions. Brown felt that when a division manager was a member of the Executive Committee, it was difficult for him to divorce himself from the divisional view and look at problems with the corporation's interest in mind.

Still another such corporate-oriented executive was Brown's friend John Pratt, who joined the Executive Committee with Brown in 1924. At that time Pratt was vice president in charge of the Accessories and Intercompany Parts Groups as well as vice president in charge of the General Service Staff—a reduced version of the original Advisory Staff. During earlier days Pratt had become schooled in the divisional excesses through his membership on the ad hoc Committee on Appropriation Request Rules (Sloan, 1964:119) and the ad hoc Inventory Committee (Brown, 1957:46). Both of these committees, along with the Financial Staff, were instrumental in designing the new corporate controls over the divisions' capital investments and inventory levels.

In summary, then, restructuring the Executive Committee with members who possessed specialized knowledge, i.e. decision-making variety, about automobile operations along with a decided corporate, rather than divisional, orientation was one approach GM's designers used to begin centralizing the firm's decision making.

THE EXECUTIVE OFFICERS

Pierre S. duPont began GM's centralization process by removing the division managers from the Executive Committee and providing the impetus for the creation of strong central Advisory and Financial Staffs. When Durant was president there was, of course, no central staff, and Pierre duPont was concerned about GM's lack of a corporate management group (Chandler and Salsbury, 1971:468). In correcting this lack of central control, Pierre formed a general staff organization and developed more adequate information flows into the central office (p. 468). Ultimately, the central staff units and not the divisions came to make the most important decisions.

The really extensive centralization at GM took place under Sloan, not under duPont. Even when head of GM's United Motors Corporation, Sloan had championed a stronger central organization, pushing first for a central research unit, then for an effective central engineering organization. Sloan talked with Raskob and Durant about coordinated financial procedures, before the central accounting office was established. And after GM's new comptroller, Turner, was appointed in 1919, Sloan gave him his ideas on interdivisional billings, capital appropriations, and uniform accounting procedures (Chandler and Salsbury, 1971:473).

Throughout much of Pierre duPont's tenure as president, Sloan felt that GM's corporate headquarters was still too weak, as the divisions often ignored performance-improvement advice. "The corporation was still too decentralized, he told Pierre and Raskob. The divisions still had too much autonomy and the general office still had too little authority to permit General Motors to get the maximum results from its combined resources" (p. 524).

Inadvertently, Sloan himself had created some of the problem. In 1920, with duPont, he had set the same salary for the heads of the operating divisions as for the president, and had set a lower salary for himself as the operating vice president to whom the division heads reported (Drucker, 1973:397). In terms of status, the division managers were on a par with Pierre duPont and considerably ahead of Sloan.

By 1923 Sloan had begun to rework GM's salary structure in order to strengthen the corporate headquarters executives' status as is shown in the "Salary" column of Figure 6-2. Only Bassett of the prosperous Buick Division earned the same salary as Sloan. By comparing the "Shares" columns of Figures 6-2 and 6-3, one can see how far Sloan had gone in restructuring GM's corporate and divisional compensation levels by 1930. In 1923, for instance, Sloan's $350,000 was not much ahead of Bassett's at $300,000 and still only three-and-one-half times those of the Oldsmobile and Oakland managers. By 1930, however, Sloan's $225,000 was far ahead of Lawrence Fisher's $100,000 (an allotment undoubtedly distorted upward because of Fisher's familial ties) and almost six-and-one-half times those of the Oldsmobile and Oakland managers.

CORPORATE EXECUTIVES

NAME	POSITION	SALARY	SHARES
A. P. Sloan, Jr.	President	$100,000	$350,000
C. S. Mott	Vice president	100,000	300,000
J. J. Raskob	Vice president	75,000	300,000
D. Brown	Vice president	40,000	150,000
C. F. Kettering	Laboratory	120,000	150,000
J. L. Pratt	Accessories manager	40,000	150,000

DIVISION HEADS (AUTO)

H. H. Bassett	Buick	$100,000	$300,000
H. H. Rice	Cadillac	75,000	100,000
W. S. Knudsen	Chevrolet	50,000	150,000
G. H. Hannum	Oakland	75,000	100,000
A. B. C. Hardy	Oldsmobile	75,000	100,000

Figure 6-2. Salaries and shares in Managers Securities Company of corporate executives and divisional heads in 1923. *Source:* Reproduced in part from a letter from Pierre S. duPont to Seward Prosser and Henry C. Bishop, November 22, 1923. GMX 30, *U.S.* v. *DuPont,* 1956:6626–6638.

More immediately, Sloan determined, "after a year's experience in the general office, that the purely advisory status of senior executive officers was no longer valid and that at least one officer in the general office should have line authority" (Chandler and Salsbury, 1971:528). Recognizing Pierre duPont's distaste for assuming aditional responsibility at GM, Sloan suggested that an *executive* vice president be appointed. In defining this new executive vice president's role, Sloan wanted "complete authority and responsibility for the administration of operations." Carrying out board as well as Finance and Executive Committee policies would then fall to the executive vice president (p. 528). But the new position was not approved, and Sloan did not gain such complete authority until May 10, 1923, when he became president.

Prior to that date, however, Sloan still made significant inroads into the divisions' spheres of authority. Consider, for example, his influence over the strategic positioning of the divisions' price ranges. Under Durant's administration, Sloan recalled that "prices were largely determined by the initiative of the different managers. . . . I remember one executive committee meeting at which one division manager said to another, 'I see you raised your price $150 the other day.' The other said 'yes,' and the first one said, 'I guess I'll do the same thing tomorrow'" (Dale, 1962:74). When Durant departed, Sloan ended the

CORPORATE EXECUTIVES

NAME	POSITION	SHARES
A. P. Sloan, Jr.	President	$225,000
D. Brown	Vice president	150,000
J. L. Pratt	Vice president	150,000
F. J. Fisher	Vice president	150,000
C. T. Fisher	Vice president	150,000
J. T. Smith	General counsel	150,000
A. Bradley	Vice president	100,000
R. H. Grant	Vice president	100,000
O. E. Hunt	Vice president	100,000
C. E. Wilson	Vice president	100,000
J. D. Mooney	Vice president	100,000
C. F. Kettering	Laboratory	70,000

DIVISION HEADS (AUTO)

L. P. Fisher	Cadillac	$100,000
W. S. Knudsen	Chevrolet	87,500
D. S. Eddins	Oldsmobile	35,000
A. R. Glancy	Oakland	35,000
W. A. Fisher	Fisher Body	28,000
P. W. Seiler	Yellow Truck & Coach	24,000

Figure 6-3. Shares in General Motors Management Corporation of corporate executives and division heads in 1930. *Source:* GTX 260, *U.S.* v. *DuPont,* 1956:3590–3596.

anarchy that had permitted a division to be "operated independently, making its own . . . price positions without relationship to the interest of the enterprise as a whole" (Sloan, 1964:59–60). The automobile price-overlap problem required immediate solution by the designers. On "April 6, 1921, the Executive Committee set up a special committee of the Advisory Staff" to recommend a more logical strategy. Sloan, as head of the Advisory Staff at that time, chaired this ad hoc corporate-dominated committee (p. 62). Somewhat later Pierre duPont, "to keep the Executive Committee out of operating detail," gave Sloan, then "vice-president in charge of operations, the authority to approve prices" (Chandler and Salsbury, 1971:527). Sloan's power was growing immensely. In the face of "tremendous opposition from within," for instance, Sloan moved Oldsmobile "into a new and more logical" market segment and "completely changed" Chevrolet's "basic policy," focusing it instead on "the mass market where Ford seemed to hold an impregnable position" (Drucker, 1973:697). Interestingly, these changes were made under the auspices of a supposedly weak corporate body, the Advisory Staff committee headed by Sloan.

The Executive Committee also established stringent rules for the introduction of new models (discussed more fully in Chapter 9). Once accepted, a design could not be changed if that action might raise the car to a higher price class (Chandler and Salsbury, 1971:519). Division managers now had their pricing dicisions thoroughly bound by the corporate headquarters.

Fixing the price of a finished automobile, in turn, *dictated* the values of many other cost-related variables for the division managers and their staffs. As later explained in the General Motors Institute manual on *Industrial Management:* "The price level which the management sets for a product places limits on the materials, tolerances, technical requirements and finishes. . . . The design engineer devotes a great portion of his time to seeing how much performance and style he can get into the product and still keep within the price, or cost, limitations" (Department of Organization and Management, 1951: Chapter 6, p. 5).

Dictating the divisions' prices was but Sloan's first step in GM's centralization. *Once president, Sloan accepted only the amount of decentralization that was forced on him by the limited channel capacity of the corporate headquarters.* During the early days this was more decentralization than Sloan desired, but still he had to move slowly. Although "there were strong temptations to step in and do things the quick way," the designers "used patience." "If," as Sloan (1924) acknowledged, "I or any of our other officials from the general office had stepped in with a sudden order, it would have halted our progress" and the men in the divisions "would have had their fears confirmed. . . . They would have left decisions altogether to us. And the work would have been too heavy for any one man or few men" (p. 138).

Just five years later Sloan could take a stronger stance toward centralization because of GM's growing headquarters staff and his mushrooming personal authority. On July 18, 1929, Sloan in addressing the Operations Committee declared himself "for new forms of co-ordination," i.e., centralized control. Sloan told the committee members that "our administration has been subject to criticism for not insisting upon changes and losing too much time in selling an idea and in not dealing promptly and effectively with weaknesses that are known to exist. . . . Resistance . . . has been greater than the man power to effect progress; therefore, progress has been slow. This . . . has got to be changed" (Sloan, 1964:171).

While forever expressing sympathy for the concept of decentralized operations, Sloan could not accept units performing less than optimally, as they often were when he did not make the decisions. Because he proceeded quite gradually his centralization activities went mostly unnoticed; indeed, his centralization policies were still being implemented fifteen years after he became president and with such subtlety that he could boast, though somewhat inaccurately, that he never gave a direct order. Consequently, Sloan was able to perpetuate the GM decentralization myth in his book, *My Years with General Motors*. (The more blatant contradictions in Sloan's book, however, led Wolff [1964:166, 168] to caution other firms to proceed most cautiously in adopting GM's so-called decentralized organizational scheme.) But clearly, Sloan had designed a decision-making atmosphere where only a fool would not know what the chief wanted and what he would reward.

To overcome the president's limited channel capacity, a formidable roadblock in any centralization plan, Sloan cut down on the number of units reporting to him and thereby reduced his load by creating the new group executive position between the GM president and the many divisions. Specifically, Sloan placed similar divisions into several operating groups (not including, of course, the Financial and Advisory Staffs) and placed a group vice president over each. In particular, both Pratt and Mott were group vice presidents, with Pratt heading Accessories and Parts Group(s) and Mott, temporarily, heading the Car and Truck Group. In addition, an Export Group was headed by James D. Mooney.

The group executives represented GM's central management and acted as advisors for the division managers of their groups (Douglass, 1954:152). In addition to supervising the work of several divisions, the group executives collectively helped set the corporation's overall policies (Chandler, 1966:165).

While the group executives were to be without line authority, it is very difficult to imagine how advice and orders were separated neatly. After all, several of the group executives were also members of the Executive Committee, the most prestigious operations unit in the firm. Would not the division managers want to heed the "advice" of such important corporate executives before it became an order? Would not deference to corporate advice become more likely as the tenure, importance, and status of the central office executives grew under Sloan's tutelage?

Even when group executives were not Executive Committee members, their activities would centralize the firm, for they *constantly* observed divisional actions and monitored performances. In contrast, the Executive Committee or Sloan could only provide more intermittent and cursory supervision. More thorough supervision resulted from reducing the number of divisions reporting to an individual or group and from grouping similar divisions together, thereby limiting the number of cognitive models needed to understand the subordinate units. In other words, since each group executive had at the most fourteen very comparable divisions under him, he was not overloaded, as Durant had been, with many multifarious units. Consequently, each division manager could expect careful inspection; that expectation alone would enhance the probability that corporate advice would be treated as corporate dictate.

Consider also the group executives' indirect centralizing influences. Mott, because of his Advisory Staff duties, was removed relatively early as the head of the critical Car and Truck Group, and the position remained unfilled from the early 1920s to the mid-1930s. Sloan, as the new president, apparently wanted these extremely important units under him directly. With the group executives monitoring the less important divisions, Sloan could devote his full capacity to the most important performance-generation units. Thus the group executives enabled Sloan to amplify his control power over all the divisions, yet concentrate on the critically important automobile divisions.

In 1933 Sloan at last relinquished direct supervision of the auto divisions, by appointing a strong executive vice president, as he had long ago suggested to Pierre duPont. In making Knudsen executive vice president, Sloan supposedly "caused the corridors of the General Motors building in Detroit to buzz with excitement" ("General Motors Coordinates," *Business Week,* October 21, 1933:16) over the breadth of Knudsen's authority. Knudsen, "as 'chief executive officer at Detroit' with 'general supervision over all car and body manufacturing operations in U.S. and Canada,'" was expected to bring to all the divisions "the policies which made Chevrolet the leader of the industry" and "Pontiac the largest selling straight 8 in the world." At long last Sloan had a man who could be trusted to operate the auto divisions as well as Sloan himself. While Sloan was loosening his personal grip on GM's internal operations (only to become temporarily more involved with finances and external relations), the grasp of the corporate office on divisional affairs did not diminish. In taking over in Detroit, Knudsen stated, "The executive offices of the corporation are in New York, and while Mr. Sloan as well as other officials have always maintained the closest possible contacts with officers of each division, I believe it is Mr. Sloan's wish that these contacts become even more closely coordinated" (p. 16). GM's ever-expanding parts and body interchangeability program demanded the greatest coordination. "The economies resulting therefrom . . . would stand a good chance of being lost in a complete decentralization of activities of the car and body companies. In his capacity as vice-president of the corporation in charge of all these activities, Mr. Knudsen can provide the necessary control over and insure the necessary cooperation between engineering and production departments of the various units" ("G.M. Reverts to Traditional Organization Plan but Centralizes Supervisory Control in Knudsen," *Automotive Industries*, October 28, 1933:536).

THE FINANCIAL STAFF

Another unit vital in the corporate headquarters' centralization effort was the Financial Staff, introduced in Chapter 5. While responsible to the Finance Committee, the Financial Staff conducted much of its detailed financial-control work in conjunction with the Executive Committee and the executive officers. Sloan, Brown, and Raskob as members on both committees and as chief operating and financial officers provided the links between these central decision units.

The Financial Staff at GM played an extremely important role in applying the DuPont financial-control procedures to the solution of GM's performance-control problems; "it was clear that there was a need to bring about better coordinated control over the corporation's growing operations" (Brown, 1957:45). The principal task of the Financial Staff was to insure that actual investment levels in each of GM's activities remained as originally allocated by the Finance Committee and did not balloon as they had under Durant.

According to Sloan (1964), the corporate executives felt that "the necessity of providing new forms of financial control was doctrine" (p. 116). Sloan's *My Years with General Motors* reveals the tight constraints imposed by the corporate office over GM's division managers with respect to the key capital-investment variables isolated in Brown's financial-control model. "Financial control at this juncture was not merely desirable," stated Sloan, "it was a necessity." He explained first "how the corporation curtailed the excessive freedom of the divisions" which jeopardized GM's survival. Though "the immediate remedies inevitably were centralizing," GM "could not afford to let its divisions continue ... the kind of mistakes they were making." The "weak divisions" were undermining the "strong ones," and the latter were operating "more for their own than the corporation's interests." While Sloan claimed that "these centralizing remedies—largely operating controls" were but "a temporary distortion in our general policy" and were later "corrected in order to return to a workable decentralization" (p. 119), no decision variables were returned to the division managers for their control. True, the "emergency" committees for controlling inventory and capital appropriations were disbanded, but only after stringent corporate controls—operated in part by the Financial Staff—made them superfluous. The alleged "key to decentralization with co-ordinated control," according to Sloan, turned out to be "a method of financial control which converted the broad principle of return on investment into one of the important working instruments for measuring the operations of the divisions" (p. 140). Supposedly, with "the means to review and judge the effectiveness of operations," Sloan and his colleagues "could safely leave the prosecution of those operations to the men in charge of them." Despite Sloan's assertion, the rate-of-return indicator simply *augmented* the controls placed over the critical variables of cash, inventory, production, and appropriations.

One of the first centralizing actions taken by the GM designers was the development of corporate-level controls for cash. Raskob first recognized the need for corporate cash-controls, and he instructed the GM treasurer, Meyer Prentis (formerly Prenskey) to develop such an administrative mechanism (Sloan, 1964:125). Under Durant's administration, "each of the many operating divisions handled their own cash" (Swayne, 1924:21), leading to an "almost unbelievable" (Sloan, 1964:122) situation of Prentis begging funds from the divisions without any idea of what GM's cash balances were or where they were located. "In 1922" GM "changed all this by setting up a consolidated cash-control system" with "depository accounts ... in some one hundred banks." Into these accounts were deposited "all incoming receipts." The central Financial Staff, in turn, administered all withdrawals, leaving the divisions "no control over cash transfers from these deposit accounts" (p. 123).

Centralized cash control assumed considerable importance because cash can be converted into a *wide variety* of uses, such as new equipment, higher inventories, increased salaries, and so on. Without a cash source the division managers' freedom of action became tightly circumscribed. Ended was an era when the division managers could ignore "the cautions, warnings, and even orders of the central office" simply because they had their "own funds for day-to-day production" and "could even borrow on their own account if the money in the division treasury did not suffice" (Chandler and Salsbury, 1971:481). "Men who decide about dollars," as Stinchcombe (1968) suggests, "are likely to be more generally powerful than men who decide about machines" (p. 166). GM's Financial Staff would look after the dollars while the division managers attended to the machines.

As Durant made no attempt to control either productive or finished inventories, he often found himself embarrassed with huge excesses. Not only did inventory levels far exceed needs, but inventories often were tied up in parts that were soon to become obsolete as design changes were made. At times capital was literally thrown away when inventories had to be written off as useless.

To eliminate such abuses, GM's planners relieved the division managers of their authority to manipulate their productive inventory levels. According to Sloan (1964), "The Finance Committee, on a temporary emergency basis, had taken control of inventories away from the operating divisions and on October 8, 1920, appointed an inventories committee, headed by Mr. Pratt, who was on Mr. Durant's staff, to bring the inventories under control" (p. 124). Brown and Prenskey (Prentis) of the Financial Staff were placed on the Inventories Committee by Pierre duPont shortly after he took over from Durant (Chandler, 1966:175). The first action Pratt's Inventories Committee took during the crisis of 1920 "was to send out, under the signature of the President of General Motors Corporation, a letter instructing all General Managers to buy nothing; to stop shipment of all purchases released—until the Inventories Committee could review the situation with each individual General Manager and decide on what material would be received and what would not be received" (Sloan, 1964:124–125). It would not take long before the divisions protested the constraint. "Some of the division managers," as Pratt testified, "felt that the Inventories Committee was tying them down too tight and didn't let them buy enough" (*U.S.* v. *DuPont*, 1956:1415).

Since Pratt's Inventories Committee was but a temporary decision unit, it recommended that in the long-run: "It would seem logical and sound in organization principle for the Vice President or Chief Executive in charge of operations to be looked to to see that the divisions effectively control inventories in accord with Finance Committee policies or good business practice" (Brown, 1957:47). As a consequence of this report, the Inventories Committee's monthly inventory and purchasing authorization procedure was handed to Sloan, at that time the operating vice president (Chandler and Salsbury, 1971:504; Sloan, 1964:127).

In considering more permanent controls Sloan (1964) realized, as Bradley had observed to him, "that the essential thing we learned from this experience was that the only way to cut back inventories—particularly in a time of declining business—is to reduce purchases and commitments for materials and supplies" (p. 125). Soon thereafter materials "already in the plant" were marked "for immediate manufacturing needs—for a quick turnover" (Wennerlund, 1924:678).

"Thus the division managers still bought the materials [a freedom that was later curtailed], but they were permitted to buy only enough at a time to make the number of cars and trucks specified in their approved production schedules" (Sloan, 1964:127). So while the division managers handled the more detailed and prosaic aspects of inventory control (i.e., what items were needed for a month's approved production), Sloan set the aggregate level of the divisions' productive inventories. (To insure that the division management made no inadvertent or willful mistakes in ordering raw material or parts, *bills of materials* were developed which gave "the quantity of each item required for one finished unit" [Wennerlund, 1924:678].) Setting the automobile divisions' monthly production schedules also "solved the more formidable problem of controlling the inventory of finished products" (Sloan, 1964:127).

Brown and his Financial Staff supplied the requisite support. This "arrangement" brought Sloan and Brown "into a continuous relationship" (p. 128). "After consulting with the vice president in charge of finance," Sloan explained, "I approved or modified the production schedule for each division" (p. 127). Not until Sloan approved the production schedule could the division managers proceed with production and make commitments for materials (p. 128).

Here again, then, Sloan and Brown constrained the division managers' actions: the production schedules automatically set parts purchases, labor employment, machine usage, and the like. So by choosing to control a few critical production variables that were connected to many other decision sets, the corporate headquarters could centralize, yet neither overload its channel capacity nor give the impression of excessive centralization. Under these circumstances the division managers acted simply as regulators of variables to insure that the desired quantities of cars were produced.

Besides developing controls over GM's working-capital variables of cash, productive inventory, and finished inventory, the GM designers established controls over fixed capital investments, such as plant and equipment. Just prior to Durant's departure, an ad hoc Committee on Appropriation Request Rules was formed and chaired by Sloan to study GM's capital-appropriations problem. Other executives on the committee were Pratt and Prentis (Sloan, 1964:119). The committee recommended new capital-appropriations control procedures. The general manager of a division could authorize small expenditures "on his own" (p. 120). Larger amounts, however, would require extensive corporate review and approval. The committee also recommended the development of "an appropriations manual . . . setting forth in detail the kind of information the divisions and subsidiaries should present" to substantiate the engineering and economic worth of a proposed expenditure (p. 121). These proposals were adopted; instructions were given to draw up the proposed appropriations manual; and A. C. Anderson, Donaldson Brown's choice to head the new Appropriations Accounting Department of the Financial Staff, developed the manual. Anderson's manual, reviewed and approved by Brown, Pierre duPont, the Executive Committee, and the Finance Committee in April of 1922, called for "an independent impartial review and checking all phases of proposed projects outside the Division or Subsidiary itself" (Chandler and Salsbury, 1971:502). The division managers could no longer set their capital-investment variables as they wanted.

In summary, since all of the above financial variables related to capital allocation within the firm, it was certainly appropriate to control them at the corporate rather than at the division level. Their corporate control, after all, was simply a more detailed extension of the aggregate resource-allocation decisions made by the GM Finance Committee when it selected the firm's product lines. Essentially, the financial controls along with the pricing limitations guaranteed that the controlled domain remained as the Finance Committee constituted it.

At the same time, the control of these variables at the corporate level enhanced GM's corporate and divisional return rates. Given its strong familiarity with Brown's financial-control model, the corporate headquarters could best control these key return-generating variables for optimum performance. Once set, moreover, these dominant variables indirectly constrained the divisional operations thereby reducing the chance that their managements would commit mistakes that might further damage performance.

THE ADVISORY STAFF

While the corporate controls over divisional finances exerted a strong centralizing influence, the greatest centralization thrust was felt in the nonfinancial areas. The GM Advisory Staff, with the full support of the Executive Committee and executive officers, played the most significant role here.

Pierre duPont began the creation of GM's huge central staff. As early as 1918 he had the DuPont Engineering Staff study the GM situation. Eric L. Bergland, later to become an important GM staff executive, conducted the survey and reported the lack of method and system at General Motors, the absence of a central engineering department, and the lack of any kind of permanent well-organized central staff (Chandler and Salsbury, 1971:468).

Under Durant, GM had no central staff; instead, it used the DuPont Company's. The DuPont Engineering Department, for instance, constructed or enlarged old automobile plants; the DuPont Personnel Office's head (John Squires) suggested centralizing the firm's employee information and personnel activities; and Pratt helped Durant plan his expansion strategy and investigated the availability of suppliers and materials (pp. 470–471). While highly impressed with Squires, Pratt, and other DuPont staff executives, Durant's only move to create a central staff was to add Kettering's Research Laboratories to the GM family. But, in reality, "Pierre and Sloan had much more to do in bringing Kettering to General Motors than did Durant" (p. 471).

Although Pierre duPont began the development of the central staff, Sloan brought the central staff to its ultimate in power. In fact, the most difficult task in building GM's powerful central headquarters was forming an effective Advisory Staff, which "required more senior executives than did the Financial Staff or the operating divisions" (p. 497). After Sloan had begun to establish the Advisory Staff, "Pierre switched Haskell and Sloan around, making Sloan vice president in charge of Operations and Haskell the head of the Advisory Staff" (p. 497). Since Haskell was ill, Mott was soon given responsibility for the

Advisory Staff in Detroit, thus bringing the divisions under closer observation than was possible from Haskell's office in New York.

The Advisory Staff's power grew over time. "At General Motors," for example, "Alfred P. Sloan, Jr., and later C. E. Wilson and Harlow Curtice acted as chiefs of staff (vice presidents or executive vice presidents in charge of staff) before becoming presidents" (Dale and Urwick, 1960:190).

One of the first nonfinancial problems Pierre duPont and Sloan confronted when they assumed command arose in product development and research utilization. Several of the divisions delivered cars with serious design faults, and their reputations suffered accordingly. If the firm was to survive, let alone prosper, these defects had to be remedied. As would happen time and time again in such situations, centralization supplied the solution. Only one division seemed safe from such corporate interference in these early days. "Circumstances had dictated that Buick, which was doing well, should be left for the time being more or less in its former, wholly decentralized state, and with its own program" (Sloan, 1964:75). Serious offenders, in contrast, demanded immediate attention. "But notwithstanding our concept of [decentralized] organization and directly contrary to it," as Sloan admitted, "expediency was permitted to centralize the affairs of other divisions. This trend was made emphatic by the decision of the top officers to impose a radical car design upon two divisions, Chevrolet and Oakland" (p. 75). The dramatic innovation to be imposed was researcher Kettering's unproven and ill-fated "copper-cooled" engine—a design vehemently opposed by the Chevrolet division manager, Karl Zimmerschied. "The Executive Committee thus made both the policy and program for these divisions on the most significant question that can come before a division, namely its engine and car design. The Executive Committee had that privilege, and in the circumstances it elected to exercise it" (p. 75).

To illustrate further the growing power of the research staff, consider next the introduction of Kettering's quick-drying lacquer. In 1921 auto manufacturers had yet to develop a durable, quick-drying paint to finish their cars; they were still using job-shop techniques developed in the days of the handmade carriage. Three severe problems resulted.

First, the poor durability of the finishes necessitated expensive repairs. It was, as Sloan testified, "the most important problem we faced" from the consumer's viewpoint. "People who bought the cars, especially the higher priced cars like the Cadillac, and paid several thousand dollars for them, could not understand why, after they had them a short time, . . . the finish would break down" (*U.S.* v. *DuPont*, 1956:1285).

Another severe problem was the huge inventories that the job-shop process entailed. During the painting of a car "something like two to four weeks went by, depending among other things on the temperature and humidity" (Sloan, 1964:235). The heavy inventories necessitated by the slow-drying paints and varnishes, of course, adversely affected working-capital requirements. "Mentally calculating the company's daily production, it was suddenly realized it took over seventeen days to paint a car—more than $20,000,000 in automobiles in warehouses waiting for the paint to dry" (Kettering and Orth, 1932:86–87)! Not only were in-process inventory levels excessive, but fixed-capital investments were huge also, for "a production of 1000 cars a day . . . required space of 18,000 cars in process . . . that is, twenty acres of covered indoor space" (Sloan, 1964:236).

Third, the repeated handling necessitated by the multiple applications of color and varnish coats demanded considerable labor. When coupled with the high inventory charges, these costs became prohibitive—except for the low-volume, high-margin makes. Inexpensive cars like Ford's Model T thus were finished with a black enamel that could be oven dried quickly and cheaply (Leslie, 1983:191). Since Sloan wanted divisions like Chevrolet and Oakland to fight the Model T with colors and closed bodies (which only aggravated the problem, given their increased surfaces), GM's research staff looked for an answer.

Kettering's Advisory Staff researchers, guided by the members of a Paint and Enamel Committee, soon arrived at a solution: a quick-drying lacquer with the trade name of Duco. Sloan (1964), in extolling the virtues of the new Duco lacquer, stressed, "Its quick drying removed the most important remaining bottleneck in mass production, and made possible an enormously accelerated rate of production of car bodies" (p. 236).

To Sloan's dismay, several of the important GM auto divisions—notably Cadillac and Buick—remained unimpressed with Duco since it had a dull, lackluster though extremely durable finish, whereas their high-quality cars typically had lustrous varnished finishes. Thus these divisions refused to trade high-luster finishes for more durability and lower in-process inventories, truly an intolerable situation for the return-conscious Sloan-Brown designers. So Mott, as chief of the Advisory Staff, "personally 'pressured' Buick and Cadillac, and secured an executive wire from Sloan to Cadillac ordering them to comply with dealers' orders when the dealers specified Duco" (Young and Quinn, 1963:192).

To justify this incursion into the supposedly decentralized affairs of Buick and Cadillac, Sloan could point to Oakland's enthusiastic endorsement of Duco lacquer. Oakland, unlike Buick and Cadillac, continued to experience serious product-acceptance problems and the new colored paint could only help improve its chances. Duco in fact did just that (Leslie, 1983:193).

But Oakland, too, could find itself opposing innovations that other divisions had proven to be valuable. In 1925, for instance, Cadillac adopted Kettering's new crankcase ventilator and oil/air filters to improve engine life. However, Oakland's manager, A. R. Glancy, "was particularly unyielding" and asserted that "if General Motors stood as a unit without these accessories, a great deal of money would be saved and no sales lost" (Leslie, 1983:199). As he often did when encountering divisional intransigence toward his improvements, Kettering appealed to Sloan to intervene. "As usual, Kettering won, and Sloan did issue an executive order" (p. 199).

Consider one more of these centralization incidents; this one again involves the irrepressible Kettering. After Pierre duPont had complained about the engine vibration in his new Cadillac (Leslie, 1983:199), Kettering embarked on a program to improve engine balancing, which until then had been poorly understood at best. "Kettering, brilliant engineer, scientist and inventor," as Ernest Seaholm, Cadillac's chief engineer, characterized him,

> was also a superb salesman, and so once he had solved the question the heat was on to incorporate it in the Cadillac car. Unintentionally, possibly, we dragged our feet. The answer had come rather late. We were in the process of releasing next year's model so why not wait another year, etc.? He just wouldn't take "no" for an answer, and while Herby Rice [the division manager] and I were in Europe for the automobile shows, he got to Mr. Sloan and they ruled it a must. (Hendry, 1983:126)

At about this time, Sloan (1924) would claim:

> Even when we make recommendations from the general corporation's office, the head of a division is at liberty to accept or reject—and once in a while he rejects. . . . We do not issue orders. I have never issued an order since I have been the operating head of this corporation. The presidency carries the power to issue orders, but I never use it, and do not intend to use it. . . . We could not follow one policy in the main, with exceptions to it once in a while. (p. 138)

Questionable disclaimers like these coupled with Sloan's sophisticated knowledge of human nature, his subtle use of organization structures, and the general confusion surrounding the terms "decentralization" and "divisionalization" helped perpetuate the GM decentralization myth. But, in reality, GM's centralization was already well under way at the beginning of 1921 when Chevrolet was instructed to adopt Kettering's copper-cooled engine. "This was virtually an order, and so far as Chevrolet was concerned the die was cast" (Sloan, 1964: 74), as it also was for the decentralization concept.

Far more significant corporate incursions into divisional affairs were to take place as the Sloan-Brown designers attempted to improve GM's performance. Kettering's Duco lacquer again provided the impetus. The day of Ford's oven-dried black baking enamel was about to end, along with more of the division manager's autonomy. The corporation would "help" style their cars.

In 1927 Sloan suggested to the other members of the Executive Committee that a special department be established within GM "to study the question of art and color combinations in General Motors products" (Sloan, 1964:269). With the automobile market moving from an engineering to a styling emphasis, art and color variables became more important to high-return performances. "The appearance of a motorcar," as Sloan (1941) later explained, "is a most important factor in the selling end of the business—perhaps the most important single factor because everybody knows that all cars will run" (p. 185).

Characteristically, Sloan's new Art and Color Section "was made a part of the corporation's general staff organization, even though it received its funds through the Fisher Body Division" (Sloan, 1964:269–270). The section, as previously mentioned, was directed by Harley Earl, who was charged with guiding production body design and conducting research and development in automobile design. As style grew in importance to automobile consumers, the Art and Color Section changed its name to the Styling Staff, grew in size, increased in power, and assumed more and more divisional functions. Sloan, ultimately, admitted: "At one time, responsibility for the styling of the cars and other products was vested in the divisions. Since then it has been found desirable to place the responsibility . . . in the Styling Staff. . . . The adoption of any particular style is now a joint responsibility of the division concerned, the Styling Staff, and the central management" (p. 430).

Eventually, the managers of the car divisions and Fisher Body lost the privilege, without special corporate-headquarters invitation, to attend the group meeting that decided their cars' specifications. In Sloan's (1964) words, "since the group concerns itself with broad corporate policy, the membership does not include the general managers of the car divisions or Fisher Body" (p. 241). Important styling variables just had to be controlled by the central headquarters staff and executives. And by 1938 divisional autonomy in styling had been restricted greatly: "Divisional originality . . . gets full play in hoods, cowling, and radiator grilles" ("General Motors," *Fortune*, 1938:156). Freedom to style hoods, cowling, and grilles—within the constraints imposed by the interchangeable chassis parts and the corporate-designed body—is not much freedom. The design restrictions were especially tight since the divisional design studios came under the jurisdiction of the corporate Styling Staff and worked under the unquestioned dominance of Harley Earl (Bayley, 1983:12–13).

Along with the centralization of styling went the GM designers' creation of larger sales, engineering, and manufacturing staffs. Because of Chevrolet's success, Sloan "wanted to spread the benefits of its management." Accordingly, he made Grant and Hunt of Chevrolet corporate vice presidents in May of 1929 "with staff responsibility for sales and engineering, respectively. At the same time, C. E. Wilson, formerly of Delco-Remy, was elected vice president for manufacturing" (Sloan, 1964:170). Until then, as Sloan put it, GM had "only a modest staff organization outside of finance, Mr. Kettering's laboratories, and the work associated with the interdivisional committees." Sloan concluded: "The election of these men therefore was a new beginning in staff activity, which would eventually . . . develop into the great staff organization we have today" (p. 170).

To see how the larger sales, engineering, and manufacturing staffs increased GM's centralization, consider the new corporate engineering group: "Under Mr. Hunt's guidance, the advanced engineering in the divisions became a corporation staff responsibility" (Sloan, 1964:255). For instance, the "special product-study groups," organized to study major engineering problems, eventually changed "from corporation task forces situated physically within the operating divisions into permanent separate organizations engaged in the continual process of research and testing in four vital areas—power development, transmission development, structure and suspension development, and the design of new types of cars" (p. 256). With the corporate designers handling all major development work and the downgraded divisional engineering staffs handling the relatively routine and repetitive problems of bringing out each year's model (p. 255), another critical control vector had been shifted to the corporate headquarters.

The depression of the 1930s brought intense pressure on the divisions to coordinate and improve their operations and added much impetus to GM's centralization thrust. "There can be no doubt," as Sloan (1964) admitted, "that, . . . some overcentralization took place and it was wrong. On the other hand, however, Messrs. Wilson, Grant, Hunt, and Bradley, all staff men, thought the contrary. Each . . . recommended some specific form of increased co-ordination" (p. 175). Rather than censuring these important staff executives for their "overcentralization" tendencies, Sloan rewarded them, appointing Bradley to the all-powerful Executive Committee on August 6, 1934, and Grant, Hunt, and Wilson on August 5, 1935.

Sloan (1964) himself best summarized the full importance of the GM staff after its many years of development under his guidance:

> Indeed, many . . . important decisions of central management are first formulated in collaboration with the staff . . . and then adopted, after discussion, by the governing committees. Consequently, the staff is the real source of many decisions. . . .
> . . . The staff contributions in . . . styling, finance, technical research, advanced engineering, personnel and labor relations, legal affairs, manufacturing, and distribution are outstanding. (pp. 431-432)

Lastly, one should note that Sloan applied the label "advisory" to GM's growing body of staff units. In the end, about all this title meant to GM's increasingly prudent divisional line managers was that they would be well-advised to accept the staff's advice.

THE INTERDIVISIONAL COMMITTEES

Probably Sloan's most subtle way of centralizing GM in the early 1920s was his development of the Interdivisional Committees. These multipurpose committees provided considerable centralization over the auto divisions because they set the values of important performance variables and were easily dominated by the corporate executives. For the most part, high corporate executives—usually Executive Committee members—both manned and chaired these committees. Corporate-headquarters executives acted as secretaries, ran the supporting staffs, and conducted the committees' extensive affairs. Divisional managers were systematically excluded from membership (except where they possessed important performance-improvement information), but lower-level divisional executives were members of the committees.

Thus, Sloan bypassed his division managers and plowed directly into the lower-level divisional affairs of the critically important car-production units. By having corporate executives and staff officers come in contact with second-level people within the auto divisions, Sloan could rest assured that corporate views would dominate and that corporate advice would be taken. Furthermore, by their superiors' meeting with their subordinates directly, the division managers' positions were eroded significantly. What might be going on in those meetings between their own chief engineers, sales managers, or purchasing agents and Sloan, Brown, Mott, or Pratt? What was said? A prudent division manager would need to be careful. Not unexpectedly, there seems to have been some severe criticism made of Sloan regarding this matter, for Sloan mounted a strong defense to justify his position. To Pierre duPont he said, "the policy of doing business exclusively with the General Manager is not only unsafe, in my judgment, but absolutely dangerous and should not be further tolerated" (Chandler and Salsbury, 1971:524).

Sloan's bypassing of the auto-division managers apparently was in response to their ignoring his Advisory Staff's advice in such areas as research and development, sales, and plant engineering and construction (p. 524). To correct these divisional shortcomings, Sloan and his trusted fellow designers on the Interdivisional Committees just had to go ahead and set these important variables themselves.

Six Interdivisional Committees were instituted: (1) General Purchasing, (2) Institutional Advertising, (3) General Technical, (4) General Sales, (5) Power and Maintenance, and (6) Works Managers. The General Purchasing Committee was the first of these units, and it established their general characteristics.

The General Purchasing Committee

There seem to have been three specific reasons behind Sloan's creation of the General Purchasing Committee. One dealt with the poor bargaining conducted by several of the divisional purchasing agents. P. S. Steenstrup, Sloan's old partner at Hyatt Roller Bearing, first drew this problem to Sloan's attention. By accident Steenstrup "ran into an old friend" (GMX 46, *U.S.* v. *DuPont,* 1956:6689) who was in the parts business and who did not know Steenstrup was associated with GM. Unsuspectingly, Steenstrup's friend "stated that it was a strange thing to him but a large corporation like the General Motors apparently had not properly coordinated their purchasing departments, and he knew of cases where one division would be buying certain articles considerably cheaper than another division within the corporation" (p. 6689). Second, Sloan hoped through the Purchasing Committee to pool purchases and thereby lower operating costs. Specifically, Sloan "believed that the company could save from $5,000,000 to $10,000,000 annually if the purchasing staff in the general office drew up contracts on items that were widely used by the corporation as a whole" (Chandler, 1966:187). A third reason behind the General Purchasing Committee stemmed from the division managers' "skepticism" over Sloan's "suggestion for more centralized purchasing" (p. 187). To overcome their "many objections," Sloan offered his division managers the palliative of participation.

The General Purchasing Committee was built around the Purchase Section of the Advisory Staff. After considering several plans for coordinating GM purchases, Sloan forwarded one to the divisions for their comments. The final plan called for a General Purchasing Committee with the Director of the Purchase Section of the Advisory Staff as its agent.

Sloan claimed—both in *U.S.* v. *DuPont* and in *My Years with General Motors*— that to meet divisional objections he "proposed the General Purchasing Committee, with a membership drawn mostly from the divisions" (1964:103). While he may have proposed divisional domination, he never finalized it. The original committee of 1922, for instance, consisted of:

> G. G. Allen, Purchasing Agent, Buick
> H. L. Barton, Director Manufacturing Section, Advisory Staff
> W. W. Cromer, Purchasing Agent, Dayton Engineering Laboratories
> D. F. Hulgrave, Purchasing Agent, Cadillac
> W. A. Kimball, Purchasing Agent, New Departure Manufacturing
> J. H. Main, Director, Purchase Section, Advisory Staff
> C. S. Mott, Vice President
> D. P. O'Keefe, Purchasing Agent, Chevrolet
> J. L. Pratt, Vice President
> Alfred P. Sloan, Jr., Vice President—Chairman
> James Lynah, Assistant Director Purchase Section, Advisory Staff—Secretary.

> (GMX 63, *U.S.* v. *DuPont,* 1956:6744)

Thus of the eleven original members, only five represented divisions. Furthermore, most of these divisional representatives had reacted favorably to Sloan's purchasing-coordination plan. Regarding Allen of Buick, for instance, Sloan wrote, "Mr. Allen seemed to be in accord with the whole program and simply pointed out, very properly so, the importance of dealing with sources of supply that were reliable and that would not go back on us in times of stress" (GMX 51, p. 6715). Cromer too felt that "there does not seem to be anything in it that conflicts with our present plan of operation" (GMX 52, p. 6717).

The 1923 salaries and Managers Securities stock allotments of the committee members, as shown in Figure 6-4, provide further perspective on the corporate-dominance issue, if salary and bonus allotments are accepted as measures of status and decision-making power.

Sloan could also draw on less tangible sources of influence, for he was rapidly becoming "the undoubted head of General Motors with moral authority within GM's general management such as has rarely been equaled" (Drucker, 1973:622). It is difficult to see how the divisional representatives on the General Purchasing Committee could escape agreeing with Sloan, Mott, Pratt, and their staff assistants. Just to be sure, however, Sloan brought Pierre duPont to the first meeting of the General Purchasing Committee "simply to give it a little atmosphere" (testimony of Sloan, *U.S.* v. *DuPont*, 1956:1061).

CORPORATE LEVEL

NAME	SALARY	SHARES
Barton	$24,000	$40,000
Main	15,000	0
Mott	100,000	300,000
Pratt	40,000	150,000
Sloan	100,000	350,000
Lynah	10,000	0

DIVISION LEVEL

Allen	$15,000	$25,000
Cromer	10,000	0
Hulgrave	?	0
Kimball	?	0
O'Keefe	12,000	15,000

Figure 6-4. Salaries and Managers Securities stock allotments of General Purchasing Committee members. *Source:* GMX 30, *U.S.* v. *DuPont*, 1956:6626–6638.

During its ten years of existence the General Purchasing Committee entered into 709 contracts (Brief for Appellees, *U.S.* v. *DuPont,* 1956:47) for the divisions, covering such important items as tires, frames, and steel. General purchase contracts for all the divisions meant that specifications among divisions had to be standardized, so subordinate committees were charged with developing common standards for innumerable items, like electrical equipment and body trim. After specifications were reviewed and approved by the General Purchasing Committee, the head of the Advisory Staff's Purchase Section was authorized to contract for the needed quantities.

Not all the divisions, of course, appreciated these corporate services. Along these lines Lynah, Sloan's hand-picked secretary of the General Purchasing Committee, testified that there "was a feeling of resentment on the part of many of the General Motors division people, particularly the purchasing agents and others who were afraid that there might be. an effort to centralize General Motors organization on the same scheme of organization under which the DuPont Company operated" (*U.S.* v. *DuPont,* 1956:1172). Since Lynah himself had been chairman of the DuPont Company's purchasing-standardization committee (p. 1075), the division staffs' centralization fears seem quite well-founded.

To circumvent the new purchasing rules, some of the division purchasing agents ignored the corporate contracts. Sloan bristled to the Executive Committee that

> certain Purchasing Agents after receiving the general contract . . . deal with the disappointed suppliers and buy independently on a lower basis.
>
> . . . as Chairman of the Committee I wrote to the Purchasing Agents of all Divisions and told them that they were not to be permitted to purchase outside of the General Purchase Committee's contracts and indicated . . . anything of that kind . . . would certainly have the disapproval of the Corporation and in a way would not be tolerated.
>
> . . . I want it distinctly understood by everybody that neither Mr. Mott nor myself nor any Central Office authority or any part of the staff is taking any position whatever in this matter. It is to be clearly understood that the Divisions themselves through their appointees acting solely on and in behalf of themselves and in no sense in the interest of the central organization, are taking a position which in their judgment is constructive and to the benefit of the stockholders of General Motors Corporation. . . .
>
> I think that the time has come when we have got to take a little stronger position of capitalizing the advantages of General Motors Corporation. I will assure you and will take full responsibility that it will have the support of the most important and most capable Divisions and the best men in those Divisions. (GTX 458, *U.S.* v. *DuPont,* 1956:4096-4098)

What must have upset Sloan here was not that some units were getting lower prices—he was all for that—but that his committee's authority was being subverted. The failure of the divisions to follow corporate guidelines was extremely dangerous at this point, for Sloan had begun to plan other centralizing interdivisional decision groups. In the same memo to the Executive Committee Sloan alluded to his plans: "This is the beginning of many activities of similar nature and unless the policy that I am developing is reversed by the Executive Committee these Committees representing the functional activities of the Divisions . . . will take positions in the interest of the Corporation which some Divisions may consider more or less arbitrary" (p. 4098). Arbitrary or not, these decisions were going to stick. Divisions were to be required, after January, 1924, "to take the percentage for which they are committed regardless of whether or not more favorable terms are offered on the open amount" (GMX 158, *U.S.* v. *DuPont*, 1956:7106).

Three important improvements resulted from the activities of the General Purchasing Committee. First, standardization of parts among all divisions was achieved on a massive scale. "The General Purchasing Committee's real and lasting success was in the area of standardization of materials" (Sloan, 1964:104). Second, the costs of parts and supplies were reduced. For the most part, savings resulted from improved divisional operations rather than coordinated monopsony buying power (though a number of quantity-discount contracts were written); after all, the "quantities of any particular product needed for one division were generally large enough to justify the supplier giving the lowest possible price to that division" (p. 103). Once the divisional operations were improved, then, there was less need for the committee, and it went out of existence in the early thirties. A third and final achievement of the Purchasing Committee was the experience it provided in coordinating "interdivisional activity" (p. 104). Sloan quickly extended this approach to the other functional areas of the divisions.

The Institutional Advertising Committee

Sloan found in 1922 through consumer studies that most people did not know of GM. He thought that the parent company should be publicized (Sloan, 1964:104) and that the divisional campaigns should be fitted to the corporate advertising. Accordingly, GM "started handling all its advertising through one agency. Before that time . . . the advertising was handled on [an] independent basis, each company concerning itself not at all with what the others were doing so far as co-operation and co-ordinated effort were concerned" (Nichols, 1923:17).

Sloan created the Institutional Advertising Committee to handle the interdivisional decision variables related to advertising. Via this committee the corporate officers came to control divisional advertising. When Oldsmobile introduced a new car after Sloan had forced it to shift price classes, for example, "the whole advertising campaign [was] a General Motors, rather than an Oldsmobile, presentation" (p. 19). Supposedly, Sloan "made a rule that if any advertising theme dealt with a particular division, it must have the approval of that division" (Sloan, 1964:104–105). But what "reasonable" divisional staff would disagree with the entire Institutional Advertising Committee? The committee's composition, as Mott's (1924) description shows, was heavy on the corporate-headquarters side, including a "vice-president of General Motors" as chairman, "the president and several [other] vice-presidents," plus "the presidents of the advertising agencies handling General Motors advertising, the director of the sales, advertising, and service section of the advisory staff and several of his aides." Non-corporate representation was limited to "the sales managers of some of the divisions" (p. 526).

The General Technical Committee

Sloan's next committee effort was the General Technical Committee. Sloan, as president, headed the committee as he had the General Purchasing Committee. Other members, according to Mott (1924) again, included "several of the vice-presidents, the president's technical advisor, the chief engineers of the principal divisions, members of the advisory staff whose work deals with technical matters, the head of the patent section and the vice-president in charge of the accessories group of plants" (p. 526). Since the General Technical Committee had more of an educational and informational flavor and less of a decision-making emphasis, "other members of the organization" were "from time to time invited to assist in the deliberations of this committee" (p. 526). (Because of its education and information roles, the committee is considered more extensively in Chapter 8.)

While the committee was information-oriented, it still exerted strong central decision-making power over the divisions. First, Sloan charged the committee with "setting up the broad specifications of the different cars to assist in . . . efforts to keep the several General Motors' cars distinct and separate products and in a proper price and cost relationship to one another" (Sloan, 1964:110). That charge inevitably circumscribed the divisions' freedom of decision. Such restrictions could grow as tight as having the General Technical Committee supervise Chevrolet's designing of the Pontiac car, which was then assigned to the Oakland division for manufacturing and marketing. The General Technical Committee also centralized GM's divisions via its extensive audit activities, which are discussed in Chapter 9.

The General Sales Committee

From the functional area of engineering, Sloan jumped to marketing by next creating the General Sales Committee. He opened the first meeting of this committee on March 6, 1924, by apologizing for centralizing yet another functional area once under division control: "While General Motors is definitely committed to a decentralized plan of operation, it is nevertheless obvious that from time to time general plans and policies beneficial to the Corporation and its stockholders, as well as to the individual divisions, can best be accomplished through concerted effort" (Sloan, 1964:111). Eventually, the chairmanship of the General Sales Committee was "given to Donaldson Brown, vice president of finance, because of the bearing of statistical and financial controls on production and sales problems. Co-ordination in sales thus extended to the Financial Staff" (p. 112). As with the other committees, a staff officer also served as secretary of the committee, in this case B. G. Koether, director of the Sales Section, who had a staff that could be "expanded if necessary." So both the Financial and Advisory Staffs played an important role in the deliberations of the General Sales Committee.

The Power and Maintenance Committee and the Works Managers Committee

Sloan's final two Interdivisional Committees were the Power and Maintenance and the Works Managers Committees. These closely related committees dealt with the efficient use of GM's plants. E. L. Bergland was the executive secretary for both. Bergland, like James Lynah, executive secretary of the Purchasing Committee, had been a DuPont Company executive. To make sure that committee activities went as he wished, Sloan was a member of both groups. Pratt headed Power and Maintenance, while Bassett chaired the Works Managers Committee.

Bassett was the only division manager to preside over an Interdivisional Committee. And if Sloan had sought a "token" division manager to chair one of these committees, he could not have found a better man (see, for example, Bassett, 1926:268-270). Within his well-managed Buick Division, Bassett had established a committee similar to the GM Works Managers Committee. "A board of factory executives, composed of the various managers and superintendents as well as the general manager and his assistants, supervise[d] the manufacturing operations of the Buick Motor Company as a whole" (Buick, 1925:39). "New methods" were "discussed at the regular meetings of this board, common problems" were "studied out, the latest ideas in factory practice and the handling of materials" were "investigated to see if they" could "profitably be applied to any or all of the factories" (p. 41). Bassett, then, could be trusted to run a corporate committee designed to bring all the divisions up to the Buick standard in efficiency. When Bassett died, rather than place another division manager in the chairmanship, Sloan chaired the Works Managers Committee himself.

POLICY GROUPS

Between 1934 and 1937 Sloan devised a series of Policy Groups to supersede the Interdivisional Committees. One of the first such decision-making bodies was the Distribution Group composed of "Sloan, Knudsen, Grant, Brown, Bradley, L. P. Fisher, and A. L. Deane" ("G.M. III," *Fortune*, 1939:105), all corporate-level executives. Sloan's Distribution Group was very active, having "met all day every other week" between 1934 and 1939. "Although the divisional sales organizations" supposedly had "been kept relatively independent," the Distribution Group promulgated "a series of resolutions and suggestions" that had "formulated a new distribution policy for G.M." (p. 105).

In keeping with Sloan's style of always downplaying his centralization moves, the Policy Groups' resolutions and suggestions were not directives backed by formal authority. Still they were not to be ignored or overlooked, for the groups "had on them the president and other executives, and those executives had line authority" (testimony of Sloan, *U.S.* v. *DuPont*, 1956:1009).

When formally established across all the divisions' functional areas in 1937, the Policy Groups culminated the GM centralization process. "They combined the top executive officers, including the president, with functional staff men," like Grant, who by then were supported by very sizable corporate staffs. They set corporate policy in the functional areas, and "the divisional managers, being charged with administration, were specifically excluded" (Sloan, 1964:182). Here, again, the corporate headquarters had come to dominate GM's decision making. About Grant, *Fortune*'s editors observed, "He is one of the four or five most influential G.M. executives, being a member of six policy groups and chairman of the Distribution Group ("G.M. III," *Fortune*, 1939:74). Grant had a sizable staff group to back his directives. It was noted at the time, for instance, that "neither Chevrolet nor any other G.M. division can now add or replace dealers in any metropolitan area unless Mr. Grant's statisticians, after applying their slide rules to a dozen considerations, say the potential will permit" (p. 105).

The division manager's choices thus were narrowed tremendously by removing many decisions from his jurisdiction completely, by constricting many others indirectly, and by prescribing many policies and goals to guide him in making the decisions left in his jurisdiction. In the latter case, the corporate head-quarters specified the desired values for many variables via policy statements and performance objectives, and the division manager and his staff insured that the actual output values indeed equaled the Sloan-Brown designers' input values. In short, the divisional management acted merely as a regulator for the corporate controllers or decision makers. "Once production objectives have been established," to cite an example presented by Donner (1967), "the divisional or subsidiary management has the responsibility for determining how these objectives can best be achieved. In what proportions will the many models and body styles be ordered and produced? What combinations of color and interior trim will be specified, and which materials must be obtained now?" (p. 32). More and more, then, the divisional managements focused on the routine and repetitive. They became restricted to making GM's high-frequency, day-to-day decisions.

The extent of centralization under the Sloan-Brown designers was reflected in the cost for central office activities, which on the average amounted to slightly "less than 1 percent of the corporation's net sales" (Sloan, 1964:431). Almost 1 percent of sales—not of profits—represented a tremendous increase from the days of Durant and his handful of assistants.

7

Organizing the Divisions

In assigning GM's less critical variables to the divisions, GM's designers achieved a horizontal decomposition that provided: (a) control diversity, (b) performance commensurability, (c) causal comparability, and (d) component decomposability. These structural characteristics greatly simplified GM's decision-making processes at both the corporate and the divisional levels. Through diversification of control among a large number of divisional decision centers, no one unit exerted an excessive impact on GM's overall performance; since each remained small and therefore relatively simple, they also did not overwhelm their management teams; hence fewer mistakes were made and less corrective action needed. By keeping the performance scales of each unit commensurable, GM's designers could readily compare their contributions to the firm's overall performance. The causal comparability or similarity in functional structure, product technology, and marketing environment of GM's components further added to the simplicity; the corporate and divisional decision makers experienced little difficulty, respectively, in switching their attention among units or in understanding their neighbors' operations. Last, and most important, the decomposability among GM's divisions allowed each unit to be treated as a separate entity. This simplicity was particularly important in the early 1920s, given the small corporate headquarters and the designers' still somewhat limited experience in automobile operations. But with a growing central unit and their staffers' increasing tenure, GM's designers became better equipped to handle more complex and intertwined decision-making situations. Their growing sophistication was already paying dividends by the late 1920s when GM added several makes to cover the strong market for middle-priced cars. Given the similarity that always existed among GM's divisions, adding cars to the line meant that price buffers shrank and common parts increased. During the depression, this trend became even more pronounced as the whole line moved toward the low-priced field. Further centralization resulted.

CONTROL DIVERSITY

Sloan's (1941) concept of management was "to divide" the organization "into as many parts as consistently can be done" (p. 135). He wanted GM "to be a collection of small or moderate-sized units" (Seltzer, 1928: 140). The Sloan-Brown designers attained (at the divisional level at least) extensive management diversity because GM employed, rather than a functional structure, a divisional scheme capable of being broken into far more decision centers; GM possessed many more products than functional activities.

Consider an example, from GM's 1928 *Annual Report,* of how the divisional structure was amenable to rather fine factorization. Specifically, the Delco-Light Company was subdivided into two subsidiaries: one "engaged in the manufacture of farm electric power and light plants and water systems," and one engaged in "manufacturing and merchandising the well-known automatic refrigeration system 'Frigidaire'" (pp. 16–17). GM's planners believed that these two operating subsidiaries would allow greater concentration on each line of products and improve operations (Green, 1933:121). Indeed, improved performance was likely to result as each product-division management team could focus solely on the tightly coupled problems of a single product line. Thus, "most of the divisions of General Motors are enormous in themselves, but separately they do not have the complication of products of highly diversified character or type" (Brown, 1927:11).

Besides improving the performance-control prospects of the various divisional managements, the diversification of control also made the GM corporate headquarters less dependent on any one divisional decision-making group. Spreading performance regulation to multiple management teams made it significantly less likely that a single team would make compounded mistakes and damage the performance of several products.

While the GM designers certainly attempted to protect the firm from divisional errors and overloads, much management diversification was lost eventually because the single central headquarters so dominated the several auto-division managements; independent engineering, production, styling, or marketing approaches could not be pursued. As long as the corporate headquarters was more competent and more in touch with the automobile market than the divisional decision makers, such centralization was valuable, on balance. Considering the divisional managements Durant had bequeathed to the Sloan-Brown group, this was long to be the case; it was especially the case during the early days when many of the divisions were so poorly run. Yet, as their centralizing grip strengthened, the designers were under more pressure to have the correct solutions or, as Sloan always harped, "to get the facts."

PERFORMANCE COMMENSURABILITY

The concept of performance commensurability was developed in Chapter 5 during the discussion of GM's strategy formulation. To measure corporate and divisional performances on a commensurable scale, GM's designers employed uniform account classifications, standardized accounting procedures, and market-referenced transfer prices. Divisional performance "reports, for example, were not usable for evaluation and comparison until they were set up on a uniform and consistent basis" (Sloan, 1964:143). Similarly, "unless a truly competitive situation can be preserved in the pricing of products, sold by one General Motors Division to another, there can be no basis upon which the performance of each individual Division can be measured" (Sloan, 1944:9). The Sloan-Brown designers also needed to maintain a set of distinct capital-resource/profit-benefit performance centers.

Capital-Resource Segregation

Sloan (1944:8) and Brown (1927:10) realized if the divisions' contributions to GM's performance were to be appraised accurately, then the resources employed by each would have to be carefully segregated from the resources of the others. Since performance was judged in terms of rate of return on invested capital, the basis (denominator) for this calculation, i.e., the invested capital, had to be isolated for each division. "It is absolutely essential," Brown (1927) emphasized, "that each unit be constituted so that it represent a self-contained business enterprise. The capital placed under its jurisdiction must be identified definitely with its own business and no other; and prices at which its products are sold must be based upon actual competitive values" (p. 10). Ultimately, the corporate headquarters used this unique capital mapping to generate and publish "a monthly ranking of all units on the basis of percentage earned on capital employed" (Swayne, 1924:22).

Brown (1927) described two methods for separating and, in turn, identifying capital with a particular division. First, "where any given plant produces a component entering into the finished product of just one of our divisions, it is deemed proper . . . that the investment . . . be placed under the jurisdiction of the consuming divisions" (p. 8). When two or more divisions bought the same product, "it is deemed desirable to place the investment in such plant and the full responsibility for its operation under the jurisdiction of a separately organized division." In this way GM's "numerous parts and accessories divisions derive their separate entity."

Using the above capital-separation rules, the GM designers established a series of distinct "investment centers" (Bradley, 1927:424). With these unique investment centers, performance commensurability was maintained. The resources employed by each product division, when combined with the resultant benefits (i.e., profits) generated by these resources, could be mapped efficiently and unquestionably onto the single rate-of-return performance scale, eliminating the need for special performance scales to account for poorly segregated capital investments.

Profit-Benefit Separation

The designers' key decision in creating performance commensurability was to decompose GM into a series of divisions. Since each division manufactured and marketed a single product line and was surrounded by input and output markets capable of evaluating its benefit contributions, the revenues, the costs, and therefore the profits could be mapped readily and uniquely onto a set of segregated performance centers. The resultant separated performances, in turn, provided accurate performance comparisons for control and correction purposes.

The divisional structure's improved performance commensurability—compared to the functional scheme—arises from its potential for close market contact. The functional department is "too far away from the market for its performance to be related to market success" (Drucker, 1946:122). Just as with the functionally decomposed firm, "the objective yardstick of competitive market position cannot be applied directly to the departments within a division. . . . They do not have a distinct marketable product the performance of which can be measured directly" (p. 116). With a functional structure the designers and decision makers, at best, can keep track only of costs incurred. Yet, "without . . . a real market in the distance—a market that is 'objective' in that it cannot be manipulated by those who are measured by it—cost accounting lacks a valid frame of reference" (p. 124). By bringing each component into close contact with its evaluating markets, the divisional structure enables the market to map the generated benefits back onto the division.

When profit benefits were referenced to capital resources, the designers developed a complete performance picture. Accordingly, the Sloan-Brown designers found that the "appraisal of multifunction divisions proved to be more precise than that of single-function departments" (Chandler, 1966:384).

CAUSAL COMPARABILITY

To improve causal comparability, the GM designers standardized account classifications and accounting procedures. The product-division decomposition scheme also enhanced comparability, and grouping divisions into similar categories under the group executives further accentuated it.

Uniform Accounting

"The development of a uniform accounting practice enabled" Sloan and his fellow planners "to analyze the internal condition of each division and to compare one division's operating performance with another's" (Sloan, 1964:143). More importantly, uniform accounting procedures permitted the creation of an ideal-type model of divisional operations, providing "guidelines" and "developing yardsticks for evaluating operating efficiency."

Product Divisions

The divisional structure also promoted causal comparability because the many divisions encompassed very similar sets of decision variables. Each GM division, for the most part, was composed basically of an engineering, a purchasing, a production, and a marketing department. Thus, only one basic model of component operation, with minor variations of course, needed to be mastered to understand generally how all the divisions operated. The single causal, or explanatory, model minimized internal variety. Accordingly, the Executive Committee was "able to deal with operating policies in respect to any division with an understanding of the characteristics of the particular business and the position of the division in its field of competition." This type of organization in turn contributed "to a high degree of centralized control" (Brown, 1927:9-10).

A functional structure, on the other hand, would have increased the variety faced by the corporate designers and decision makers, for the functional departments differed greatly and required multiple explanatory models. Engineering is quite different from purchasing, production, and marketing, and a thorough knowledge of product engineering provides little insight into the other functional areas.

The divisional structure, moreover, lends itself readily to performance improvement because the similar divisions can be used for planned experiments or serendipitous discoveries. The introduction of Duco lacquer at the Oakland division, for example, showed Sloan that the market would accept the somewhat dull, lackluster finish of the new quick-drying lacquer. Armed with this knowledge, Sloan was able to badger the Buick, Cadillac, and Fisher Body Divisions to accept this important inventory-reducing improvement (GTX 386, *U.S.* v. *DuPont*, 1956:3942). Oakland, in this instance then, was used as a test unit for the other divisions. Similarly, "when G.M. introduced knee action [front suspension] on its cars, it put the cross-lever or exposed-coil-spring type on all except Chevrolet and Pontiac, which were given the enclosed Dubonnet type. The corporation wanted to make sure it wasn't missing any bets" ("General Motors II," *Fortune*, 1939:104). Eventually, the cross-lever suspension proved cheaper, and all divisions were switched to it.

Divisional Groupings

The final way in which the Sloan-Brown planners improved causal comparability was to group similar divisions together under the previously mentioned group executives. Under Durant the divisional groupings were confused, some units being within United Motors, others being part of Chevrolet or Buick, and others reporting directly to Durant himself (Chandler, 1966:164).

Specifically to improve interdivisional comparability, Sloan "grouped together those operations which had a common relationship, and . . . placed over each such group for co-ordinating purposes . . . a Group Executive" (Sloan, 1941:136). Sloan's early divisional grouping consisted of Accessories, Parts, and Car divisions. According to Sloan's "Study of Organization," these new divisional groupings placed divisions with similar problems together (GMX 1, *U.S.* v. *DuPont*, 1956:6540). The Accessories Group, for example, faced "common problems such as sales and advertising policies, competitive conditions, proper placing of the Corporation's capital to effect the best results to the Corporation as a whole and many other questions of a commercial nature which do not enter into the operations within the Corporation" (p. 6542). The output of the

Parts manufacturing units, on the other hand, was "largely to be consumed within the Corporation's own divisions, involving therefore no commercial relations with outside interests" (p. 6543). Thus prices had to be set under the "jurisdiction of the President," as "the divisions in this group [were] production or manufacturing operations, pure and simple" (p. 6544). Moreover, "the conditions so far as control [were] similar in each." The Car operations were grouped together because they manufactured and marketed "complete motor cars—purchasing part of the component parts from outside sources; part from other divisions of the Corporation and manufacturing part with their own facilities. These divisions in general [sold] none of their products except incidentally to other divisions. . . . Their products pass entirely outside of the Corporation's control" (p. 6541). In summary, Sloan grouped divisions according to the following criteria:

1. Operations having both commercial and manufacturing problems and that distribute part of their products outside the Corporation itself.
2. Operations which from a manufacturing standpoint present similar problems and have no outside commercial relations.
3. Operations which might be more effectively grouped geographically. (p. 6540)

By 1924, the Intercompany Parts divisions were grouped with the Car and Truck divisions, perhaps in an attempt to centralize their activities under Sloan, the president. This overload and other grouping inconsistencies were corrected when Sloan realigned the divisional groupings in January 1925 (Sloan, 1964:115) so that the Car and Truck Group contained only divisions producing for the finished-automobile and truck markets; the Intercompany Parts Group, parts divisions selling more than 40 percent of their output to other GM divisions (Chandler, 1966:164); the Accessories Group, parts divisions selling over 60 percent of their products to outside customers (p. 164); and, finally, the Export Group, consisting of overseas manufacturing and marketing divisions.

Since each of the divisional operating groups was composed of strongly similar units, Sloan and his group executives could isolate the causes behind performances through detailed interdivisional comparisons. Furthermore, close contact with the divisions helped the corporate-headquarters groups disseminate this performance-improvement information rapidly to the similar units. In this way, successful experience was put to use quickly throughout the divisions of each group.

COMPONENT DECOMPOSABILITY

Causal comparability is useful to designers and decision makers only if the units under investigation are loosely coupled (otherwise the various performances would be hopelessly intertwined). Component decomposability, more generally, permits each unit to be treated as a relatively separate entity with little regard for the others. GM's designers sought a loosely coupled factorization—especially in the early days of the small corporate staff—because they could not hope to deal with the firm in its full complexity.

Resource-segregation and benefit-separation techniques aided component decomposability. The designers further enhanced GM's inherently loose horizontal couplings by: (a) retaining the product-division structure, (b) using the transfer-price-setting parts markets to separate the supplying and consuming divisions, (c) standardizing the transferred parts, (d) employing inventory buffers of parts, (e) separating the divisions geographically, and (f) requiring their market specialization and segmentation.

Product Divisions

To visualize the loose coupling advantages of the divisional decomposition, imagine the converse. If the many GM products had been handled by one massive engineering department, one huge purchasing component, one large production facility, and one immense sales unit, confusion would have reigned along with unexplained poor performances. "With an executive in charge of all manufacturing operations, another in charge of all sales, and with central purchasing and engineering departments," as Brown (1927) cautioned, "unrecognized inefficiency in manufacture may throw an impossible task upon a very efficient sales department. On the other hand, an inefficient sales department may make it impossible to capitalize a highly economical manufactured product" (p. 10). Furthermore, GM's planners had quickly realized that "the company's activities were too large, too numerous, too varied, and too scattered" to be managed either easily or successfully through a "functionally departmentalized organization" (Chandler, 1977:460).

Indeed, the DuPont Company experienced considerable organizational chaos before switching from a functional department structure to a product division structure in 1921 (Chandler, 1966). GM's retention of Durant's product-division structure can be traced to the DuPont source via two different routes.

While still with DuPont, Donaldson Brown belonged to an important subcommittee on organization created by the DuPont Executive Committee toward the end of World War I. After several months of "studying the problem, consulting with industrialists in other companies and conferring" (Brown, 1957:37), Brown's committee recommended grouping the related products of one industry into self-contained units, each under one divisional manager with authority for all operations and responsibility for profits. Beyond coordinating purchasing, production, and sales, the committee stressed that this organizational structure "affords more direct and logical control of the investment of Working Capital" (Chandler, 1966:121).

What was needed then was "general managership" (Brown, 1927:10) over each product line. General, or product-division, management was "essential, not only to gain proper coordination of functional activities, but also from the standpoint of having some one in sufficiently close touch with the details as to be competent to size up the effectiveness and efficiency of departmental management." Thus with product divisions, divisional managers could be expected to devote most of their time to controlling and regulating the variables assigned to them, without having to deflect much capacity to deal with the managers of other components. Furthermore, with responsibility limited to only one product line, each division's staff could be expected to handle the tight couplings among their various functional departments. What could not be done at the corporate level with the functional structure could now be easily accomplished at the component level with divisional decomposition. So when Brown joined GM, he came with a belief in the performance-control potential of the product-division approach.

Sloan, who had been in contact with members of the DuPont subcommittee on organization, was the ultimate source for GM's retention of the product-division decomposition scheme (Dale, 1960:253). In January 1922 Sloan formalized his preference for a divisional structure in a report to Pierre duPont and Raskob; the report, in turn, was disseminated widely within GM (Chandler and Salsbury, 1971:527). Later Sloan would explain: "We . . . set up each of our various operations as an integral unit, complete in itself. . . . We would place in charge of each unit an executive responsible, and solely responsible, for his complete activity" (Nevins and Hill, 1963:325).

Transfer Prices

Just as external couplings had been weakened by the creation of the parts markets and the institution of the multiple-supply-source policy, internal couplings were loosened too. The parts markets and their transfer prices became the decoupling mechanisms.

The designers achieved this separation along two dimensions. First, if the purchasing divisions did not like the transfer prices charged to them by the parts and accessories divisions they were free to enter the surrounding markets to fill their needs (Raskob, 1927:131). "Our internal supplying divisions," as Sloan (1964) detailed, "must be fully competitive in price, quality, and service; if they are not, the purchasing divisions are free to buy from outside sources" (p. 432). Thus the automobile divisions were not saddled with inefficient internal suppliers. Their purchasing freedom permitted them to sever constricting couplings. Second, the designers employed the markets' prices as reference points in interdivisional bargaining. Negotiating procedures were strictly standardized around market prices. Furthermore, the competitive market prices could deactivate any special monopoly or monopsony power variables controlled by the interacting divisions.

"For sales to outside customers," Sloan (1964) recognized "that the market would determine the actual price, and if this yielded a desirable return, the business in question might justify expansion" (p. 49). Such market-generated prices, in turn, would dictate the prices that the selling divisions should charge other GM divisions. Similarly, "the independent purchaser buying product[s] from any of our divisions is assured that prices to it are exactly in line with prices charged our own car divisions" (Brown, 1927:8). As Knudsen (1927) put it when describing his Chevrolet Division's internal GM purchases, they were made "in the regular way at competitive prices." The items included: "large purchases of electrical equipment from the Remy Division, wood wheels and steering wheels from the Jaxon Wheel and Inland Manufacturing Divisions, steering gears from Muncie Products, ball bearings from the New Departure Division [and] radiators from Harrison Division" (p. 68).

"Where there are no substantial sales outside, such as would establish a competitive basis, the buying division determines the competitive picture—at times partial requirements are actually purchased from outside sources so as to perfect the competitive situation" (Brown, 1927:8). Finally, for situations where it was not possible to use transfer prices generated through the market mechanism, Sloan "recommended that the starting point should be cost plus some predetermined rate of return, but only as a guide." And "to avoid the possibility of protecting a supplying division which might be a high-cost producer," he "recommended a number of steps involving analysis of the operation" (Sloan, 1964:49). The surrounding parts markets along with their transfer prices, then, aided in providing a good boundary definition among GM's interacting divisions.

Parts Standardization

Standardizing the many parts that went into GM's finished autos weakened both internal and external couplings. In general, standardization "is of great importance, because the ability to produce large quantities of parts, each one just like the other or sufficiently alike, within a predetermined allowance of inaccuracy, is the foundation of mass production, as we understand that term today" (Sloan, 1941:40).

First, reducing the number of different parts and materials to a limited range of standardized items greatly reduced the variety of interdivisional and environmental interactions required. "In December, 1919, the several divisions of the General Motors Corporation were using a total of 13,355 different parts in their products. A few months later the same divisions were using a total of 2,099 different parts in the same products" (Baird, 1923:334). Similarly, "steels have been reduced from more than 100 different analyses to 18, nonferrous metals have been reduced in approximately the same numerical ratio, and standards have been determined for oils, greases, petroleum products, and other expense items of importance" (p. 336). Correspondingly, the (unnecessary) internal variety of each division decreased. "The first division of the Corporation to standardize its parts," for example, "reduced the number of bins in its stock rooms from 1,640 to 78" (p. 334).

Second, parts standardization erected "walls of constancy" that further blocked the flow of variety and thus decoupled connections both between the divisions and between the divisions and the firm's environment.

> That word "interchangeable" is the sum and substance of all standardization. It is the key to mass production. Fabrication has become a matter of timed operations rather than of individual ingenuity in getting mismatched pieces together. There are specified tolerances that are rigidly adhered to. Pieces are no longer antagonistic. They are interchangeable—they fit. (Kettering and Orth, 1932:100)

The auto divisions and parts divisions or external parts suppliers still remained linked, of course, but the linkages now were fixed by the *unvarying* parts standards. Price and quantity could vary with market changes, yet the specifications for the item transferred remained constants, i.e., unvarying parameters. Operations could be conducted without concern for parts specification changes.

Parts standardization also decoupled the auto divisions from the parts divisions by establishing common markets that suppliers and buyers could enter freely. Such accessible common markets, in turn, could serve as inventory and production buffers as well as consumption sinks for the parts divisions' excess capacity. Thus GM's parts and auto divisions operated relatively independently of each other by having the interunit markets absorb and provide slack production capacity.

While GM's standardization of parts decoupled many of the divisions' high-frequency interactions, the requirement that all auto divisions purchase identical parts removed more of their autonomy. "When two or more divisions use common components," as Sloan (1964) admitted, "the independence of each division is limited to the extent that there must be a common program between them" (p. 181). Since someone "must co-ordinate such a program, ... more questions come into the policy area" of the corporation from the "administrative area" of the divisions. In short, the corporate office became involved in many internal and external interactions that were once the exclusive prerogative of the auto divisions. At first the Factory Division of the Advisory Staff conducted these centralizing activities. The initial standardization program was limited to such minor items as "bolts, nuts, screws, washers, pins, keys, and other small articles of hardware" which could be "standardized without affecting the individuality of design of the different cars" (Baird, 1923:334). When the standardization effort came under the direction of the General Purchasing Committee, however, it became a significant threat to divisional autonomy.

Inventory Buffers

The designers also employed inventory buffers to decouple the parts and accessories divisions from the auto-assembly divisions. During the slow winter season before the greatly increased spring production schedules, for example, the parts divisions accumulated inventory (*Automotive Industries,* December 10, 1927:870). Similarly, the auto divisions carried their own inventories to insure steady production. Pratt explained: "The banks of material, or, in other words, stocks of materials at our various car divisions' plants ahead of production must vary with the volume of production. When our schedules are increasing, the bank of material is increased; when schedules are decreasing, the bank of material is decreased" (GTX 364, *U.S.* v. *DuPont,* 1956:3893).

Geographic Separation

Besides separating the divisions by the use of product divisions, transfer prices, standard parts, and inventory buffers, the Sloan-Brown designers also attempted to break their geographic connections. As a consequence, most GM divisions came to have a complete set of functional activities and departments in a unique geographical locality. Geographic separation decoupled the divisions by severing connections in their environment. The primary environmental connections cut by the geographic separation of facilities were transportation and labor ties.

Inadvertently, Durant had started this geographical separation of divisional facilities by accumulating a series of complete companies which, for the most part, just happened to be located in different localities. He did, however, overload the Flint, Michigan, area with GM facilities (Chrysler, 1937:158). Since Flint was Buick's home, Durant brought many supporting parts firms, e.g., Mott's axle manufactory, to Flint. "A dependable supply of parts," as Sloan (1941) explained, "might well make the difference between success and failure. Distance added uncertainty. It was natural, therefore, for the industry to correlate all its manufacturing within a rather narrow geographical area. It was a practical urge, but underneath it was a very definite economic justification" (p. 49). Yet the extreme concentration of industry in one locality could also produce dangerously uneconomic coupling effects. In the Flint area Durant had soon strained transportation facilities and created a tremendous labor shortage, forcing GM to import workers as well as to expend capital on building workers' housing. (During the depression, when Flint's labor shortage shifted to a surplus, GM's performance was to suffer even more because of Durant's excessive concentration in Flint.)

The Sloan-Brown group, in contrast to Durant, recognized the value of moderate geographical facility decoupling, and in the early days of the reorganization they pursued it as an explicit policy. Raskob, for example, wrote in 1927, that the divisions' independence was accentuated by the manufacture of cars in different cities:

> The Cadillac and LaSalle cars are made in plants in Detroit, the Pontiac and Oakland in plants at Pontiac, the Olds at Lansing and the Buick at Flint. The Chevrolet, being a high-quantity-production car, is assembled in twenty-three principal plants throughout the world. . . . It may be added that the thirteen accessories divisions are operated as so many independent factories, also in widely separated localities. (p. 131)

By dispersing its facilities GM was able to avoid many of the labor and transportation problems that Ford encountered with his massive River Rouge Plant. Moreover, GM was not forced to invest its capital in railroads and steamships, which the GM management had little expertise in operating.

In relocating a portion of Chevrolet's engine and axle production facilities from Flint to Buffalo in 1937, GM pursued an ongoing policy launched after the 1935 Chevrolet strike in Toledo ("G.M. and Ford Are Moving Out," *Business Week*, April 3, 1937:16). The Buffalo move enabled Chevrolet "to get away from labor-torn Flint," where labor had been able to tie up production by closing down the Flint engine factory. Buffalo, like Grand Rapids (where GM had located a big stamping plant in 1936) had a large percentage of home owners, portending labor stability. In addition, "of five cities studied, Buffalo got the new plant because of its lake transportation facilities."

Sloan, in the 1935 *Annual Report* to GM's stockholders, declared that GM "believes that the soundest policy, both economically and socially, is to distribute" production "among as many different communities as is practically possible." Sloan also said that spreading expansion among many cities eased housing and other social service demands. Furthermore, during periods of economic distress, "the burden can be better carried because of the greater number of communities among which it is divided." In addition, producing parts in different cities gave GM more sources of supply; that, of course, lessened the impact of a strike at a particular plant. Geographic separation of facilities, then, helped to decouple the GM divisions from each other as well as to decouple the GM divisions from the external environment.

One major exception in the designers' attempt to separate GM's divisions geographically was the location of the Fisher Body Division's plants adjacent to the auto divisions' assembly plants. In this way, bodies could be "delivered to the line clean and fresh, without any chance of transportation damage," and "the stocks . . . carried in their least bulky form" (Knudsen, 1927:68). Yet the Fisher Body and the auto assembly plants maintained strict separation on other dimensions, not even pooling their paint purchases ("General Motors II," *Fortune,* 1939:45).

Market Specialization and Segmentation

The GM designers also attempted to specialize the parts divisions into unique markets and to segment the auto divisions into differentiated markets. The Parts and Accessories Group divisions (as well as such subsidiaries as GMAC) were relatively loosely coupled among themselves because the GM designers kept these units in noncompetitive markets. If the parts divisions had competed in the same market, volume would have been split among several units, most likely with damaging effects on each unit's return on capital. After all, "it is an axiom of economics that costs are reduced and trade created by specialization and the division of labor" (Sloan, 1964:432). Early on, Brown (1927) noted, "There are numerous divisions manufacturing parts and accessories for automobiles, each one a highly specialized business and, in most cases, selling product[s] in competitive markets as well as to our own car divisions" (p. 3).

Decoupling the parts divisions required little effort because Durant had accumulated a group of parts units with little market overlap, that is, a *single* parts producer for each of the major automobile components, such as radiators, springs, and ignitions. Consequently, Sloan and his colleagues only needed to insure that the parts-producing divisions remained specialized and did not stray into each other's markets.

On the other hand, the finished-automobile market was difficult to segment and keep segmented because strong environmental pressures were constantly tending to bunch the auto divisions together. In times of prosperity, these units moved toward the middle- and high-price ranges *en masse,* and during periods of depression they migrated toward the low-price range. With the deep and prolonged depression of the 1930s, many of the divisions ended up in the lower price classes. Even worse, many dealerships had to be consolidated to generate sufficient volume.

In the early 1920s the Sloan-Brown group achieved the greatest decoupling among the auto divisions, spreading the coverage and reducing the overlap of their divisions. With Sloan's new variety marketing plan came the ability to treat each auto division independently. "Each of our units now has to make a product aimed at a particular group of consumers," Sloan (1924) explained, "the groups being graded according to certain strategic price classes of the product, ranging from below $1,000 to above $3,000. None of our products now competes with any other within one price class" (p. 138). But even with the new price classes, the auto divisions still offered substitutable products in that they supplied the same overall market: finished automobiles. Consequently, the auto divisions always remained somewhat coupled. (What the designers lost in decoupling advantages, however, was offset to a degree by improved comparability.)

The early extensive decoupling did not last long. Sloan himself argued for the new Pontiac car to be introduced by the Oakland Division in 1925. The Executive Committee questioned the wisdom of selling "an automobile that would take away customers from both Chevrolet and Oldsmobile." But Sloan convinced them that "it will be better that we take business from our own Divisions than have competitors do so" (Chandler and Salsbury, 1971:557). Sloan, then, was willing to accept higher internal variety (i.e., tighter coupling) to be assured GM's product offerings matched the variety demanded by the automobile marketing environment. While the Pontiac addition shrank the $100 price buffer between the Chevrolet and Oldsmobile, Pontiac did not overlap into the other price ranges, for it was priced at $825, a full $50 above the highest-price Chevy and $50 below the cheapest Oldsmobile. So, with strong prosperity GM's wide-band price buffers were compromised slightly to match the environment better. The introduction of the La Salle in 1927 provides a second example of this phenomenon. The La Salle's mass production by the Cadillac Division for the $2000 price class—at that time the very top of the Buick price range—increased the line's variety to cover a $1000 gap, with only minor costs to the loosely coupled automobile divisions' price separations.

During the same period of prosperity GM also tried to reinstitute buffers in a line of cars whose boundaries occasionally grew fuzzy. Thus, the Oakland Division, whose models overlapped the high-priced Oldsmobile and the low-priced Buick, focused completely on the new Pontiacs (Chandler, 1964:152). (And given Oakland's terrible reputation for an oil-burning engine and poor running gear, the brand-name switch was most helpful to GM.) The 1920s, then, were a period when GM's designers increased the firm's internal variety to match the environment yet attempted to minimize internal complexity by maintaining some price-gap decoupling.

The Sloan-Brown group was considerably less successful in maintaining price-gap decoupling during the severe depression of the 1930s. In Knudsen's words: "The business first stopped in the higher priced field—the Cadillacs, the Buicks, and the Chryslers. . . . The low-priced cars hung on the longest, but finally they started to slide, too, and the only thing the automobile business could do was to follow the trend of dropping income, try to hit for the lowest priced field, and give the best possible car for the money so we could tempt people to go back in and buy some more cars" (Beasley, 1947:151). Note the damage to Sloan's loosely coupled price structure by the end of the depression: four of GM's cars are bunched together within a price range measuring "about $300 from the cheapest Chevrolet to the cheapest Buick, and less than $130 from the most expensive Chevrolet to the cheapest Buick; while the fifth, Cadillac-LaSalle, is languishing for lack of access to this range. 'A car for every purse and purpose' obviously no longer describes the G.M. lines" ("General Motors," *Fortune*, 1938:156).

The luxury of treating the auto units as independent, or decoupled, components was gone, and each auto division had to be treated simultaneously with every other auto-producing unit. GM's internal complexity had grown considerably. Sloan, in his testimony during *U.S.* v. *DuPont*, (1956) summarized his altered administrative approach:

> Going back to 1927, two or three years minus or plus that year, . . . it was my practice to go up to the divisions, say to the Buick Division in Flint, [and] . . . go over all the problems of that particular division, particularly the engineering program, and of course . . . the distribution program and others, and everything was there . . . so far as that particular divisional operation was concerned. When we pass to 1937, the conditions were entirely different. . . . a much more [significant] degree of coordination became necessary. In other words, if, in 1937, I had gone up to Buick as I did in 1927, I couldn't have dealt with it in that [separated] way, because I would have had to consider the impact of the program on the other car operations. (pp. 998–999)

An additional reason for increasing the headquarters' direction of the auto divisions then stemmed from the fact that the pressures on price forced drastic internal economies, which could be achieved only through central control. Alerted by Bradley's econometric predictions (see Chapter 10) to expect an economic downturn (Cray, 1980:269), Sloan got an early start on the consolidation process. "Shortly before the stock-market crash," he "addressed a general letter to the organization noting the end of expansion and promulgating a new policy of economy for the corporation" (Sloan, 1964:171). Sloan instructed

GM's decision makers "economy in operation must now be the key-note rather than expansion of plant and equipment" (p. 172).

Thus "the depression years were hard on rugged individualism in General Motors, as elsewhere." More specifically, they "brought the family closer together. There were sales consolidations (Buick-Olds-Pontiac Sales Co.) and manufacturing consolidations (Chevrolet-Pontiac). Parts manufacturing units at distant points ... doubled up with Buick at Flint, where reduced volume made room available" ("General Motors Coordinates," *Business Week*, October 21, 1933:16). Sloan (1964) added, "The most difficult economies to get were in commercial or selling expense and here we took the most drastic measures of reorganization" (p. 177). In fact, Sloan removed the sales activities from Buick, Oldsmobile, and Pontiac and centralized them under the corporate sales officer, R. H. Grant. (As the depression eased, these activities were again returned to the auto divisions which—as is shown in the next chapter—now had more depression-oriented, i.e., Chevrolet-trained, decision makers.)

Sloan appointed Knudsen to the post of executive vice president to make sure that the auto divisions adopted Chevrolet's low-cost manufacturing methods. Further, "to economize we co-ordinated to a greater extent our work in purchasing, design, production, and selling, and some of these changes were of lasting value" (Sloan, 1964:177).

The interchangeable-body program, initiated by the corporate stylists, probably had the strongest centralizing effects of any of Sloan's depression-motivated cost-reduction programs. By standardizing bodies among divisions the designers dictated many other divisional decisions. The body of an automobile is such a major component that once its many variables are fixed, many other related variables are automatically specified in turn. Engine design, running-gear design, and even production-facility layout became set. Parts standardization, this time of a major component, again tightly coupled the divisions to the central headquarters.

8

Training the Decision Makers

Besides defining the firm's goal, formulating its strategy, and organizing its structure, GM's designers needed to train numerous decision makers to implement their policies. The designers on GM's Finance and Executive Committees simply could not control the firm's performance without considerable help. Donaldson Brown (1927), for instance, observed, "Obviously, it is humanly and physically impossible for the executive committee of General Motors to maintain the same kind of intimate contact with the details of its business as would be practicable in the case of a very much less diversified business" (p. 7).

Brown was aware, though, that even without the requisite channel capacity the GM designers had assumed an obligation to the GM stockholders, as the system's clients, to provide adequate performance control, for Brown (1927) acknowledged that "still the responsibility to stockholders is exactly the same" as it is with a smaller organization "and the proper organization of control has been forced by absolute necessity. Otherwise the business were better split . . . so that the stockholders of each unit respectively could elect a board of directors capable of assuming the usual responsibilities" (pp. 7–8). Thus, a method was needed by which the designers' own limited power to control GM's performance directly could be *amplified*.

Sloan (1924) in particular realized the designers' decision-making capacity needed to be amplified considerably. Specifically, he wrote: "Every executive has to recognize sooner or later that he himself cannot do everything that needs to be done. Until he recognizes this, he is only an individual, with an individual's power, but after he recognizes it, he becomes, for the first time, an executive, with control of multiple powers" (p. 137). Sloan elaborated: "My office force is very small. That means that we do not do much routine work with details. They never get up to us. I work fairly hard, but it is on exceptions [less repetitive matters] or construction [design], not on routine or petty details" (p. 195).

Consequently, many regulators, i.e. decision makers, were needed to provide more detailed, routine, and repetitive performance control. "After a while we thoroughly grasped the principle that the work must be done by other men," said Sloan (1924:137). In particular, the new openings in GM's expanded corporate headquarters and the vacated positions in its revamped divisions would require many additional executives. Moreover, those executives already with the firm would require extensive training, as they tended to know more about the production of cars than the production of profits. For new and old executives alike, Sloan and his colleagues "had to find a way to arouse their initiative" (p. 137). In reality, Sloan and the other designers went considerably further than just arousing "initiative." Rather they devised extensive value orientation, factual education, and executive selection procedures.

Ultimately, Sloan was after the massive control amplification that could result only from simultaneous—not one-by-one—orientation and education schemes. Strong divisional similarity, obtained from performance commensurability and causal comparability, permitted the simultaneous training of many decision makers. Accordingly, Sloan (1924) emphasized the training of *groups* of managers, e.g., "the purchasing agents of [the] five principal units that manufacture cars" or the corporate and divisional decision makers occupying the 70 or 80 managerial positions on "which the corporation must rely for the successful management of its business" (p. 140). Mass orientation and education programs greatly amplified the control of the designers on GM's Finance and Executive Committees because this small group of individuals could deal with comparatively few variables, i.e. have their control vectors shrunk, yet still achieve *detailed* and *continuous* direction of even the most remote aspects of GM's operations. Further amplification in performance control could be obtained through the careful selection of corporate and divisional decision makers who either possessed the desired value and factual premises for decision making or were predisposed to attain them quickly.

So if GM's designers could not control the firm's performance directly, they would control it indirectly by creating a large decision-making corps to act in their stead. Once properly trained, these surrogates could be expected to decide matters just as the designers would have if they had been able to occupy all the positions themselves. And by infusing GM's corporate and divisional decision makers with the prescribed concepts, the designers could achieve the divisional and corporate performances that would have eluded them without help. The designers thus amplified their control by creating in their own image a large group of proxy decision makers.

VALUE ORIENTATION

Most of the orientation programs developed at GM were designed to create in their participants a realization that: (a) the clients to be served by the organization were the stockholders, and (b) the measure of how well the stockholders were served was the long-term return on their invested capital. The solution to the GM employee-orientation problem was quite simple: just make the more important employees serving the clients, clients also. The GM designers (Brown, 1927, 1957; Raskob, 1927; Sloan, 1941, 1964) described the approach as an attempt to create a close, enduring partnership between the GM management and the GM stockholders.

Specifically, a variety of different programs was designed to facilitate the partnership between decision makers and clients. These programs were tailored on the basis of the employee's position in the GM hierarchy. Raskob (1927), the member of the Finance and Executive Committees closely associated with the dominant DuPont interests, described the GM hierarchy (in order of responsibility) as:

1. The senior executives of the Corporation, including all its divisions and subsidiary companies.
2. Junior executives occupying positions of responsibility and importance.
3. Other junior executives such as the heads of departments and other salaried employees occupying positions of importance, under direction.

Finally, come the foremen, minor executives, and the main body of employees. (p. 132)

Noting that GM's policy admitted the three numbered classes of executives as partners with the GM stockholders, Raskob allowed that the "main body" of employees also were to have "an opportunity to save, with a view to investment in the corporation's securities, thereby admitting them to partnership" (p. 132).

Since the heavy responsibility for GM's decision making rested "upon the major executives in classes 1 and 2," Raskob reported it was "believed to be in the best interests of the stockholders that these responsible men should have a definite financial share in the business" (p. 132). And for that purpose the Managers Securities Company was formed in 1923.

To enable "the third class of executives," that is, the heads of departments and other important salaried employees, to "share as partners in the results they helped to achieve" (Raskob, 1927:133), the Bonus Plan was adopted in 1918. Through this program, annual awards of General Motors stock were made "for conspicuous and meritorious service" by all employees whose salaries were

$5000 or more. The divisions received Bonus Plan allotments in proportion to their contributions to the "prosperity of the corporation as a whole."

Because GM's incentive programs were meant to develop "managerial ability" and because it was "quite possible that such ability may exist among the foremen and employees at large," GM's plan of "every employee a partner" was extended to provide every worker an opportunity.

Thus, there were several specialized orientation-through-partnership programs operating at GM during the period under study: (a) the Managers Securities Company, (b) the General Motors Management Corporation, which succeeded the Managers Securities Company in 1930, (c) the GM Bonus Plan, (d) the Savings and Investment Plan, and, finally, (e) the Housing Plan. As should be expected, the partnership interests open to employees diminished in value while the number of employees eligible increased rapidly with descent down the firm's hierarchy.

Orientation via Managers Securities Company

Brown designed the Managers Securities Company on instruction from Pierre duPont and Raskob. It was Raskob, according to Brown (1957:43), who was the prime mover behind this scheme. Raskob had two objectives in mind: (1) to divest a portion of the DuPont Company's GM stock without losing voting control and (2) to orient GM's more important executives toward accepting the stockholder as system client. At the time, Managers Securities Company was viewed as "the most far-reaching effort ever made by a large corporation to provide a complete identification of the interests of management and stockholders through having the managers become large stockholders themselves" ("General Motors Corp.—To Make Partners...," *Commercial and Financial Chronicle*, November 3, 1923:1998).

Raskob's (1927) thinking paid homage to other industrialists who had employed able executives to help them control "the complex activities of a great corporation" that were "far beyond the detailed direction of any one man, however capable" (p. 130). For example: "The far-sighted iron-master, Carnegie," recognized "that there was a limit to the inducement offered by salary." Consequently, his managers became "his partners so that their combined ability would continue to be concentrated more effectively upon the development of his steel business" (p. 130). So Raskob too was aware that the amplification of control needed by a large firm like GM was possible only if the men regulating the corporate and divisional performances were properly motivated to pursue the stockholders' interests. Most importantly, Raskob felt that the decision makers could be oriented to pursue the stockholders' interests by making them partners in the firm's ownership.

In a 1923 statement to the GM stockholders, Sloan (1924) expressed the same view:

> There are approximately seventy managerial positions upon . . . which the corporation must rely for the successful management of its business. . . . Your directors believe that the men occupying these positions can secure a better, broader and more sympathetic understanding of the stockholders' interest, and, therefore, of the interests of the corporation as a whole, if they can be attracted into partnership with the stockholders through becoming substantial stockholders themselves. (p. 140)

Seventeen years later Sloan's perspective remained unchanged. "I believe strongly that those managing our great industrial enterprises should have a real stake in the business. . . . It is essential that corporate executives be placed as far as possible in the same position as the joint owners of a private business" (Sloan, 1941:153–154). Still later Sloan (1964) wrote: "Managers Securities Company created a top management team with a heavy personal stake in the success of the corporation as a whole. Managers Securities Company was certainly a great individual incentive" (pp. 414–415).

Via Managers Securities, the GM planning group made sure that corporate and division executives who did not originally have extensive GM stockholdings acquired them rapidly (Seltzer, 1928:221). So, these seventy to eighty participating executives too could be expected to treat the stockholder as the firm's client. In other words, Managers Securities Company focused the corporate and division executives' attention on the stockholders' well-being, thereby sensitizing the decision makers to the clients' needs.

Brown (1957), as the Managers Securities Company designer, stated that "executives throughout the far-flung operating divisions of the Corporation, through their equity holdings of Company stock, came to recognize that their own welfare was tied up with the welfare of all GM stockholders" (p. 43). Going a step further and discussing the benefits of this program to the duPont interests—the major GM stockholders and the driving force behind the program— Brown (1927) wrote:

> The duPonts were far-seeing enough to make a part of their holdings in General Motors available for . . . the Managers Securities Company, having confidence . . . that in the long run they would gain handsomely [from] ownership management. The incentive thus supplied is a vital force operating in the interest of all stockholders, since stockholders at large inevitably enjoy a proportionate benefit of whatever is beneficial to the managing group. (pp. 14–15)

Several safeguards were built into the Managers Securities Company orientation scheme to assure that the top GM managers would not displace the stockholders as the system's client. The safeguards can be sorted into two categories: those designed specifically to protect the duPont interests and those designed generally to protect all stockholders.

The duPonts feared that GM might be taken over again by Durant or other powerful interests. So while they wanted to divest some of the overinvestment made during the Durant debacle, they did not want to give up a voting control of these shares. The reluctance was probably heightened because many of the recipients could be former Durant men. In short, the DuPont Company "wished to give them earnings but not votes" (Green, 1933:287).

The duPonts accomplished their objective by creating General Motors Securities Company. Thus, the GM executives (via their Managers Securities stock) were limited to a 30 percent ownership while E. I. du Pont de Nemours & Company held the remaining 70 percent of stock (p. 288). "In such a situation, the holders of the 30% interest could vote on such matters as were presented to the stockholders of the General Motors Securities Company, but in no possible way could they extend their influence to the General Motors Corporation. Any interest under 50% was powerless to exert any influence in the management of the securities company and, through it, in the control of the operating motor car producer" (p. 288). Indeed, DuPont Company voted Managers Securities Company stock as it saw fit until 1938 when the duPont voting control reverted to 23 percent, which was what it actually owned. Via General Motors Securities Company, then, the duPonts assured themselves that they would remain GM's principal client group and that the firm's managers would serve their interests.

The duPont group also perpetuated their dominance by vesting the power to approve or disapprove membership in Managers Securities Company with the GM Finance Committee, which they, of course, dominated until 1937. (The Finance Committee, in turn, "gave the task to a small subcommittee consisting of Seward Prosser, Arthur G. Bishop, and Pierre as chairman" [Chandler and Salsbury, 1971:542].) "Actually, the decisions as to allocation of the stock were left almost entirely to Pierre and Sloan, with Brown probably giving the most additional advice" (p. 542). Pierre duPont prepared the first list of executives to be considered; Sloan, exerting far more selectivity, pared Pierre's list from 150 to approximately 80 executives (GMX 30, *U.S.* v. *DuPont*, 1956:6629–6638). While the Finance Committee could alter Sloan's selections, it accepted, as did Pierre, "the judgment of the corporation's chief operating executive" (Chandler and Salsbury, 1971:542).

Still other safeguards were built into the Managers Securities Company plan to protect the interests of the stockholders at large. These safeguards were: (a) a minimum return on investment before GM would make its contracted payments to Managers Securities, (b) a review of the individual performances of the executives who were stockholders in the Managers Securities Company,

(c) an emphasis on broad corporate rather than narrow divisional contributions, and (d) a restriction on the participants that prevented them from diversifying away their risk. (The latter safeguard will be discussed below in conjunction with the GM Bonus Plan.)

While it is true that the GM corporate and divisional executives participating in this plan were all made millionaires (Brown, 1957:43), they benefited only after an adequate investment return had been earned for the stockholders. Under GM's contract with the Managers Securities Company, it agreed "to pay that company annually . . . from 1923 to 1930, both inclusive, 5 per cent of its excess net earnings over and above 7 per cent on the net capital employed" (Brown, 1927:13). Therefore, the participating executives benefited from the plan only after they had earned a 7 percent return for the stockholders; their benefits, in addition, increased proportionately the further they raised the return on investment figure above the 7 percent level. "If the business was successful," Sloan (1964) recalled, "they would be in a position to become substantial owners of stock" (p. 410). Given the disasters of 1920 and 1921, the participants were not likely to think successful performances could be attained without considerable effort.

The stockholders were protected not only by the minimum-earnings safeguard but also by the performance reviews that Brown and Sloan designed into the orientation scheme. In contrast to the minimum-earning requirement that affected the executives as a group, these other performance-review procedures were individual. (Individualized reviews were an extension of a philosophy Sloan had developed at Hyatt Roller Bearing, where "more than nine out of ten of his men were employed on piecework; the more they produced, the more they earned" [Douglass, 1954:141].) Specifically, the annual review determined whether the participation and performance of each executive "was out of line compared with other executives, including those not in the plan" (Sloan, 1964:412). If an executive performed poorly or resigned, "General Motors held an irrevocable option to repurchase all or part of any executive's holdings." On the other hand, should an executive's performance excel, Sloan, as president, "could recommend an additional allotment." Since the stakes were extremely high (more than a million dollars worth of stock for each executive), it seems safe to assume that the decision makers involved were duly concerned with achieving adequate performances.

As was written at the time, "the yearly review" of the stockholders in Managers Securities Company made "the plan dynamic with plenty of opportunity for subordinates to move up to the key job" ("General Motors Has Plan to Hold Its Key Men," *Printers' Ink*, November 15, 1923:36). One such subordinate who moved up in financial status and remuneration was A. R. Glancy, who headed the team that developed the new Pontiac replacement for the Oakland car. Glancy had not been awarded stock initially, but Pontiac's success earned him some feelers from Studebaker. GM promptly countered via the Managers partnership offer (Chandler and Salsbury, 1971:581, 706).

By the same token, even the highest executives could expect to be reviewed, with negative outcomes if their attention wandered. In 1925, for example, Pierre duPont, as Chairman of the Finance Committee, "proposed that Sloan's allotment . . . be increased from 350,000 shares to 450,000 shares; while Mott's be reduced simultaneously from 300,000 to 200,000" because "Mott was taking a less active part in top management" (Chandler and Salsbury, 1971:580). Accordingly, Mott (who was as deeply involved in the affairs of Flint, Michigan, as he was in GM's activities) had his allocation reduced (p. 706). Just the threat of such reviews could serve to keep most decision makers' attentions focused on GM and the stockholders' interest.

To insure adequate overall performance, Sloan and the Finance Committee evaluated the division executive's performance in terms of his corporate-wide contribution, not just his more parochial divisional performance (Sloan, 1964:415). Coupled with the strong emphasis on the corporate rate-of-return figure inherent in the Managers Securities Plan, this evaluation procedure oriented the divisional executives toward protecting their own as well as the other divisions' performances. So whereas the division executive "could conduct the affairs of his unit with little regard for the other units of the group," GM could not "have one unit reap advantage at the expense of another" (Pound, 1934:397–398). In addition, the chances that performance-improvement discoveries would be transferred among divisions was increased. A divisional executive who had developed an improvement applicable not only to his division but to others also, would benefit from both the rate-of-return improvement of his own unit as well as from the corporate-wide use of the innovation. In this way, "the benefits obtained by one unit" were "made available to the others, so that each" advanced "not only its own but the common good of all, which, of course," was "the Corporation itself. These benefits of counsel, cooperation, and teamwork" were "necessary for effective control" (p. 398).

The system-wide orientation was further enhanced by the Sloan-Brown group's cancellation of the individualized contracts Durant had made with his important division managers. These contracts granted the manager a share of his unit's profit and thus "exaggerated the self-interest of each division at the expense . . . of the corporation itself. It was even possible for a division manager to act contrary to the interests of the corporation in his efforts to maximize his own division's profits" (Sloan, 1964:409). As Brown (1927) explained, "Every man charged with responsibilities and vested with authority must be brought to realize that his function is tributary to the accomplishment of a central motive" (p. 4).

Managers Securities Company's initial allotments, suggested by Pierre duPont and pared by Sloan, provided thorough coverage of the more important positions, especially in the critical auto divisions. In addition to allotments to division managers, Sloan awarded partnerships to divisional assistant general managers, factory managers, factory assistants and managers of subfactories, sales managers, sale branch managers and district managers, chief engineers, assistant engineers, purchasing agents, and traffic managers. Corporate executives and staff officers also were well covered, especially in the finance and accounting areas.

Exactly how significant was Managers Securities Company to the participating executives? According to *Barron's* ("General Motors Corp. Shares Profits," *Barron's*, March 3, 1930:15), the GM payments to the Managers Securities Company for the years 1923 to 1929 were

1923	$ 1,876,119
1924	1,140,190
1925	4,633,535
1926	8,274,099
1927	10,488,072
1928	12,408,595
1929	10,000,000
Total	$48,820,610

Managers Securities Company also earned substantial dividends on its 4,218,750 shares of GM common stock, and GM stock appreciated in value tremendously from 1923 to 1930. In March of 1930 *Barron's* claimed the Managers Securities Company had received about $250,000,000 in income from GM and in equity appreciation. Participating executives, on average, had received $3,000,000 apiece (Reeves, 1927:65), a substantial return for their combined investment of only $5,000,000 in the company. On an unaveraged basis the minimum participation, which required a cash payment of $25,000 in 1923, was "worth more than $1,000,000" in 1928 (Raskob, 1928:104).

Whether associated with the corporate headquarters or with the various divisions, executives who were such substantial GM stockholders themselves were thoroughly oriented to the interests of the GM stockholders. Even those executives who were not members could hardly fail to become stockholder-oriented after observing such monumental benefits. Hearing of such significant financial and status rewards, most nonparticipants surely must have sought to behave so that they too would be admitted to GM's most select managerial group.

Orientation via General Motors Management Corporation

Not surprisingly, the designers decided to extend the Managers Securities Plan after its initial expiration date. So, "with the termination of the contractual relationship with the Managers Securities Company as of December 31, 1929, the advantages of the plan were so manifest that the General Motors Management Corporation" (Pound, 1934:401) was organized in March, 1930. The GM Management Corporation again was established to provide the firm's "executives with an opportunity to increase their ownership interest in General Motors and to provide added incentive" (Sloan, 1964:415).

Probably because of the success of the original plan, the GM Management Corporation was expanded to include about 250 executives (GTX 260, *U.S.* v. *DuPont*, 1956:3589–3596). Another difference between the two orientation schemes was that instead of receiving its stock from the duPonts as the Managers Securities Company had, the GM Management Corporation's 1,375,000 shares of GM stock was accumulated, over three years, by GM itself (Sloan, 1964:416).

Otherwise, the two orientation schemes were quite similar. The GM Management Corporation, for instance, set aside a block of GM common stock which was "to be paid for by the participants by an initial, partial cash payment and by the application of their participation in supplemental compensation for a number of years in the future" (Sloan, 1964:415–416). The GM Management Corporation was to continue until 1937 the basic orientation scheme begun in 1923 with the establishment of the Managers Securities Company.

Orientation via the Bonus Plan

In 1919—while Durant was still president—a GM Bonus Plan was established for both senior and junior executives. As Pierre duPont explained in 1918: "I am a great believer in paying reasonable salaries with very substantial added compensation for successful results. This is the plan under which we work and under which our record has been made" (DuPont Exhibit [DPX] 60, *U.S.* v. *DuPont*, 1956:5706).

Sloan, too, supported the plan ardently. As the head of GM's United Motors Corporation Sloan encountered severe motivational problems. In a 1918 letter to Haskell he expressed a great interest in the Bonus Plan as a means to deal with the "astounding amount of unrest everywhere, even among those that you would hardly think it possible" (GTX 27, *U.S.* v. *DuPont*, 1956:6614). Sloan later testified: "I was having difficulty in keeping my organization [United Motors] satisfied. . . . I am a great believer in incentive to get the most out of an organization, and I think the nearer we can come to the old proprietor form of management, the better off we are" (*U.S.* v. *DuPont,* 1956:1215).

In 1923 one-half of the payments of the Bonus Plan were diverted to the Managers Securities Company to pay for the GM stock presented by the duPonts; the other half of the Bonus Plan money was paid to the junior executives.

Payment of Bonus Plan awards to Managers Securities Company and GM Management Corporation made the most senior executives almost completely dependent on GM's design group. If Bonus Plan payments of stock had been made to the executives over a period of a few years, as they were prior to 1923, the participants would not have accumulated such a fantastic stake in their continuing success and in the *long-term* success of GM. Instead, each year's bonus would have been theirs within a few years and wholly independent of their performance in subsequent years; furthermore, they would have been able to divest themselves more readily of their GM stock, thereby quickly diversifying away their personal stake in GM's continued success. But by accumulating the participant executives' yearly bonus earnings in these plans and by subjecting each man's entire holdings to a yearly performance review, Sloan, Raskob, and the duPonts held much more leverage than if the review could alter only a single year's bonus. In carrying this line of reasoning to its conclusion, then, the executives participating in the plans could be Raskob's millionaires (Brown, 1957:43) given that Sloan, the duPonts, and Raskob thought their becoming so was valuable to GM stock. As Raskob (1928) himself summarized the designers' position, "The welfare of an executive ought to be bound up with the corporation which he helps to manage" (p. 106).

While there were procedures for involuntary withdrawal from the plans, there were no established ways to withdraw voluntarily. The extreme financial dependency that could result for a GM executive is illustrated by the difficulties faced by Raskob himself when he left GM after a political disagreement with Sloan in the late 1920s. Raskob faced sizable obstacles to converting his $19,960,000 worth of Managers Securities shares into marketable GM shares. In a letter to the duPont-dominated General Motors Securities Company he explained that when he had been devoting his "best efforts to the General Motors Corporation" he remained "perfectly willing to have practically" his "entire fortune invested" in GM. But when "no longer a partner engaged with those responsible for shaping the policies and carrying on the management of the Corporation," he became anxious "to reduce the large investment." He concluded, "I sincerely hope that this will receive most sympathetic consideration as certainly it is not constructive to be forced to remain in a partnership of which I am no longer a part" (GTX 262, *U.S.* v. *DuPont,* 1956:3598–3599). Participants with less status and power than Raskob must have been even more fearful for their fortunes.

After 1923, when the Managers Securities Plan participants were removed from Bonus Plan coverage, the Bonus Plan centered on GM's lower-level executives. The GM Bonus Plan thus provides yet another example of the attempt to keep the stockholder in the prominent position at GM. In contrast to the Managers Securities value orientation scheme, the Bonus Plan covered a much wider group of employees, but with the same purpose. "Junior executives, heads of departments, and other employees occupying important positions," in short, "the senior executives of the future, obtain a partnership interest in the financial success of the institution as a whole through . . . the Bonus Plan" (Brown, 1927:15). Raskob (1927:133) also saw dual purposes in the GM Bonus Plan for lower-level executives, that is, to create an identity of interest with the stockholders and to stimulate desire for continued advancement.

Sloan (1964), in addition, recognized that the Bonus Plan developed "a tremendous incentive among employees not yet eligible for bonus awards to become eligible." One top executive recalled to him "I well remember the thrill . . . when I was first awarded a bonus—the feeling of having made the team and the determination to continue to advance in the organization" (p. 426). The plan also capitalized on psychological motivators, or as the executive continued, "The potential rewards of the Bonus Plan to ego satisfaction generate a tremendous driving force within the Corporation" (p. 426).

Sloan particularly appreciated the *great flexibility* the Bonus Plan afforded GM's planners in adjusting rewards and penalties (both financial and ego) to performance. While salary penalties were difficult to implement, "a substantial reduction in a man's bonus that runs counter to the trend . . . constitutes a severe penalty—and one of which the individual concerned is very much aware" (Sloan, 1964:427). On the upside, the Bonus Plan also provided much greater flexibility than did the salary structure. A raise for superior performance could upset the entire salary stratification and would commit GM indefinitely. In contrast, the bonus enabled the reward to be tailored to the period of unusually high performance (p. 427). Hence, an executive's "bonus award may fluctuate widely" (p. 408).

Further, the Bonus Plan with its annual awards permitted the designers to exert increasing leverage as participants advanced to more sensitive decision-making positions. Sloan (1964) noted that the bonus tended "to increase in a kind of geometric (rather than arithmetic) progression" (p. 426) as a man advanced, thereby making it an increasingly effective stimulus as long as the man stayed with the corporation.

Thus the designers had built powerful controls with an extremely *high variety of motivational settings* to orient GM's executive personnel toward high long-term performance. And the motivational impact reached all executives whether they were Bonus Plan participants or not.

In actual operation the Bonus Plan possessed many safeguards similar to those used in the Managers Securities Company Plan. First, the total amount of the bonus available for payment was the same 5 percent of profits after a 7 percent return on investment had been achieved. The Bonus Plan also was based on reviews of individual decision-maker performance and was not a profit-sharing plan to which employees were regularly entitled. As Sloan (1964) later articulated: "Each man must earn the right to be considered for a bonus award each year by his own effort. . . . The knowledge that his contribution is weighed periodically, and a price put on it, acts as an incentive for each executive at all times" (p. 408).

The Bonus Plan review procedure paralleled that of the Managers Securities Company. The bonus was allotted, at year's end, to the various divisions in proportion to their contributions to the corporation's prosperity as a whole. Each division's general manager then recommended for the Finance Committee's approval or modification how the allotment should be distributed within his division (Raskob, 1927:133). Thus individual performances were judged on corporate rather than purely divisional contributions. Furthermore, benefits were tied directly to GM's long-term corporate earnings and common-stock value since the Bonus Fund was invested in General Motors common stock (Brown, 1927:15).

Again as with the Managers Securities Plan, the Bonus Plan payments to participating executives were made in GM stock: "One-fourth of the stock so awarded is delivered at the time of award, and the balance in three equal annual installments, provided the employee remains in the service of the Corporation" (Brown, 1927:15). If the beneficiary left GM of his own free will or was dismissed due to unsatisfactory performance, the unpaid balance reverted to the bonus fund (USFTC, 1939:544). Even the junior executives occupying comparatively insensitive managerial positions, then, were required to keep their bonuses in GM stock and encouraged to remain (successfully) with the corporation. In 1933, *Fortune*'s editors could write, "Perhaps the most generous bestower of management bonuses is General Motors, yet General Motors is also high among the corporations whose managements show a keen sense of the responsibilities of their position" ("Corporate Management," *Fortune*, 1933:102). Not surprisingly, six years later they added that "the bonus plan . . . ties G.M.'s best men to their jobs by strings of withheld stock" ("General Motors IV," *Fortune*, 1939:150). So "with a heavy stock interest in the corporation," GM's top managers were "more conscious of the identity between their interests and those of the shareholders than they would be if they were professional managers only" (Sloan, 1964:408).

The decision makers' identification with the stockholders' interests coupled with their concern for losing their extensive stockholdings freed the designers from having to provide detailed, daily control of GM's multitudinous decision variables. Such a task could not possibly have been performed directly by

Sloan, Brown, Raskob, et al., but they could conduct yearly, post-control performance reviews (via the Finance Committee) without overtaxing their channel capacity. Thus, GM's designers significantly amplified their limited control powers.

Orientation via the Savings and Investment Plan and the Housing Plan

Both the Savings and Investment Plan and the Housing Plan were directed primarily toward the worker level. While this level was not extremely critical to GM's success, especially in the early days, it did receive value-orientation design attention.

The designers established the Savings and Investment Plan in 1919. "Any employee, after having been in the service of the Corporation for three months," could "place 20% of his wages from the date of employment, not to exceed $300" (Sloan, 1930:2-3) in any year. "For each dollar thus paid into the Savings Fund, the Corporation pays 50¢ into the Investment Fund.... Money paid in by the Corporation is invested in common stock of General Motors Corporation for the account of the employee" (p. 3).

The designers connected the Savings and Investment Plan to the Housing Plan by permitting an employee to apply his payments in the Savings Funds to the purchase of a home and "at the same time receive the full benefits from the Investment Fund at maturity" (Sloan, 1930:3). Sloan, in justifying this program, reminded GM's stockholders "that home-owning employees . . . are happier and better employees and citizens and to the degree that this can be accomplished there results a more efficient and therefore more prosperous institution as well as a better community" (p. 3). And, of course, employees with an equity to preserve would be more cautious and thus more controllable. Moreover, as capital owners themselves they would be more likely to empathize with the positions of other capital owners, e.g. GM's stockholders.

FACTUAL EDUCATION

While well-oriented decision makers could be safely left alone on a daily basis, they could not be left unattended for extended periods without further design effort. Decision makers also had to be educated with respect to the factual premises of decisions to achieve an amplification of control for the clients and designers.

Several types of education were necessary. For example, since Brown's financial-control model was a radical innovation for the GM auto executives, these engineering- and production-trained decision makers needed instruction on its effective use. Second, the executives in charge of the divisions along with their functional staffs required on-going education on the best methods for improving performance in the still rapidly changing auto industry. Finally, junior executives had to be trained so that they were ready to assume positions of higher responsibility.

Education via Brown's Financial-Control Model

Several attempts were made to orient and to educate simultaneously. One of these dual orientation and education attempts originated with the Managers Securities Company. "At the close of each year" Sloan "held a Managers Securities Company shareholders meeting, attended by all the participating executives, . . . to review the results of the year just ended." Here was Sloan's "chance to emphasize the mutuality of interest between its executive shareholders and the General Motors shareholders" (Sloan, 1964:415).

While Sloan emphasized the orientation task at these meetings, Brown concentrated on the educational aspect. This effort was but one lesson in an on-going instructional program. "My task, in those early years," Brown (1957) reminisced, "was to find some means whereby the significant accounting facts and financial considerations affecting the business would be revealed, . . . understood and appreciated by those . . . responsible for the various operations; management at all levels" had to "be made more aware of the constructive opportunities . . . for serving better the long-term interests of the stockholders" (p. 61). With respect to Managers Securities' "all-day meetings," Brown recalled that "comprehensive statements" displayed "how those common interests" between executives and stockholders "were served by effective control of capital expenditures, of inventories and receivables, efficiencies in manufacture, sales and distribution, and in product-appeal to the consuming public" (Sloan, 1964:415).

Probably because of the infrequency of the Managers Securities Company stockholders' meetings, Sloan and Brown used a more frequent technique to instruct the divisional decision makers, namely the divisional rate-of-return analysis form previously presented in Figure 4-1.

In discussing the use of this form—prepared monthly by the division staffs—Sloan (1964) emphasized that "the early return-on-investment form . . . was the first step in educating our operating personnel in the meaning and importance of rate of return as a standard of performance" (p. 143). While Sloan was concerned primarily with orienting the divisional decision makers to the measure of performance which reflected the clients' well-being, Brown again exhibited more concern for its instructional thrust. Brown (1957), for example, stressed that "this form was designed with a view not only of eliciting the data desired but also of planting in the minds of the reporting divisions the significance of the various factors involved." Brown wanted to impart "a greater consciousness . . . and an appreciation as to how improvement in overall results could be achieved by suitable attention to each element of investment, cost or expense. This method of reporting contributed importantly to a better understanding of the end results sought. (pp. 48–49)

Education via the Interdivisional Committees

The divisional decision makers also needed basic instruction in the functional areas to improve performance. Purchasing agents, for example, had to move from Brown's financial-control model to specific purchasing practices that would increase inventory turnover and thereby boost the rate of return on investment. As was noted in Chapter 6, the Interdivisional Committees were important centralized decision-making bodies. Another function was to insure that the auto divisions, primarily, but also the parts divisions, had properly trained decision makers. So as Sloan's orientation schemes focused thinking on the stockholders' interests and Brown's financial-control model directed thinking toward the return rate and the critical variables that influenced it, the Interdivisional Committees promulgated those variable settings that improved performance.

Soon Sloan (1924) could boast:

> Our buyers are learning purchasing. For instance, the fundamental principles of inventory control cannot be too often studied or too intelligently applied. Our ratio of stock on hand to production is constantly going down, without interfering with continuity of manufacture. The saving on investment is considerable, [as] is the reduced risk of loss of capital tied up in inventory if prices for material go down. . . .
> Our buyers are learning to buy for the needs of production, rather than to buy for a speculative profit—another fundamental principle. (pp. 194–195)

The General Purchasing Committee provided the means for Sloan's teaching. Excerpts from one committee meeting, for just one example, show that "ways and means of reducing inventory to the minimum consistent with uninterrupted production schedules were discussed" (GMX 125, *U.S.* v. *DuPont,*

1956:6935). Since the General Purchasing Committee's minutes were regularly distributed to all purchasing agents, Sloan's educational efforts reached far beyond those individuals represented on the committee. Furthermore, the committee's provision for revolving members meant that a selected purchasing agent could be brought in when needed, as Sloan (1924) put it, "giving him a bigger view" (p. 140).

Similarly, the General Technical Committee provided education for the divisions' engineers. In fact, the General Technical Committee's "most important role was that of a study group" or "seminar" (Sloan, 1964:109). Meetings featured a general discussion of "one or two papers on a specific engineering problem or device." Members brought back "to their divisions . . . a broader understanding of new developments and current problems of automotive engineering" and a knowledge of the activities of "their associates in other areas of the corporation."

Education via General Motors Institute

Not all of the designers' educational efforts were directed toward current decision makers; the designers also allocated resources to the education of future managers via the General Motors Institute (GMI).

GMI was established in 1926 on urging from Harry Bassett, head of the prospering Buick Division and member of the Executive Committee (Gronseth, 1936:405). GMI, according to Sloan, gave employees an opportunity "to train for greater responsibilities in the corporation, its divisions and subsidiary companies" (Faurote, 1930:428). Overall, the Institute "provided an agency for types of training which would best be conducted for the Corporation as a whole through a central program" (Pound, 1934:410).

GMI's educational programs were divided into two categories, one for full-time students and the other for employees taking courses in their spare time. Cooperative engineering, technical trades, and automobile-service courses were available to full-time students. Spare-time course work included "job training, semi-skilled trades, accounting, industrial engineering, foremanship, management, and practically every phase of the automobile industry" (Dibble, 1926:181).

Of particular interest, here, is the Cooperative Engineering Course. The Cooperative Engineering Course, according to Albert Sobey, GMI's head, was organized as a source of future executives who were broadly trained in engineering and management and who received or had experience in the divisions' plants. Specifically, "the training for future executive-material was organized in two separate programs, one for the young men of promise already employed in the [GM] industries and the other for the young men of the right type coming out of high school" (Sobey, 1927b:242).

The programs for the 500 students in both tracks were quite similar in content. Sobey (1927a) classed the subjects under two headings: "(a) fundamental engineering subjects that are common to all college engineering courses, and (b) specialized training in the principles of industrial organization and management" (p. 555). The management course offerings of the last two years reveal the GM planners' thinking on what constituted a sound managerial education. They included such topics as: (a) psychology, applied psychology, labor problems, and incentives; (b) economics, accounting, and costs and costs control; (c) industrial management and management problems; (d) manufacturing methods, salesmanship, and business law; and, finally, (e) a very *avant garde* course in statistics, business cycles, and barometers.

It would seem that these courses were exactly what future GM division and corporate executives required to be effective decision makers. They needed, for instance, to understand the use of incentives. Psychology was important, for as Sloan knew, "The human problem is far more delicate and difficult to handle than any production . . . distribution . . . engineering or financial problem" (Forbes, 1924:759). Going still further, Sloan was noted to hold that "personality and a knowledge of the theory and practice of psychology constitute at least 50 percent of the material requisites of success in any executive and 75 percent in a motor executive. Fine minds, without a knowledge of how to make other fine minds work best, are just so much waste material" (Benson, 1928:130). Similarly, GM's future decision makers needed to understand accounting procedures, and how they could be used to measure and control financial performance. Most importantly, they needed to know how business cycles influenced GM's performance, particularly since GM produces a consumer durable, subject to wild demand fluctuations. An important way to counter the disruptive effects of these demand fluctuations, as was explained in Chapter 2, is to anticipate their occurrence through the use of predictive, or feedforward, information: thus the probable explanation for a GMI course on business-cycle barometers.

In sum, the GMI course offerings show the designers' concern for providing effective performance control at GM through the use of properly educated—or designed—decision makers. In passing, it is interesting to note that Edward N. Cole earned his Bachelor of Science degree from GMI in 1933; Cole attained the GM presidency in 1967.

SELECTION

In addition to orienting and educating their decision makers, the designers could select executives who already possessed the orientation and education needed for effective performance control. Since such executives had the correct performance-control focus, they could be used immediately. If executives could not be located with exactly the correct orientation and education (certainly a likely possibility considering the scarcity of executive talent in the developing automobile industry and, more importantly, the novelty of Brown's financial-control model), at least the designers could select division executives who would readily accept orientation and education.

Selection of the New Division Managers

The Sloan administration's particular care in orienting and educating its division managers as well as in reviewing their efforts and results came on the heels of the Durant management's marked avoidance of such procedures. "Mr. Durant," according to Pound (1934), "relied on trusted lieutenants and trusted some of them too far, their capacities sometimes being unequal to their responsibilities" (p. 199). As Durant's assistant, Pratt worked with these men in trying to improve their operations and concluded that they were largely "impractical men" (testimony of Pratt, *U.S.* v. *DuPont* 1956:1406). He added that Durant "would like the men that would come up as some of the automobile men had by working in the shop, and really producing something with their own hands. They had done it by gut and try, and some of them did a very good job; a lot of them didn't" (p. 1406).

When Sloan and Pierre duPont first assumed command of GM, then, they found themselves in a bad position with respect to the division managers. With no corporate office in existence, immediate and thorough orientation and education were not yet possible. Furthermore, Durant's division managers were decidedly independent men, wholly unsuited to accepting and learning Sloan's carefully crafted control procedures. As Cray (1980) suggested, Durant's men "pulled long faces whenever Sloan talked about rational managerial controls" (p. 189). Not surprisingly, then, these divisional heads, "accustomed as they were to the Durant procedures, . . . anticipated difficulties for themselves under duPont management" (Pound, 1934:194). Indeed, personnel changes were called for as well as "a wholesale realignment of relations between the plants and the central offices. Whether that could be done seemed for a time doubtful; the situation was so tense, for instance, that in the case of one new manager there was grave doubt whether he might be allowed access to his office without interference from men who were being replaced" (p. 194).

But the opposition soon collapsed, and the recalcitrant division managers of Cadillac, Olds, Oakland, and Chevrolet were dismissed quickly (testimony of Sloan, *U.S.* v. *DuPont*, 1956:988–989). In Sloan's words, "drawing money from the corporation that was not properly authorized" (Chandler, 1966:170) underlay the dismissal of R. H. Collins, manager of the Cadillac Division, and E. Ver Linden of the Oldsmobile Motor Division. Collins had withdrawn $491,000 for his personal compensation for the year 1920, an amount the corporate headquarters disputed (DPX 64, *U.S.* v. *DuPont*, 1956:5714). Ver Linden too had withdrawn without authority "cash against his own disputed calculation of 1920 compensation and also cash against [Bonus Plan] certificates of stock" (DPX 64, p. 5716).

Unauthorized withdrawals also seem to have played a part in the dismissal of Oakland division manager, Fred W. Warner (Chandler and Salsbury, 1971:499). Other factors, in particular a $3.7 million inventory error resulting from an astonishing overstatement of the Oakland Division's physical inventory in 1919, also influenced Warner's dismissal (p. 503).

Of the above managers, "Pierre was not concerned about the loss of Warner and Ver Linden, for their divisions were badly managed, but he may have been sorry to lose Collins, who was an able executive" (p. 499). Cray (1980) concluded that "Collins' loss was keenly felt; General Motors needed salesmen, and Collins was that" (p. 193). But even when they were able, overindependent executives were not to be tolerated by GM's designers.

At Chevrolet Fred Hohensee quickly left. Pratt testified that "Mr. Hohensee was a great driving force, but he had no technical education. His great forte . . . was his ability to get work done, but he didn't care too much how well it was done" (*U.S.* v. *DuPont*, 1956:1407). As a result, the Chevrolet car was the poorest product of GM's auto divisions. It was so defective, in fact, that a consulting firm even suggested that the whole division be scrapped (Sloan, 1941:139).

But Pierre duPont and Sloan demurred and instead launched a salvage effort. Karl Z. Zimmerschied, an engineer loyal to GM and not Durant, was appointed as the new division manager. To no avail, he worked to improve the car and the division's performance. Since little progress was made, Zimmerschied—an appointee of Pierre himself—was replaced within a year and a half. Another problem with Zimmerschied's tenure at Chevrolet was that he opposed duPont (and the Executive Committee) on the use of Kettering's new copper-cooled engine for the Chevrolet. DuPont, "convinced that this new motor would be useful," moved "Zimmerschied out of the division" (Crabb, 1969:392) and returned him to his previous position as assistant to the president. Zimmerschied left GM shortly, supposedly after a "nervous breakdown" ("General Motors II," *Fortune*, 1939:39). "Chevrolet's Zimmerschied," put more harshly, "had rebelled until he suffered a physical breakdown and was hospitalized" (Cray, 1980:209). GM's new regime meant "business." You could stand in its way if you wished, but it would not turn aside as Durant did, and in the end it would move on even though its course might be temporarily askew as it was with the copper-cooled engine. The corporate juggernaut had started to roll.

DuPont himself took over as the Chevrolet Division manager, but he soon gave the job to Knudsen, an ex-Ford executive who "had studied the 'copper-cooled' motor and thought it had production possibilities" (Crabb, 1969:392). So Knudsen was in early agreement with Pierre duPont regarding the use of the Kettering-designed engine, considered at that time to be a potential means for competing with Ford in his high-volume market (Chandler and Salsbury, 1971:529). (The engine, however, was discarded eventually as unworkable in a production automobile.) Over the longer run Knudsen had other characteristics cherished by GM's designers. "Bill Knudsen," as Brown (1957) put it, "had a marked appreciation of the requirements of coordinated control, and of the need for administration to conform to basic policies" dictated by the corporate headquarters. "He and I worked wonderfully in harness" (p. 103).

To fill the posts in the other three auto divisions, Pierre—on suggestion from Sloan—appointed Alexander Hardy to Oldsmobile, George Hannum to Oakland, and Herbert Rice to Cadillac (testimony of Sloan, *U.S.* v. *DuPont*, 1956:989). Hardy, like Zimmerschied, had spent some time in the corporate headquarters (pp. 969, 989). Specifically, "Alexander B. C. Hardy moved from Chevrolet to the advisory staff's purchasing section and, having proved himself loyal, finally to Oldsmobile as general manager" (Cray, 1980:193). Before heading Oakland, Hannum had been general manager of one of the accessories divisions where he worked under Sloan (testimony of Sloan, *U.S.* v. *DuPont*, 1956:989). Undoubtedly, he was familiar with Sloan's approaches and expectations. And prior to taking over Cadillac, Rice was "in the central organization. In fact, he was treasurer of General Motors" (p. 989). Rice, interestingly then, had more financial than production experience. Given the planners' emphasis on financial performance, this appointment was not at all surprising. As a group, "these men proved to be competent, though not exceptional divisional managers. Yet they quickly came to have as much confidence in Pierre as they once had in Durant. Their attitudes, in turn, helped to persuade the divisional personnel to support the new management wholeheartedly" (Chandler and Salsbury, 1971:499).

Thus the managers of the major divisions as a group were now more amenable to the Sloan-Brown controls. As Chandler (1966) noted, "The most important result of the departure of [the] strong and independent managers, and this result seems to have been quite unintentional, was to make much easier the regaining of control over these divisions" (p. 170). While Chandler claimed this result was "quite unintentional" it is hard to believe it was unforeseen by Sloan and duPont. Since at least two of these "strong and independent" managers were fired for violating the new cash-control procedures, it seems unlikely that they would be replaced with others equally independent and uninterested in the new accounting and statistical controls. Further, it seems Sloan and Pierre duPont surely would have anticipated that the replacements would know exactly why their predecessors were fired and behave accordingly. In fact, Pierre wrote to Lammot duPont, then a DuPont Company vice president, "With the change in management at Cadillac, Oakland and Olds, I believe you should be able to sell substantially all of the paint, varnish and fabrikoid products needed; especially is this true of Cadillac" (GTX 421, *U.S.* v. *DuPont,* 1956:4012). Finally, it should be noted that Hardy at Olds, Hannum at Oakland, and Rice at Cadillac were all allotted additional reserves of Managers Securities Company stock, which would be extremely valuable if successful performances were achieved. Here was added motivational leverage in Sloan's hands.

Nor would the lessons of these dismissals be lost on those executives remaining with the divisions. When Collins left Cadillac, for instance, he took with him the chief engineer and three principal department heads. Ernest Seaholm, who stayed on eventually to become Cadillac's new chief engineer, observed: "You can imagine things looked dark. It was catastrophic, like suddenly losing your backbone. All the leaders from the top down were gone and those of us left were confused and anxious" (Hendry, 1983:125). It seems safe to conclude that the "confused and anxious" would not be expected to resist effectively Sloan's centralization efforts. Here then was the fertile and broken ground from which "a new organization took shape" (p. 125).

In sum, Sloan and Pierre duPont seem to have selected their division managers for technical knowledge and a cooperative attitude and not for brilliance or insight or intuition (Sloan, 1964:433). Brilliance, after all, fostered independence and often required the design of specialized jobs, or roles. And as Drucker (1973) noted, "Alfred P. Sloan, Jr., the architect of General Motors, was adamant that jobs had to be impersonal and task-focused" (p. 411). He made only one exception to accommodate Charles Kettering, GM's brilliant inventor.

Selection of Other Important GM Decision Makers

The only auto division manager to survive the Sloan reorganization, *Harry Bassett,* was probably GM's most able and cooperative division head. By far, Bassett's Buick Division was the top GM performer. The high-precision, mass-production Remington Arms Company first employed Bassett. "Harry showed [an] aptitude for figures, and the treasurer annexed him as his assistant" (Forbes and Foster, 1926:5). Eventually, Bassett's close "analyses of the different operations enabled him to make suggestions for improving methods and cutting costs" (p. 6). Mott hired Bassett away from Remington to be his assistant superintendent with Weston-Mott Company (Young and Quinn, 1963:29), and he soon considered Bassett to be one of his "right-hand men" (Sloan, 1941:50). When Weston-Mott merged with Durant's GM, Bassett became the assistant general manager of Buick. In addition to working under Walter Chrysler at Buick, Bassett had known Sloan when he was still selling his Hyatt Roller Bearings. As the Buick manager Bassett ardently supported such Sloan philosophies as the testing of automobiles and the annual model change. With respect to the latter innovation, Bassett said: "I believe in striving to make one's product better every year. . . . It's important not only to make it better, but to make it look better" (Forbes and Foster, 1926:3). About GM, Bassett felt: "Its Managers' Securities Plan makes us feel that we are working in the interest not only of the corporation but of ourselves. Under the leadership of such men as Mr. Pierre duPont and President Sloan, we are all inspired to do our level best" (p. 12).

During the early design days, not surprisingly, managers with experience at the DuPont Company were often selected for GM's important decision-making positions. Such executives were predisposed to accept the stockholders as clients and to understand and solve the performance-control problems which influenced the stockholders' well-being. *Alfred R. Glancy*, a former DuPont engineer, came to GM in 1920 ("General Motors," *Fortune*, 1938:155). Glancy, as has been mentioned, helped develop the Oakland Division's new Pontiac car, whose success resulted in the dropping of the Oakland make and the renaming of the division. Accordingly, Glancy succeeded George Hannum in 1925 as the division manager. At DuPont *E. F. Johnson* had become a good friend of Pratt's and, according to Brown (1957), "had demonstrated marked managerial ability as general manager of the Old Hickory Plant" (p. 36). Johnson also directed DuPont's important Development Department in Wilmington (Chandler and Salsbury, 1971:498). At GM Johnson first worked under Mott to head the Intercompany Parts Divisions. Later Johnson came to be the assistant group executive under Pratt when he directed both the Intercompany Parts and the Accessories Group Divisions. *Eric Bergland* became involved in GM affairs as early as 1918 when he wrote a report, eventually presented to Pierre duPont, critical of GM's lack "of a central engineering department, in fact, of any sort of permanent, well-organized central staff" (Chandler and Salsbury, 1971:468). After joining GM Bergland worked as the executive secretary of the Works Managers and the Power and Maintenance Committees and, accordingly, headed the staff functions involved with these production-oriented committees. Similarly, upon leaving DuPont *James Lynah* came to head GM's central purchase-contract investigation and standards staff. Lynah also became the executive secretary of the General Purchasing Committee.

Throughout the design process, managers from GM's own Parts and Accessories Divisions were often selected for important corporate and divisional roles. As Pratt testified, "Sloan knew the boys in the accessory divisions, and when he had a big job he wanted to do in the car divisions, he reached down in the accessories divisions and got them, and I had to find somebody else" (*U.S.* v. *DuPont*, 1956:1421). Thus the Accessories and Parts Divisions provided a training ground for the vitally important auto divisions or the corporate-headquarters positions. For instance, *James D. Mooney* became the Export Group executive in 1921 after having worked for Sloan at Hyatt and serving as president of Remy Electric ("General Motors," *Fortune,* 1938:167). Mooney was interested in organization and authored several books on the subject. *Harlow Curtice* started at GM in 1914 as a bookkeeper with GM's AC Spark Plug Co. of Flint. "A year later, at twenty-one, he was named controller; at thirty-five he was president" and at forty became general manager of the Buick Division (Sheenan, 1956:11). During the late 1930s Curtice rejuvenated Buick's faltering sales. In 1952 Curtice became GM's president and, true to form, he had "some slight leaning toward centralization" (p. 19). Curtice, for instance, spent much time on styling GM's cars, but it was in the crucial matter of scheduling that Curtice most actively and consistently intervened (p. 14). *C. E. Wilson,* also destined to become GM's top executive, joined GM in 1919 as chief engineer and sales manager of the automotive section of Remy Electric, then a GM subsidiary (Forbes, 1948:448). Later he became Remy's chief engineer and helped put its "operations on a sound financial basis" (p. 449). After becoming division manager in 1925, Wilson was selected as special assistant to Sloan, for whom he conducted many centralizing functions, like the corporate-wide purchasing of tires, commodities, etc. Wilson also served as the vice president of the Accessories Group and "staff advisor on all manufacturing" ("General Motors," *Fortune,* 1938:45) to Knudsen when he was president. *R. H. Grant* was a salesman at National Cash Register Company (and worked with Kettering there) before joining GM. After heading the Delco Light unit he was promoted to Chevrolet, where he guided "Chevrolet's sales through the twenties, and so became the top salesman in the United States" (Sloan, 1964:97). The Knudsen-Hunt-Grant Chevrolet team overcame Ford's dominance of the automobile industry. Grant, like Hunt and Knudsen, was transferred to the corporate office. "A new executive group moved up in rank and into the general corporate area where their influence would affect the entire corporation" (p. 170). Grant instilled "most of Chevrolet's major selling methods in the other car divisions," and "Grant-trained Chevrolet salesmen turned up in charge of sales" at Pontiac, Oldsmobile, Buick, and Cadillac as well ("G.M. III," *Fortune,* 1939:78).

Finally, the managers picked for important jobs often had accounting and financial backgrounds. In many cases this training was quite extensive. *Marvin E. Coyle,* for instance, succeeded Knudsen as the general manager of Chevrolet when Knudsen became executive vice president in charge of all car, truck, and body manufacturing operations. Coyle started with GM in 1911 in the then minuscule central accounting office, moved to Chevy in 1917, and eventually became Chevy's comptroller under Knudsen. To this financial-control-oriented executive the precious Chevrolet Division was entrusted during the dismal depression of the 1930s. Indeed, the *Fortune* editors characterized Coyle as "thoroughly an 'organization man' " who managed Chevrolet "on so strict and clockwork a basis that the Coyle blood is commonly supposed to run with decimal points instead of corpuscles" ("General Motors II," *Fortune,* 1939:42).

In sum, the designers attempted to provide managers who willingly accepted corporate views and directives. As Frederic Donner, another important GM executive, put it: "General Motors attaches the highest importance to the selection of its managers both in the United States and overseas. . . . They must know the policy limits of their authority. They must recognize when a decision might breach these limits and require a policy determination" (Donner, 1967:30). And most important, decision makers were selected for positions or roles in the organization structure; positions or roles were *not* created to accommodate specific individuals. With the high comparability that existed among the many GM divisions, the GM decision maker became a standardized commodity capable of high interchangeability. The standardization, of course, made for easy orientation, education, and selection, and consequently provided the designers with a considerable amplification of control.

THE AMPLIFICATION OF PERFORMANCE CONTROL: SOME EVIDENCE

Did the division managers and their staffs indeed operate as amplifiers who provided constant and detailed performance regulation for the various GM divisions? The data available to answer this question can be separated into two classes: (1) information on the amplified control of the causal variables, such as inventory level or turnover, and (2) information on the amplified control of the overall return-rate variable.

Amplification of Control over Causal Variables

Evidence that the divisional managers and their staffs accepted the designers' concepts and, in turn, amplified control is found in statements by several important divisional executives. The general manager of the Delco-Remy Division, C. E. Wilson, wrote: "Our bogey for inventory turnover is 12. In other words, our average turnover, figuring the cost of sales with relation to the value of the raw material, work in process, finished stock, is 12 to 1. That is many times better than it used to be under our old system of purchasing and production" ("Group Bonus Is Important Factor in Delco-Remy Production," *Automotive Industries*, December 10, 1927:870). W. W. Clark (1926), G. H. Koskey (1926), and C. B. Durham (1927), all in important positions at the Buick Division, also wrote about the importance of inventory control. Durham (1927) stressed, "The less money you tie up in your business, provided only that you do not handicap yourself by inadequate facilities or too short stocks, the more money you should make" (p. 30). Durham went on to mention that an important area of investment reduction was that of working capital: "We felt, back in 1920, that we were doing a pretty good job of turning over our working capital. But if we represent our 1920 rate of turnover by 1, then our 1926 rate was 10. This year we shall do considerably better" (p. 30). Going still further, Durham added these points:

- Nowhere in our plant is there space for a day's supply of any finished part except frames. . . .
- On purchased parts, one day's supply comes in some time during the day before it will be used. Raw material stock comes in much the same way. . . .
- It used to take 18 days from the time a wheel entered the wheel paint-shop until it was ready to use. Now, within 4 hours of the time a wheel enters the paint-shop it is on the automobile. (p. 56)

Amplification of Control over Rate of Return

Articles by members of the Delco-Light staff (Fordham and Tingley, 1923, 1924a, 1924b, 1924c; Tingley, 1924) show that the division managers and their staffs amplified control over the ultimate rate-of-return variable. Delco-Light, in which GM held the stock but did not own the physical assets directly, was "a subsidiary of large and growing importance" (Bradley, 1926d:4).

The budget chart presented in Tingley's (1924:384) article, "Visualizing Budgetary Control," provides an excellent summary of the entire series of articles and illustrates how the general manager, comptroller, and various functional staffs (i.e., engineering, sales, finance, and manufacturing) produced a target return rate on invested capital over the course of a year. The description of the general (or division) manager's role reveals clearly how the designers' control over rate of return was amplified. Specifically, to regulate the return on investment, he was to review his operations monthly and, should actual operating results indicate poor performance, to take corrective action. He also was to report "estimated operations and profits and loss to the board of directors with a statement of his plans to secure the desired profit" (Tingley, 1924:386). Finally, the general manager had "but one aim for the entire business—a substantial profit on the money invested" (p. 383). Thus, "every policy established, every piece of equipment purchased, every rearrangement of the machinery, every detail of the production system, every financial transaction must be carefully weighed and performed only if it will yield a profit. Even such an altruistic part of the business as the first aid or medical department should be continued only if it can show a profit" (Fordham and Tingley, 1923:719).

Certainly this sentiment paralleled closely Sloan's feelings regarding safety glass (see Chapter 4). "Accidents or no accidents," as he later responded to Lammot duPont, "my concern in this problem is a matter of profit and loss." Perhaps this concert of value premises on noneconomic matters is why Sloan (1941) could later say, "After all, what any one individual can accomplish is not great, but through the power of organization the effect of a few may be multiplied almost indefinitely, especially if the few have the capacity for real leadership" (p. 155). Sloan's concept of leadership, or "industrial statesmanship" (p. 145) as he called it, will be discussed more fully in Chapter 11.

9

Coordinating the
Divisions and Headquarters

Besides providing for the on-going training of GM's many corporate and division executives, the Sloan-Brown designers also had to link these decision makers together. That is, they had to build an internal information network that connected, or recomposed, the various components into a unified body capable of coordinated decision making. Accordingly, Sloan "contrived to provide" GM "with a composite brain commensurate with its size. His achievement may be summed up ·as one of intercommunication, getting all the facts before all the people concerned" ("Alfred P. Sloan, Jr.: Chairman," *Fortune*, 1938:114).

The loose horizontal coupling provided by the designers (see Chapter 7) required a minimum of divisional intercommunications. In contrast, because the designers tightly coupled the divisions, especially the auto-producing units, to the corporate headquarters, considerable vertical communication was required. The corporate headquarters needed vertical linkages to instruct the divisions and receive divisional performance data so new instructions could be formulated.

LINKING THE DIVISIONS

The loose horizontal couplings that were built into the firm's structure were designed not only to reduce the cognitive complexity of the control problem, but also to lower the amount of communication needed to solve the coordination problem.

A functional structure with its tight horizontal couplings would have necessitated many massive high-frequency communication links between components. GM's divisional structure, on the other hand, permitted each divisional staff to control its unit's performance with little need for frequent coordination with other units. While the couplings were tight within a given division, they were relatively loose and infrequent across the division's boundaries. Thus a particular unit needed to observe few of the others' decisions and very limited interdivisional communications were required.

During the 1920s most communications dealt with the transfer of parts or accessories among the product divisions. The standardization of these parts not only decoupled much variety flowing among units, but also facilitated communication regarding these transfers. For example, GM's universal numbering system for its standardized parts eliminated much confusion. "With the universal numbering system," as Alex Taub, engineer of the Advisory Staff's Factory Division, put it, "we all speak one language. Prior to that time each firm spoke a different foreign language" (Baird, 1923:335). So "the greatest benefit from the universal number system" resulted from "eliminating the 'Babel' of numbers heretofore existing" (p. 336). Thus the combination of massive parts standardization and the universal numbering system eased the problem and costs of communications among GM's many divisions.

Other than these facilitation efforts, the corporate designers stayed out of the higher-frequency interdivisional communications until the late 1930s. "Each division of General Motors had handled its own inter-city traffic, some by telephone, others by telegraph and a few by teletype" (Warner, 1940:51). In 1937, however, GM launched a study of "an interplant communication network" and concluded "that the coordination of its divisional communication activities would provide faster and more economical service." That is, "the corporation" found "that centralization of its communication facilities would result in marked efficiency as well as increased speed through the use of common channels" (pp. 50–51). Eventually, GM's centralized communication net connected over 500 General Motors facilities across 88 cities, making it the world's largest privately operated communication linkup (p. 52). The corporate headquarters' new Communication Section ran the operation. From its Detroit headquarters this unit was connected, by private lines, to New York, Philadelphia, Dayton, Chicago, Flint, Pontiac, Lansing, and Grand Rapids. The New York and Flint connections allowed for simultaneous message transmission. Intermediate relay points were used to connect with other cities—Atlanta, St. Louis, Pittsburgh, Baltimore, Trenton, N.J., Janesville, Wis., Milwaukee, and Kokomo, Ind. (pp. 52-53). These outlying points served as message collection centers; therefore a message from Miami went to Atlanta and then on to Detroit via Dayton. "Clearing messages from the South through Dayton save[d] materially in wire tolls because the leased teletype wire from Dayton to Detroit [was] more economical for a heavy volume of traffic than sending the messages individually from Atlanta to the points of destination" (p. 54). At the end of 1939, 80 percent of the 45,000 messages per month handled by the centralized communication network dealt with manufacturing operations, i.e., interdivisional and interplant parts transfers, but many sales points were being integrated into the net so the sales message traffic was expected to increase.

LINKING THE DIVISIONS
TO THE CORPORATE HEADQUARTERS

The Sloan-Brown designers created vertical connections within the firm through the: (a) Interdivisional Committees, (b) Policy Groups, (c) Advisory Staff, (d) Financial Staff, (e) Executive Officers (f) Operations Committee, (g) Executive Committee, and (h) Sloan Meetings.

The Interdivisional Committees

Sloan's Interdivisional Committees were an extremely important aspect of GM's communication network. Sloan (1927) himself pointed out that "co-ordination is effected through what we call Inter-Divisional Relations Committees where those interested in the same functions of the important divisions meet together and discuss their own problems as well as the same problems from the standpoint of Corporation policy. For instance, our Purchasing Agents meet together in the form of a General Purchasing Committee, presided over by a Vice-President of the Corporation" (pp. 11-12). As *Fortune*'s editors reported years later, Sloan's "zeal" was "not so much to 'run' G.M. as to arrange and activate its committees;" his "science" was "not only automobiles but management itself" ("General Motors," *Fortune*, 1938:178).

Sloan's institution of the Interdivisional Committees brought the divisional staffs of the critically important auto divisions into close proximity with the top corporate officers, notably Sloan and his fellow Executive Committee members. Close proximity, in turn, led to close observation. And with the intimate contact promoted by the Interdivisional Committees, the Sloan-Brown designers could be assured the important auto divisions were using the best methods isolated by the Advisory Staff.

One of Sloan's motives in creating the Interdivisional Committees was to improve communication links between the corporate headquarters' and the divisional staffs following the major dispute over Kettering's copper-cooled engine. After this bitter fight, Kettering and his men were simply not talking to the divisional engineers anymore, a gap that could not be tolerated by Sloan, who was bent on improving GM's products. The functional committees provided the communication vehicle where the disparate views could be brought together and reconciled. Accordingly, they had their own resources—secretaries, offices, and apparently budgets (Chandler, 1966:188). Their representatives came from the similar functional departments of the various automobile divisions and sometimes from other operating units. Corporate-staff department heads soon moved from mere committee membership to become the Interdivisional Committees' secretaries (p. 188), thereby occupying focal positions in the communication network. Sloan served on all the committees, chairing at various times the General Technical, the Works Managers, and the Purchasing. At least one other Executive Committee member sat on each committee. Brown chaired General Sales, for instance, and Pratt headed the General Purchasing Committee. "The committees, therefore, provided a systematic and regular means by which the line, staff, and general officers could meet monthly or even more often to exchange information and to consider common problems" (pp. 188–199). The General Technical Committee, for example, united GM's research scientists, divisional engineers, and corporate executives (Sloan, 1964:253).

The Interdivisional Committees greatly improved the low-frequency lateral communications among the extremely similar auto divisions. In short, they periodically made "available to each Division the experience of all members of the family" (Bradley, 1926d:5). Thus performance-improvement information generated in one division could be transmitted to another to improve its performance too. "There have been in General Motors," as Sloan put it in 1924, "certain products turned out which are excellent and some which are only fair. Every General Motors product is going to be better than the best is now. This can and will be brought about by concentrating the brainiest men, those who have produced the excellent products, upon improving the products which most need improving" (Forbes, 1924:760). A bit later Brown (1927) stressed: "There are certain general policies which, if good for one division, are good for all." For example, fundamental engineering developments needed to be "brought to light" and have their "adaptability determined," as did manufacturing and sales methods and policies (p. 11). The Interdivisional Committees served to crystalize important corporation policies, make them effective, and facilitate "the adoption of engineering improvements and operating methods." Or, as Raskob (1927) detailed, "the General Technical Committee" made sure "that new engineering developments of a fundamental character" were disseminated among the divisions. "The Works Managers Committee" considered "the application of new, economical manufacturing methods . . . developed in one plant, to others." And "the General Sales Committee" acted "in the same capacity with regard to sales and advertising policies" (p. 132).

The presence of Sloan and at least one other Executive Committee member on each Interdivisional Committee was vital to facilitate lateral communication. First, the top corporate executives' participation testified to the importance of the committees' communications tasks. Second, their presence created a favorable communication attitude in the senders and receivers of performance-improvement information. Since the divisions competed to some extent, high-performing divisional staff members might be unwilling to share performance-improvement knowledge discovered by their divisions. As Sloan well knew, the right "atmosphere" had to be created to foster communication. In the presence of top corporate officers divisional executives would be assured recognition for their discoveries and corporate-wide performance contributions. Similarly, the divisional executives receiving the improvement data would be more likely to use it, for their futures would be determined to a significant degree by the top executives. In Sloan's words: "Those in charge must first be brought into a receptive frame of mind. They must be 'sold' on the advisability and importance of supplementing their own knowledge and ability by accepting the best counsel others can bring to their assistance" (Forbes, 1924:760). Thus the designers bridged several important lateral communication gaps and corrected "the lack of any extensive interchange of information among the divisions" (Sloan, 1964:253).

Before the creation of the Interdivisional Committees, the corporate officers had problems insuring the effectiveness of their downward communications. The divisional engineers, for example, distrusted Kettering's research group (Chandler and Salsbury, 1971:525). Similar difficulties occurred with sales. Although the corporate staff under Norval Hawkins developed valuable programs on advertising, trade-in policy, sales methods, car maintenance and repair, and dealer organizations, the divisions ignored this advice. Even an occasional effort by Pierre duPont and the Executive Committee could not change the situation (p. 526). But when the divisions' functional specialists joined Sloan in regular Interdivisional Committee meetings, their attitudes changed; they began to accept what the corporate headquarters officers were saying. Even Kettering admitted to Pierre duPont: "The Technical Committee meetings are a fine thing and . . . are going to do much toward developing the spirit of cooperation in each of the divisions" (p. 548).

The Interdivisional Committees also became important conduits for upward flowing communications. Under Sloan's direction, for instance, the General Technical Committee established the GM Proving Ground for testing GM's (as well as competitors') cars. "When all the charts have been consulted and the chief engineer finds that his car is trailing the field in the several particulars he knows at once that something has to be done about it" (Condit, 1926:11-12). And with the technical audit reports receiving close scrutiny by such Technical Committee members as Sloan, Mott, and Kettering, division engineers would be expected to take any corporate-office advice seriously.

Even though membership on the Interdivisional Committees was generally limited to the auto divisions, the audit activities conducted by the committees and their respective staffs covered all divisions. Consider for example one extensive investigation made under the auspices of the General Purchasing Committee: "Secretary reported that representatives from Purchase Contract Investigations Section had called upon" no less than 17 divisions and plants "to check . . . participation in general contracts, and to receive suggestions and criticisms for use in renewing these contracts" and in developing future general contracts. "The results of the survey indicated that in general participation was satisfactory, and co-operation whole-hearted. A few misunderstandings were cleared up." Visits were to be made at regular intervals as the contact was found "most helpful" (GMX 163, *U.S.* v. *DuPont,* 1956:7143). Along with providing considerable data for the corporate headquarters on the matter of divisional compliance with performance-improvement methods, such extensive audits surely had a chilling effect on divisional autonomy. What divisional purchasing agent would want to be cited for noncompliance with a General Purchasing Committee contract?

Prior to the late 1920s, in sum, relatively little coordination was needed "between a program proposed by one [automobile] division and one that might be put forward by any other car divisions" (Sloan, 1964:184). Under these circumstances, the infrequent meetings and loose contacts of the Interdivisional Committees were adequate. GM's interdivisional purchasing contracts, for instance, were negotiated for extended periods of time; and, once finalized, they did not demand continuous, detailed attention.

The Policy Groups

By the late 1920s, however, Sloan and his colleagues were poised to nudge their narrowly diversified, divisionally organized firm—with its matrix-like Interdivisional Committees—toward a slightly more functionally oriented structure. This modification in the organizational form would help them recoup still more of the scale economies lost with GM's early adoption of the easily managed divisional structure. Sloan's experiments with the Chevrolet-Oakland-Pontiac, Oldsmobile-Viking, Buick-Marquette, and Cadillac-La Salle joint ventures had shown that significant cost savings were attainable through pooled volume, that is, increased consolidation. Furthermore, the considerable experience the designers had gained with the automobile divisions' operations during the 1920s allowed them to push for increased interdivisional coordination. By 1930, then, they had acquired the requisite capacity to manage the more complex decision-making situation. The depression, of course, added considerable impetus and justification for further integration; reduced sales meant that volume had to be pooled if profits were to be sustained and dividends maintained. It was not long before GM's five automobile divisions shared three common bodies.

Once the Sloan-Brown designers created the interchangeable-body program and further centralized the styling and engineering functions, it became impossible to treat the divisions independently. "In other words, a divisional product program, instead of being integral in itself," became "deeply involved with the product programs of many of the other divisions." Consequently, GM's annual models had to be created from the "corporate point of view." And throughout the lengthy development period, "detailed contact" had to "be continually maintained between the engineering departments of all the car divisions, the Styling Staff, the Fisher Body Division, and more or less the accessory divisions" since they all were "commonly concerned with and working together on a single problem" (Sloan, 1964:184).

These tight interdivisional and interlevel relations required a new organizational mechanism to coordinate all the units involved. Sloan's new Policy Groups in the various functional areas, aided by the associated corporate staffs, provided the requisite recomposition device. With respect to the common-body program, for instance, the corporate engineering staff collaborated with the divisions "to effect the necessary co-ordination." Sloan's "co-ordination agency" that addressed "the problems involved" was the Engineering Policy Group (p. 184). The evolution of GM's corporate-level Policy Groups from the intermediate level Interdivisional Committees was complete, and the firm could now better coordinate its activities to meet the depression's exigencies.

However, the designers did not always seek to facilitate communications among the divisions. For example, the designers did not want the auto divisions to communicate about styling matters (Sloan, 1964:272). After styling was centralized within the corporate headquarters, each division was allocated a separate space. Harley Earl "kept a competitive creativity going among his staff by having a locked door policy for each of the five divisional studios so that none knew what the others were doing. Only Earl had the key" (Bayley, 1983:73). With interchangeable bodies (from engine firewall to rear bumper) the cars could easily come to look alike and not provide the product variety demanded by the automobile markets. Internal communications thus had to be blocked, on occasion, to insure an external variety match.

The Advisory Staff

The communications work of the Interdivisional Committees and then the Policy Groups was connected closely with the corresponding units of Sloan's Advisory Staff. Thus, one head of the Sales Section, B. G. Koether, also served as secretary of the General Sales Committee; James Lynah was both secretary of the Purchasing Committee and head of the Purchase Section; and Harry Crane, Sloan's technical advisor, was also secretary of the General Technical Committee (Chandler, 1966:188).

Accordingly, the decisions reached by the Interdivisional Committees cleared "through various central office staff organizations, maintained so as to perfect the flow of information back and forth and to facilitate the orderly consideration of common problems of important policy and procedure" (Brown, 1927:12). With respect to the Policy Groups, Sloan (1944) explained, "Serving each committee is a central office staff of specialists which works under the jurisdiction of the committee and is charged with the responsibility of gathering information, conducting tests and preparing exhaustive reports and analyses designed to throw light on the problems discussed by the committees. . . . Their entire efforts are devoted to *finding the facts*" (p. 10).

In general, the Advisory Staff's work within the GM headquarters was extensive and its relationship with the Executive Committee was close. According to a 1924 article by Mott entitled "Organizing a Great Industrial," numerous sections of the Advisory Staff kept the Executive Committee updated via monthly reports that summarized their activities.

The various Advisory Staff units were also intensely involved in fostering lateral communication within the firm. Among the functional departments of the various divisions, they promoted the exchange of performance-improvement information related to sales, production, accounting, engineering, and research (Chandler, 1966:185). In this way the Advisory Staff fashioned the communication towers and lines that leaped the boundaries erected by the adoption of the divisional structure and linked the product-specialized divisions.

It is important to consider briefly the lateral communication loads placed on the Advisory Staff (as well as on the Interdivisional Committees) in transmitting performance-improvement information. While the divisional structure certainly increased these communications loads, it should be noted that the comparable divisions needed to be linked only infrequently; performance improvement information is not available frequently since it takes lengthy periods to discover and then verify. In contrast, had a functional structure been adopted by GM's planners it would have necessitated considerable high-frequency communication among the functional components just to conduct and coordinate each day's business. Thus, the planners' early selection of the divisional structure kept the lateral transmission infrequent and the communication loads manageable. (They only modified this simplified approach with the richer connections of the Policy Groups after they had gained a decade of managerial experience in automotive operations.)

The Advisory Staff also helped the corporate line executives with their downward communication by providing the divisional managers "with expert advice and assistance which was used to improve current operations and to formulate future plans" (Chandler, 1966:185). Extreme care seems to have been taken by staff units in their communications with the divisional managements to improve effectiveness. Kettering (1928) for example, in describing the communications of his laboratories, wrote: "Our technical-data section . . . receives and edits all reports before they go from the laboratories to any one else in the company. This section is . . . a public-relations section for the laboratories, with its public confined to the General Motors Corporation" (p. 738). In short, it interprets "the laboratories to the company . . . in concise, plain, understandable English."

The Advisory Staff provided still another important communication function for the designers. Since the corporate officers (i.e., the group vice-presidents, the president, and the Executive Committee members) constituted a very small group, they, of course, had a quite limited channel capacity for making detailed checks on the particular methods used by the many divisions. The Advisory Staff increased the corporate group's channel capacity, for the Staff checked on the divisions to "see how various policies were being followed" (Chandler, 1966:185). The "steady flow of information on operating activities" that they provided to the corporate group was, "like statistical data provided by Brown's financial staff, . . . relatively free of any divisional bias" (p. 185). Moreover, the corporate executives were now less dependent on the divisions, no longer having to base their decisions on information developed by the divisions' functional departments (p. 383). The Advisory Staff, then, provided a reliable local source of information for the corporate executives (Drucker, 1946:55) and better information for developing the divisions' improvement plans.

The Advisory Staff's power is clearly illustrated by the strong new product "auditing" authority that the Executive Committee gave to the Staff's Research Section on March 15, 1921. Specifically, the Executive Committee directed the Research Section of the Advisory Staff to review the preliminary sketches of the divisions' new car designs, to test the corresponding experimental models, to report on the results, and to recommend changes. Should the Research Section and a particular division not be able to reconcile their differences, "the Operating Vice President" was to "advise the Operating Division as to what changes, if any, in his judgment are essential to protect the Corporation's interests" (DPX 58, *U.S.* v. *DuPont*, 1956:5699). Ultimately the full Executive Committee received the standardized reports "on all special tests as well as on the regular 25,000-mile one" (Chandler and Salsbury, 1971:519).

The copper-cooled engine fiasco at GM further enhanced the Advisory Staff's power. Kettering's development of this engine and the divisions' unwillingness to use it created extreme problems for Pierre duPont and Sloan. To increase control over the divisions, Sloan endorsed Raskob's suggestion that the Advisory Staff be given the same audit role as the Financial Staff. If competent managements appreciated financial auditing, as Sloan argued, the division should also welcome checks on "the quality of its product, the efficiency of its Sales and Engineering Departments." Thereafter, it would not just be "a privilege but a duty of the staff sections of Dayton and Detroit to familiarize themselves with the respective engineering, sales, construction and purchasing sections of each division" (Chandler and Salsbury, 1971:526). "Thus staff sales executives reviewed marketing policies, controls, and procedures with the sales managers of the many divisions; those on the manufacturing staff did the same with divisional production managers; and so with automobile design, advertising, and other comparable activities" (Chandler, 1977:462). While the Advisory Staff supposedly was to remain in an advisory role only (like the auditor), a prudent division manager or his staff could hardly ignore such advice.

The Financial Staff

In contrast to the Advisory Staff, which concentrated on discovering the causes behind divisional performances, the Financial Staff emphasized overall performance measurement and reporting. They determined whether the reporting and control procedures still worked properly, saw to it that the corporate office and the operating managers received information promptly, and fine-tuned the statistical and information controls (Chandler, 1966:184).

Besides operating the overall performance-measuring apparatus via its central headquarters personnel and divisional comptrollers, the Financial Staff operated important (and related) information networks associated with the cash, inventory, and capital-investment variables. In their attempt to reduce the cash balances needed to run GM, for instance, the Sloan-Brown designers made several important communications-design changes. The first, and most important, change was their pioneering use of the Federal Reserve telegraph network for transferring funds among GM's various bank accounts (Pound,

1934:201-202). According to GM's Swayne (1924), "the service thus rendered the 71 divisions and subdivisions resemble[d] very closely that performed by the Federal Reserve System for the national banks of the country" (p. 22). Another communication improvement the GM designers made to reduce GM's cash balances was the interdivisional billing system—previously introduced in Chapter 5—for which the Financial Staff served as a "clearing house." With this approach, "Cadillac pays Hyatt for bearings with a white slip. Hyatt bills Cadillac; Cadillac acknowledges to Hyatt on a white slip; Hyatt mails the slip to the [GM] comptroller, who charges Cadillac and credits Hyatt and thus the settlement is made without the use of cash" (Swayne, 1924:22).

As a further communication improvement, the corporate-headquarters group designed into the new cash-control methods a *daily review procedure.* One objective was to insure, on a short-term basis, that excess cash was not held, for it earned a very low return. Sloan (1964) alluded to an important by-product of the corporate review, namely the careful monitoring of divisional daily activities: "We began calculating a month ahead what our cash would be each day . . . , taking into account the sales schedule, payrolls, payments for materials, and the like. Against this projected curve we compared each day the corporation's actual cash balances." A divergence would signal the need to find the cause "and to take corrective action at the appropriate level of operations" (p. 123). The GM treasurer, then, could monitor the divisions by comparing daily cash receipts and disbursements with his expectations. Moreover, the division managers had to know that their activities were under constant and careful scrutiny.

A similar situation developed with respect to inventory control. At first, it was Pratt's Inventories Committee that reviewed the divisions' inventory levels and production schedules. Once "the flow of incoming material was stopped," as Sloan (1964) explained, "each General Manager submitted a monthly budget" that estimated the next four months' sales and production requirements. After the Inventories Committee had discussed these budgets "with the General managers," the Committee released "material for one month's production at a time" and therby "gained control over runaway inventories, reduced them, and conserved cash" (p. 125). With the demise of the Inventories Committee, Sloan took over reviewing and approving the divisions' production schedules. As noted in Chapter 6, Brown and his Financial Staff supplied Sloan with the requisite support.

The Financial Staff also received extensive data on capital appropriations: "The divisions were to make monthly reports of construction in progress. . . . Proper records were to be kept of expenditures and approvals for expenditures, and uniform treatment was to be given to appropriation requests throughout the corporation" (Sloan, 1964:121). Anderson's Financial Staff unit, as could be expected, was to receive the divisional reports and assure that funds were spent as intended. By April 1921, GM's allocation procedures for long-term, permanent investment were as precise and effective as DuPont's (Chandler and Salsbury, 1971:502).

The Group Executives

The group executives provided still another communication linkage between the corporate headquarters and the divisional components. From the earliest days, those positions had been conceived of by Sloan as primarily communication vehicles, especially between the divisions and the Executive and Finance Committees. Brown (1957) emphasized that these executives were "to keep in intimate contact with the operations in their respective jurisdictions, to serve in an advisory capacity, and to bring about common understanding of purposes and policies affecting divisional operations." That is, the group executives served as "the top echelon liaison officers in the 'two-way flow' of authority and ideas essential to proper coordinated control of corporation activities" (p. 67).

Each group executive supervised a set of highly comparable divisions. Consequently, he could determine relatively easily the causes of good and bad performances. He could then disseminate the causes of good performance to all his several divisions and, thereby, improve the entire group's performance. Essentially, then, the group executive's communication role was quite similar to the Advisory Staff's. The group executive, however, in contrast to the Advisory Staff units' promotion of lateral communications, dealt with a smaller set of more comparable components and with less-specialized functional problems.

Besides being an important point of contact between the corporate headquarters and the divisions, the group executives simultaneously provided close ongoing control themselves over the divisional performances (Chandler, 1966:165). For example, James D. Mooney (1924), the group executive in charge of the Export Group, explained that he exerted his performance-control function by means of a number of reports from the various units under his direction. Specifically, each unit sent "a complete monthly report showing assets, liabilities, reserves, income, sales inventories, etc., as well as a detailed sheet showing analysis of profits, turnover, and return on investments" (p. 31). This performance information was supplemented with data on: (a) "total production for the month of the different car lines and the percentage which goes to the various export organizations" (p. 31), (b) "the total automobile exports divided into commercial and passenger cars, and divided also into the various countries to which these automobiles are sent" (p. 31), and (c) "local conditions and any items of interest which affect export sales" (p. 28).

According to Mooney, some of these reports were "conclusively informative" on their own, while others were "of value only when compared to like data from other sources, or the same data for previous periods of time" (p. 28). Once

Mooney had studied the reports he turned them over to his staff for further analytical work. "Graphic control charts" were prepared to monitor the Export Group's sales volume and market share "with reference to the business as a whole." Additional "sets of curves" showed the individual status of the various members of the Export Group, including their turnover, percentage of profit, and return on investment. In sum, "all these control curves and charts" were "watched carefully for a sudden break in the wrong direction. When the break" was "sharp and . . . of sufficient importance, one of the executive staff" would "be assigned to make a complete study and to determine means of remedying the weakness" (p. 30). Clearly then, Mooney, as a group executive, paid close attention to the performance of his group, focusing especially on the maintenance of high performances among all export units and investigating poor performances thoroughly. In short, he was a vital link in the vertical control network established by the Sloan-Brown designers to insure GM's high performance.

The Office of the President

To establish direct communications between the president's office and the various divisions, Pierre duPont, accompanied by Sloan, spent many of his early days at GM on the road, visiting and evaluating the various divisions inherited from Durant (Sloan, 1964:56). After Sloan assumed the presidency he continued his inspection trips to supplement GM's expanded reporting procedures with some firsthand reconnaisance. Particular attention was paid to the critical car and truck divisions. "In the 1920s," as Brown (1965) put it, "Mr. Sloan made frequent visits to the car divisions with the purpose of discussing all angles having to do with problems of coordinated consideration, and planting in the minds of division managers the concepts relating to . . . stockholder interest" (p. 333). In addition, the managers of the car and truck divisions were summoned to Detroit once or twice a month for meetings, often attended also by the specialized corporate staffs (Dale, 1956:55). Once formalized under Sloan, the frequent visits were to bring the divisions under closer scrutiny and to make them less free to do whatever they wished. After the initial crisis of 1921 ended and Sloan became president, he spent "the greater part of every other week in Detroit" (Sloan, 1941:173). From Detroit he often inspected GM's facilities throughout the area. The divisional privacy enjoyed under Durant, who hardly ventured out of GM's New York financial headquarters (Chandler and Salsbury, 1971:444), was eroded still further.

To supplement these visits plus the many regular committee meetings, Sloan and Brown used the division return rate as a summarizing performance indicator. By employing a measure that summarized the contribution of all decisions made by the division management, GM's designers were able to monitor and improve divisional performance while not overloading the channel capacity of the small presidential office. "Financial control as worked out by General Motors," according to Sloan (1964), "gave the corporation a review of operations that reduced the need to administer operations from the top. Central-office management was able to know whether the decentralized management was operating well or poorly and had a factual basis for judgment regarding the future of any particular part of the business" (p. 148). Hence, the overall rate-of-return indicator assured Sloan that even the more detailed and routine divisional decisions would be made correctly.

Sloan and Brown used both yearly and monthly formalized performance reviews to gauge the divisional performances. The yearly performance reviews, conducted as an integral part of the Managers Securities Plan and the GM Bonus Plan, can be characterized as heavily postcontrol in nature. While they were essentially evaluations of completed performance, literally beyond improvement, they were also to some extent precontrol performance-improvement efforts: Sloan attempted to reorient and Brown attempted to reeducate the divisional staffs to pursue the clients' interests more successfully in the next period of operations.

The monthly budgetary reviews can be considered as primarily current control in nature, designed to adjust a poorly performing division's characteristics so that it would perform better during the coming periods, which in this case was the next several months. To make these monthly performance reviews work, each division manager submitted monthly reports. "The central financial office" took the data and placed it on forms "to provide the standard basis for measuring divisional performance in terms of return on investment." In turn, "each division manager received this form." And if the results were unsatisfactory, Sloan "or some other general executive would confer with the divisional manager about the corrective action." In addition, when Sloan made his routine calls on the divisions, he "carried a little black book" that systematically "exposed the facts with which to judge whether the divisions were operating in line with expectations as reflected in prior performance or in their budgets" (Sloan, 1964:142).

The precontrol nature of the corporate headquarters' review of divisional performances was furthered by combining the data on the past month's actual performance with budget data on future months' expected performances. Early in the tenure of the GM designers, divisions were expected to forecast sales so that in-process inventories could be controlled properly. Sloan (1964) explained, "The instrumentality of control . . . became the divisional four-month forecast of expected business, which came to me as vice president in charge of operations" (p. 127). Later the designers increased the scope of the budget forecasts to include plant investment, working capital, outstanding inventory commitments, estimated sales, production, and earnings. The expanded forecasts were to be in Sloan's "hands on the twenty-fifth of each month. They covered the current month and each of the three ensuing months" (p. 127). Still, these forms were not overextensive, so they were not likely to overload Sloan's office.

The Operations Committee

While the less important units could be left to the care of the group executives, like Mooney and Pratt, Sloan himself kept a close observation on the auto divisions. Besides his many personal inspection trips and his periodic performance reviews, Sloan used the Operations Committee to keep the managers of these important components in close proximity. Unlike the Interdivisional Committees, the Operations Committee was concerned with the divisions' total performances rather than their operations in any specific functional area.

When Sloan reorganized the Executive Committee and excluded most of the division managers from it, he established the Operations Committee. Sloan placed on the committee "all the general operating officers on the Executive Committee and the general managers of the principal divisions, thus making it the major point of regular contact between the two types of executives" (Sloan, 1964:113). Sloan had created then another set of communication links between the controllers in the corporate headquarters and the regulators of the important auto divisions, for "in a large enterprise some means is necessary to bring about a common understanding" (p. 113).

Sloan constituted the Operations Committee to achieve sufficient participation of the critical divisions yet still allow corporate domination. In 1927 the Operations Committee, for example, was composed of seven Executive Committee members, five others from the corporate level, and five division managers (GMX 3, U.S. v. DuPont, 1956:6562). Even if one discounts the committee's executive secretary and assistant and includes Lawrence Fisher with the division managers, they still were significantly outnumbered by the corporate-level members. So, with the Operations Committee, then, Sloan had still another corporate-dominated group that could observe the division managers of the most critical units as well as provide them with extensive operating instructions.

The Operations Committee thus provided an important vehicle for lateral and downward communication of performance-improvement information. In this vein, Sloan could use the Operations Committee to pit division managers against each other so that they would comply with performance-improvement suggestions. In 1924, for example, George Hannum, the Oakland division manager, strongly supported Sloan's desire to paint GM's cars with the new Duco lacquer. Bassett at Buick, however, feared the dull Duco finish's potentially negative sales impact. Bassett obviously needed to be motivated, and

Sloan made "a note to discuss this matter at the next meeting of our Operations Committee with a view perhaps of developing a little more atmosphere in the mind of Mr. Bassett." (GTX 394, *U.S.* v. *DuPont,* 1956:3947). A reasonable man, like Bassett, could not fail to see the light, accept the performance-improvement information, and make the changes. In fact, Bassett shifted abruptly from opposition to wanting to convert Buick's entire production to the new Duco lacquer.

While downward and lateral communications were important, the Operations Committee's upward communications were still more important to GM's high performance because they permitted Sloan and the other corporate executives to exert a final performance review function over the divisions. In particular, the Operations Committee met at least monthly, and according to Sloan (1964), "would receive a full set of data on the performance of the corporation and would review that performance" (p. 113). In addition, "the Operations Committee, including the general [divisional] managers, appraised the performance of the divisions" (p. 114). With the important divisional managers close at hand, such postcontrol review by the Operations Committee easily became precontrol adjustment aimed at improved future performance. Finally, the very threat of the monthly performance review probably kept the division managers as well as the corporate decision makers operating correctly in their regulation of current operations.

The Executive Committee

Further up in the GM hierarchy, the Executive Committee also provided crucial linkages. For instance, consider the Executive Committee's role vis à vis the critically important Fisher Body Division. Fred, Lawrence P., and Charles T. Fisher were all added to the Executive Committee. In this regard Pierre duPont wrote: "Interesting . . . members of the Fisher family directly in General Motors will have a very beneficial effect in breaking up a line of separation of the two companies' interests that has not been altogether wholesome. From lack of knowledge, the two sides have tended to criticize each other, without good result. Hereafter the Fishers will better understand General Motors problems and difficulties and, I think, General Motors men will better appreciate the Fisher problems" (GMX 32, *U.S.* v. *DuPont,* 1956:6658).

The Executive Committee also established vital connections to the most important of GM's auto divisions, first to Buick and Cadillac and later to Chevrolet. In 1924 Buick and Cadillac were the most efficiently managed GM divisions (Pound, 1934:204). At Buick, with its high volume and high unit profit, Bassett was carrying on the sound manufacturing and marketing techniques of Walter Chrysler, who had built the car's reputation for quality along with an excellent dealer distribution network. By 1927 Buick was "running at its capacity of over a thousand cars a day, with 5000 [fewer] employees than in 1923 when production reached the figure of 925 cars per day" (Clark, 1926:85). Similarly, Cadillac was continuing the high-quality production policies of its innovative, perfectionist founder, Henry Leland. After Lawrence Fisher's Executive Committee appointment, he became Cadillac division manager and was specifically charged with improving the Cadillac car's appearance to match its high-quality engineering. Sloan was particularly anxious to follow Cadillac's development, for he was well aware that the right styling adjustments could help GM capture an increasing share of the then static market. With Harry Bassett and Lawrence Fisher on the Executive Committee Sloan had access to important performance-improvement information for upgrading the other auto divisions and could keep these most critical units' managers in close proximity. In short, extremely tight control over GM's most vital auto divisions could be had.

William S. Knudsen's appointment to the Executive Committee provided further performance-improvement information and the opportunity to bring a division rapidly growing in importance into intimate contact with the top corporate executives. Knudsen joined the Executive Committee on May 9, 1929, when his Chevrolet division was on the rise and had outsold Ford in 1927 and 1928 (Pound, 1934:230; Rukeyser, 1927:373).

Executive Committee contact was by no means limited to the finished-auto divisions. Via the *summarizing* transfer prices charged to the auto divisions, the Executive Committee monitored the performances of the parts and accessories divisions. For instance, in October, 1921, the Committee requested each division to report in detail why it had purchased items (outside) that could have been bought inside GM (Chandler, 1966:174). Several years later, the group executive (usually also on the Executive Committee) had to approve all outside purchases before they could be made. A manager of a purchasing division who found the prices charged by a selling division out of line with the open market was to notify the Executive Committee, which then had the price differences investigated (Chandler and Salsbury, 1971:500). With this procedure, then, the Executive Committee could use the auto divisions to monitor the performance of the many and far-flung parts and accessories units, thereby bringing the entire divisional level under close corporate observation without information overload. Again the mere threat of such scrutiny would no doubt motivate the supply divisions to perform well.

The Executive Committee, as GM's pinnacle internal policy-setting unit, maintained contacts besides those with the divisions (Sloan, 1964:102). In 1924, for instance, the Executive Committee included seven members associated with operations (Sloan, Mott, Pratt, Bassett, and three Fisher brothers), two concerned with finance (Raskob and Brown), and one concentrated on external relations (Pierre duPont).

In sum, the Executive Committee occupied the focal position in GM's early internal communication network. Sloan, its chairman, was the undisputed master at arranging and linking GM's extensive committee structure. The Interdivisional Committees brought together corporate staff officers and their divisional counterparts along with corporate line executives and "gave a measure of co-ordination to the functions of purchasing, engineering, sales, and the like" (Sloan, 1964:114). Moving upward, the Operations Committee linked the divisions' general managers with the Executive Committee members—the principal line and staff decision makers of the corporate headquarters. While GM's Interdivisional Committees focused on reviewing functional specifics, the Operations Committee concentrated on judging the divisions' monthly and quarterly overall performances. The Executive Committee, in turn, set corporate-wide policy, reviewed long-run capital-allocation needs, and made investment recommendations to the Finance Committee—which controlled the firm's strategy via its capital-budgeting purse strings. "On the operating side," however, "the Executive Committee was supreme. Its chairman was the president and chief executive officer of the corporation; and he had all the authority he needed to carry out established policy" (p. 114).

When interlocking the above committees for coordinated decision making, Sloan himself filled the major linking-pin role. In 1927, for instance, Sloan chaired two Interdivisional Committees and was a member of the other two. Furthermore, he chaired the Operations and Executive Committees and occupied a seat on the pinnacle Finance Committee. Brown chaired the Interdivisional Committee on sales and joined Sloan on the Operations, Executive, and Finance Committees. Mott sat on each of the Interdivisional Committees as well as on the Operations and Executive Committees. Pierre duPont, Raskob, Pratt, Fred Fisher, Charles Fisher, and Lawrence Fisher also served on multiple committees and thereby helped knit GM into a well-coordinated decision making body. Here, then, was a thoroughly recomposed, i.e. unified, firm. "This was the new General Motors scheme of management from which developments down to this day, through much evolution, have been derived" (Sloan, 1964:114).

The Sloan Meetings

As one of his evolutionary developments Sloan instituted "special meetings to create common understanding" (Drucker, 1946:60). These "Sloan meetings" (p. 61) were started around 1936 and held twice a year in Detroit under Sloan's chairmanship. "The results of the various divisions . . . and reasons for success and failure" were discussed, as were the divisions' and central management's suggestions for improvements. "Unplanned but effective personal contacts" were also "established between central management and divisional personnel." Since several hundred people attended the meetings as regulars and an equal number in rotation, "practically every senior employee—beginning perhaps at the level of plant superintendent—[gained] an opportunity to see the business as a whole, to see his place in it and to familiarize himself with the basic policies and the program of the company" (p. 61). Through the Sloan meetings, then, a broad recomposition of the firm's many decision makers was attempted.

Still Sloan was not satisfied, for the "Sloan meetings" were "too large to establish the personal contact" desired "between central office and divisional personnel" (p. 61). So the Detroit meetings were supplemented with smaller meetings at various production centers where central-management representatives met for several days with local divisional executives, including "all the people . . . invited to the 'Sloan meetings' and a number of lesser employees from the local plants and offices" (p. 61).

Not only did these meetings add to GM's formal communication linkages, but they helped foster the development of an informal network. By establishing a conducive atmosphere and the initial contacts, Sloan could rest assured that GM's many executives would on their own initiative fill out the firm's internal communication network wherever necessary. This supplemental activity was required because a complete communications layout was well beyond the capacity of GM's design team in spite of Sloan's considerable accomplishments in this respect.

It is interesting to note, in conclusion, that Sloan convened these meetings at a time when GM's management was practically under siege by strong external forces: notably, government and labor. Under these pressures, as will be shown in Chapter 11, Sloan directed his attention inward and concentrated more on GM's internal workings.

10

Synchronizing the Firm and Environment

Synchronizing the firm with its environment entails a circular communication/ decision process in which the enterprise both receives and transmits information. To address the task of synchronizing GM with its immediate economic environment, therefore, Sloan and his codesigners, primarily Brown and Bradley, established the requisite communication channels. Sloan (1964), for example, "had some consumer studies made in 1922, and . . . found that people throughout the United States, except at the corner of Wall and Broad Streets, didn't know anything about General Motors" (p. 104). In response, Sloan had the Finance Committee approve a plan submitted by the advertising agency of Barton, Durstine, and Osborn (later BBDO) to publicize the parent company.

Though the communication flows the designers sought to synchronize were admittedly two-way, GM's outgoing flows played a less obvious yet important role in its success. The extensive information GM collected on consumer preferences, for example, was usually transmitted back to the environment not as verbal or written copy but as product characteristics: the shape of a fender or hood, the color of paint or upholstery, the number of cylinders in an engine, the type of transmission, etc. GM's planners, of course, did not overlook the more obvious outward communication flows. But determining product characteristics demanded by consumers was more important to GM's planners than finding the best advertising campaign to sell a less desirable product. In often making GM the industry's leading advertiser, the firm's top executives knew full well that they could not ignore advertising, as Ford did periodically; even so, advertising would not move a poor product in the long run. There was far too much meaningful product competition in the then evolving automotive industry for any such simple solution to be successful.

The GM designers' purpose for building the environmental connections—whether incoming or outgoing—was to coordinate the firm's activities with the other units and groups in the automobile-distribution channel. On GM's input side there were the raw-material and parts suppliers and workers; on the output side were the dealers, competitors, and consumers. As the automobile purchasers occupied the critical position in the distribution channel, the designers eventually geared most of GM's synchronization efforts as well as its internal operations to consumer demand. Considerable effort was focused on generating feedforward data capable of predicting future customer desires.

SYNCHRONIZATION WITH THE INPUT MARKETS

Because of GM's great purchasing power and its multiple internal and external sources of supply, comparatively little time was spent in synchronizing the firm with its raw-material and parts suppliers. Yet GM's performance depended partly on the performance of its suppliers; hence, to attend to information transfer across this environmental boundary, the Sloan-Brown designers established the Purchase Contract Investigation Section under the direction of the General Purchasing Committee. This unit collected and digested data on the divisions' requirements for purchases in large quantities (Mott, 1924:525). Moreover, its staff literally investigated the productive facilities and financial positions of GM's various suppliers to insure that they could fulfill the purchase prices, quality specifications, and delivery dates established in the GM contracts. Maintenance of production at GM was far too critical to leave such important variables unobserved (GMX 158, *U.S.* v. *DuPont*, 1956:7106).

GM's provision of extensive standardized parts information (discussed in Chapter 5) permitted raw material and parts suppliers to align their products with GM specifications. The designers also expected these suppliers to adjust their deliveries rapidly to the firm's varying needs. GM's growing purchasing power allowed it to enter "requirements contracts," which meant it did not have to buy specified quantities but only what it actually needed to maintain *current* production. Thus GM passed the variety emanating from its output front straight through to its input side and did not have to absorb it by accumulating costly inventory. "All our suppliers," as Sloan reminded Lammot duPont in 1926, are required "to adjust their schedules in line with our schedules. . . . In other words, the supplier always takes the obligation to supply what material is needed for the manufacturer's requirements and accepts increases or decreases in schedules as the demand on the . . . manufacturer's products may necessitate" (GTX 365, *U.S.* v. *DuPont,* 1956:3895).

Suppliers, naturally, desired feedforward information on GM's forthcoming production schedules (GTX 366:3898). But Sloan did not want to release the firm's actual production schedules, preferring to release its buying schedules which indicated *exactly* what GM wanted. As he testified, "There is no direct relationship between the production schedule and the incoming materials released by the purchasing department to the various suppliers" (p. 1315). For instance, materials were requisitioned to create a "float," that is, a buffer inventory to protect the planned production rate. Depending on the material, the "float" could vary greatly. For steel, which involved a lengthy manufacturing process, the "float" had to be bigger. "Not only that," Sloan emphasized, "in December and January, we have to build up large inventories . . . in the hands of our dealers" as consumers typically do not purchase cars "until April and May; while we are doing that, we have to increase the 'float' to protect . . . the higher production schedules" (p. 1315). Similarly, as production dropped during the spring months, so did the "float." Finally, Sloan testified that "when we come to the liquidation of a model, we have to reduce the incoming materials . . . much below the . . . production schedule that is forecast." When the model year ended, the "float" had to go too. So there was only "a limited relationship" between GM's production schedules and the purchasing schedules necessary "to support their production" (p. 1316).

GM's purchasing agents were instructed, also, to vary their inventory buffers, or float, with changes observed in the distribution network. "So many days' supply of this material, so many days' supply of that material, and so on, depending upon the condition and location of sources of supply, transportation conditions, etc." (Bradley, 1926d:7). Timing was important, too. "If," as Sloan (1926) cited, "railroad service should suddenly become slower, then all of our commitments would have to be made further in advance." Factors affecting the periods between placing the order and receiving the material were "all carefully watched" (p. 997). To help the divisions monitor changes in the transportation system, the corporate headquarters' Advisory Staff included a Traffic Section (Mott, 1924:526).

In part, the suppliers' problem in synchronizing their activities with GM's stemmed from the weak contact many had with the divisions. In Pratt's words, "The trouble here may largely be due to our policy of making General Motors contracts rather than allowing each Division to contract for its own requirements" (GTX 368, *U.S.* v. *DuPont,* 1956:3909). Indeed, GM's planners had tried to promote contact between the divisions and their suppliers by limiting the coverage of the general contracts. The General Purchasing Committee Minutes of January 2, 1924, noted, "for the purpose" of keeping the divisions "in touch with sources·of supply" and "in a position to feel out the market at any time and . . . encourage best possible service on part of contractor," "general contracts would . . . cover not more than 80% of General Motors requirements of any item" (GTX 158:7106). Problems, however, still arose. But while centralizing contract writing in the corporate office weakened the synchronizing communication links between GM's divisions—the users of the parts and supplies—and the suppliers, GM designers felt it was the suppliers' and to some extent the divisions' problem. Pratt, for instance, stated, "As far as I know all of our suppliers under General Motors contracts realize that General Motors contracts do not in any sense relieve them of the responsibility of keeping in close touch with the various Divisions in order to render proper service as to deliveries and requirements" (GTX 368:3909). Sloan placed the responsibility on the divisions, for he wrote to Lammot duPont, "I feel very keenly that our divisions should cooperate closely with their suppliers and that modifications either up or down of our shipping schedules should be minimized to the fullest possible extent, otherwise it tends to increase the cost which in turn we must pay" (GTX 367:3906).

GM's designers, then, devoted minimal time to perfecting the outgoing communication channels needed by other firms to synchronize their operations with GM's. Such neglect was tolerable simply because GM was an extremely desirable account with its huge volume. Moreover, the divisions could work out the specific, high-frequency details of shipping schedules once their purchasing agents and division managers were trained to minimize in-process inventories without interrupting the production schedules. Accordingly, the suppliers and the divisions were to provide much of the detailed synchronization communication for GM's input interface.

The selection of specific suppliers and the setting of specifications and prices, of course, were not given to the GM divisions for many important inputs; the corporate headquarters' units, not the GM divisions, negotiated supply sources, prices, and specifications. Thus, the corporate headquarters set these long-term market relationships while the divisions handled the shortterm, high-frequency, less critical interactions.

The GM corporate headquarters also monitored other markets that supplied the firm with requisite input factors. The Advisory Staff's Industrial Relations Section kept track of labor conditions and helped the divisions correlate and standardize all matters pertaining to industrial relations (Mott, 1924:526). As labor was often in short supply when the industry was still expanding, the Modern Housing Section operated the divisions' various housing projects for employees (p. 524). A Real Estate Section acted as consultant as well as agent for the purchase, sale, or lease of properties by the corporation or divisions (p. 524). An Office Building Section of the Advisory Staff operated the General

Motors Building Corporation. And a Power and Construction Section was directed by a consulting engineer, whose services were at the disposal of the divisions and included the investigation of power rates, the study of fuel supply and markets, the analysis of building-construction costs, and the collection of data on boilers, ovens, heat-treating plants, and allied equipment.

SYNCHRONIZATION WITH THE OUTPUT MARKET

While the GM designers generally cared relatively little about the firm's connections with its input markets, they were greatly concerned for GM's output market. Extensive communication channels were built to synchronize GM with its dealers and, still more importantly, its consuming public.

In fact, GM's input side as well as its internal operations were kept in close coordination with the output market. "Working capital items," Bradley (1926d) wrote, "should be directly proportionate to the volume of business. For example, raw material on hand should be in direct proportion to the manufacturing requirements. . . . Work in process should be in direct proportion to the requirements of finished production. . . . Finished product should be in direct proportion to sales requirements" (p. 7). Brown (1929c) explained the designers' preoccupation with the final automobile market in the following way: "In a multi-stage industry, where there are numerous intermediate steps between producer and ultimate consumer, stabilization must depend largely upon . . . ultimate consumer demand. After that, with properly applied methods of forecasting and planning, synchronization becomes a matter of one stage keeping step with another" (p. 300). Sloan (1941) too stressed that "production must be determined by retail sales—no longer by the ambition for big figures—and statistics must be available to chart the course" (p. 201).

The effort of the GM designers to synchronize the firm with its dealers was based primarily on past, or feedback, information. In contrast, the efforts of the GM designers to synchronize the firm with the final consumers was based more on predictive, or feedforward, information.

Synchronization with the Dealers

The principal variables of concern to the designers in their synchronization efforts with GM's dealers were the finished-auto divisions' production rates.

Close corporate scrutiny was directed toward these variables because huge capital investments could be accumulated rapidly in unused inventories of finished cars. Eventually such an inventory accumulation would force a costly production cutback or complete stoppage to clear the backlog. With excessive inventories and idled capital, ROI would drop below acceptable levels.

Even after Sloan had removed the divisions' authority to set their own production schedules (as was described in Chapter 6), the division managers still possessed certain dangerous control powers. In particular, they were to prepare the four-month forecasts which were submitted to Sloan and Brown for approval, but for which the corporate office group lacked any actual production data to judge their reasonableness. This situation troubled Sloan, so at the beginning of 1921 he "asked the division managers" to report "their actual unit production and sales at the factory for ten-day periods" (Sloan, 1964:129). In addition, the division managers were "to report, at the end of each month," their unfilled orders and finished inventories along with "how many cars they estimated their dealers had on hand." Even these divisional ten-day and monthly reports did not satisfy Sloan, primarily because they too were generated by the divisions.

To help prevent the divisions from accumulating excessive inventory investments, Sloan in 1921 "made a historic deal with R. L. Polk & Co. whereby, for about $50,000 a year, Polk gathered monthly registration figures by makes in thirty-one states, thus giving sales and production managers everywhere their first authentic measurement of the market on a national scale" ("G.M. III," *Fortune*, 1939:78). These data helped somewhat to keep divisional production rates synchronized with the actual dealer sales levels.

But even with the division managers' added reports and the R. L. Polk data, effective production and finished-inventory control was still difficult. The division managers' ten-day reports, for example, provided information on inventory being shipped to dealers but only weak and delayed estimates of the inventories held by dealers. Moreover, the Polk feedback data on registrations, with their inherent lags, did not lend themselves to an instantaneous determination of the dealers' sales and stocks of finished and used cars. "The big gap in our information system at headquarters and in the divisions," Sloan (1964) recalled "was at the retail level. . . . We were not in touch with the actual retail market." The resultant "time lag could be dangerous. It became, in fact, the source of a new crisis" (p. 129).

It was Brown who became convinced, from an analysis of R. L. Polk's registration statistics, that the automobile divisions, particularly Chevrolet, were overproducing (Brown, 1957:52). "Signs of distress began to appear early in 1924," for compared to the previous year's reporting period GM's "production had increased about 50 per cent while sales to the ultimate customer had declined about 4 per cent. Here the time lag entered," for Sloan did not receive "these figures until the first week in March 1924" (Sloan, 1964:130). Without current data on inventory levels, Brown was unable to counter the divisions' arguments that inventories were in proper relation to demand. But "in May 1924," as Sloan remembered, "Mr. Brown and I made a trip into the field to discuss distribution problems with the dealers." They indeed found plenty. "In St. Louis, my first stop, in Kansas City, and again in Los Angeles, I stood in the dealers' lots and saw the inventories parked in rows. The figure man in this instance was right and the salesmen were wrong. Everywhere the inventories were excessive" (p. 131). Thus, new inventory control procedures were needed beyond those installed after the 1921 inventory crisis.

To synchronize production with demand accurately, the designers faced a two-part problem. They had both to limit the divisions via a better sales-forecasting method and to obtain more current information from the dealers. Regarding the first of these issues, Sloan (1964) wrote: "We were in a better position than the divisions to forecast total industry demand and total sales of our products for the entire model year," so "we took the first actual step in putting limits on the divisions in the spring of 1924." Specifically, Brown and Sloan developed an " 'index volume,' that is, the volume to be regarded as the pointer for a twelve-month period" (p. 134).

Ultimately, the Divisional Indices, or yearly volume forecast, were based on the previous three years' actual sales experience, which covered the basic automobile boom-depression demand cycle (Pound, 1934:216), and on "the general business outlook for the coming year" (Sloan, 1964:134). These indices were "of extreme importance," as Grant (1931) stressed, "because upon their accuracy depends whether or not we set our budgets properly. We make intensive studies, in addition to using our observation and common sense in arriving at these indices" (p. 197).

Because this corporate-level prediction "could easily be upset by actual developments in the market, there was further need for a corrective device" (Sloan, 1964:135). So after the 1924 inventory crisis, the designers secured less-delayed feedback information by requiring all GM dealers to submit both to the divisions and to the corporate headquarters inventory reports every ten days. These reports covered the sales of new and used cars to customers and the number of new and used cars in stock. "The divisions and the headquarters staff were then able to take corrective action and make new forecasts with greater accuracy" (p. 136). So "instead of attempting to lay down a hard-and-fast production program a year ahead and to stick to it regardless," the corporation followed "the policy of keeping production at all times under control and in correct alignment with the indicated annual retail demand, and with the minimum accumulation of finished product in the hands of dealers" (Bradley, 1926c:160).

Brown's Financial Staff assumed the primary responsibility for developing the divisions' production forecasts and for constantly adjusting these forecasted volumes to current demand conditions. In fact, Brown's duties as head of the Financial Staff "included continual analysis of the 10-day reports. When individual forecasts and associated production schedules seemed to be out of line with basic trends," he "would contact the operating division concerned" (Brown, 1957:55). As GM president Sloan too kept a close eye on current production levels and the dealers' ten-day reports. He mentioned in 1926 that if "all of our dealers should show a greater accumulation of unsold cars than normal in, say, January, then we immediately curtail production schedules.... If sales are below the expected, within 10 days of the slump we know it and are ready to scale down our manufacturing schedules if the trend continues" (Sloan, 1926:996). Brown (1929b) summarized the new feedback-control system over GM's production rate in the following terms:

> It is nothing more nor less than a system of control whereby production, purchase of materials and the employment of labor and of capital are coordinated with the sales requirement. Flexibility is a prime requisite so that there may be a quick response and ... adjustment throughout the system to ... changes.... The focal point is the sales outlet. The flow at this point must be gauged and every other activity must be coordinated with it. (p. 181)

Sloan, as has been previously mentioned, supplemented the formalized data coming from the Polk summaries and the ten-day reports with evaluative impressions obtained directly from visits to many of GM's dealers (Sloan, 1964:283). As these meetings were designed "to gain an objective feel of dealer reaction relating to product, factory relationships, potential of the area served by the individual dealer, and any suggestions looking to betterment of the situation from all angles," "no representative of any car division was included" (Brown, 1965:332–333). Thus, this intelligence could not be distorted by the divisional staffs to serve their short-run purposes, and it could be used safely to evaluate the long-run impact of divisional decisions. (The divisions, of course, could make their own evaluative surveys; Knudsen and Grant, for instance, made similar visits to their Chevrolet Divisions' dealers [Beasley, 1947:133].) Sloan used these personal visits, then, to shortcut the great distance between the top corporate executives and the dealers who made the final contact with the customer (Dale, 1956:47). As Sloan explained in 1928:

> I have personally visited, with many of my associates, practically every city in the United States.... It has taken weeks of the hardest kind of work and continual travel to accomplish.... On these trips I visit from five to ten dealers a day. I meet them in their own places of business, talk with them across their own desks, and solicit from them suggestions and criticisms....

I go out from the standpoint of general policies and get the facts in a very personal way, without the intermediary . . . organization which is apt to overlook the most important points and inject its own personal view-point. . . . I make careful notes. . . . When I get back home I study this information . . . for the benefit of the company. (Benson, 1928:130)

Sloan was concerned in these visits with getting data on such factors as the character of the product, the trend of the consumer demand, and the dealers' viewpoints regarding the future of the automobile market (Pound, 1934:344). Specifically, "he wanted to hear about all the criticisms of General Motors products . . . and wanted to find out . . . how to produce a better car which would be easier to sell" (Douglass, 1954:159). Besides meeting with the dealer, he often would call on the local banker and the newspaper editor to get at "the roots of General Motors merchandising in a personal way" (p. 160). Later, when GM's foreign operations assumed greater importance, Sloan also visited overseas dealers. For example, "in October 1929 Sloan went to Russelsheim to meet the dealers who handled Opel cars in Germany" (Douglass, 1954:164).

From his extensive dealer visits, Sloan concluded "that the dealer was perhaps the weakest link in the General Motors chain" (Douglass, 1954:160). Sloan was particularly concerned about the dealers' lack of sound accounting procedures and set about to organize the Motors Accounting Company. As the name implied, the purpose of this service subsidiary was to assist dealers in setting up proper accounting systems. " 'Accounting,' . . . here, is meant to convey an entirely different meaning" than "ordinarily associated with that term—a more modern and more scientific interpretation, which is not merely an historical record of the past, but which develops the probable results of the immediate future and enables alterations in operating policies to be made in advance" so that "a reasonable profit at the end of the period becomes far more certain" (*GM Annual Report,* 1929:11).

Motors Accounting personnel developed a standardized accounting system that could be applied to all dealer operations; in addition, through Motors Accounting GM sent a staff into the field to help install the information system (Sloan, 1964:287). An auditing procedure was developed too, whereby a dealer's accounts were checked periodically (p. 287). In describing the latter program Grant wrote in 1931 that GM employed "some 600 field men" to audit the accounts of dealers representing "about 75 percent of our total volume. . . . In spite of the times, in spite of our desire for economy, in spite of everything, this movement is so constructive that we are going to spend $1,250,000" on it "again next year" (p. 268). As the 1930s depression deepened, GM's dealers became more reluctant to pay for Motors Accounting audit work. Consequently, the GM designers continued the work at the firm's expense, albeit at a reduced and more cost-conscious level. The designers modified the Motors Accounting auditing procedure to sample a cross-section of the dealers' operations rather than to attempt complete coverage of GM's many dealers. In addition, GM received monthly financial statements from a large percentage of its dealers (p. 288).

Besides providing the dealers with accounting support, Motors Accounting Company was to give the corporate headquarters rate-of-return performance information on the many GM dealerships and instruct the dealers themselves on the importance of and methods for achieving high return rates (Sloan, 1929:92).

Dealer return-rate performance data would indicate whether the inventories at the dealer level were controlled properly, i.e., in strict correlation with demand requirements. According to Brown (1957:53–55), the designers adopted universal and standard accounting practices for the dealers so that the GM headquarters could determine if the divisions were overloading the dealers with inventory. While division managers might be tempted to improve their own units' short-run performances by "dumping" inventory on the dealers, Sloan and Brown worried about the high mortality of GM dealerships over the long run. They felt the firm could not supply the capital to operate such a large number of dealerships (Brown, 1957:55) nor the channel capacity to control so many individualized trading situations (Sloan, 1964:282). Therefore, the data on dealer profitability were used "to identify situations where there might be undue influences occasioned by over-supply, leading to forced movement of vehicles at retail, with consequent loss to the dealers" (Brown, 1957:55).

Another objective of this "big and expensive effort" was to enable the divisions and headquarters "to look through the whole distribution system, dealer by dealer and group by group, and determine just where the weaknesses were and what should be done about them" (Sloan, 1964:288). The Chevrolet Division had provided the impetus for these dealer-improvement efforts, for "Chevrolet itself inaugurated a standard accounting system for all [its] dealers in 1923" ("G.M. III," Fortune, 1939:78). Later, when Sloan promoted Chevrolet's Grant to the principal corporate sales position in GM, "the accounting system became compulsory and elaborate; it enabled the sales managers to set up bogeys in every department of the dealer's business—used-car, new-car, parts, service, etc.—and gave direction to their constant stream of canned demonstrations, convention shows, lantern-slide lessons, salesmen's and dealers' manuals and other educational stimuli" (p. 78).

In many cases the dealer could make improvements without direct intervention from GM's decision makers. For example, the dealer had a "control sheet" which told him how many cars he had to sell to break even (Grant, 1931:268). "Furthermore, the dealer himself could not only judge his own complicated business intelligently, but also compare his operations, item by item, with group averages. Often soft spots would be discovered in time to make corrections before harm was done" (Sloan, 1964:288). Thus Sloan and his fellow designers felt confident that they could "lay before" the dealer "the bogies of expense and facts and figures" which would serve "as sign posts down a straight road to a definite and satisfactory profit" (Sloan, 1929:96). Accordingly, the GM designers expected "to stabilize" the distributing organization so that "the failure of any dealer to make on the average a return on the capital he is employing or the loss on the part of any dealer of his capital *will, in general, be a thing of the past*" (p. 92).

Earlier, in Chapter 5, it was pointed out that while GM avoided substantial involvement in the retailing of its cars, the firm did lend capital via its Motors Holding Corporation to a selected group of dealers to insure thorough coverage of the retail market. Motors Holding Corporation also provided GM's decision makers with an improved understanding of the dealers' operations. Through Motors Holding Corporation, as Grant (1931) explained, "We can get a much better 'feel' of the situation, know what our policies ought to be, and do the big job we are after much better than if we are only dealing with the other man's money" (p. 268). Similarly, Sloan (1964) wrote many years later that the Motors Holding operation had given General Motors "a clearer and more sympathetic knowledge of dealers' problems" (p. 290), plus "a better knowledge of the retail market and consumer preferences" (p. 290). Brown (1957) felt that "this intimate contact by the corporation with problems of dealer management . . . contributed to the evolutionary formation of basic distribution policies by GM. Motors Holding dealers serve as a sort of proving ground" (p. 57). Here, then, GM had strung another important communication linkage with the economic environment.

Synchronization with the Consumers

Even though GM did not deal directly with individual automobile customers per se, Sloan still wanted to leap over the dealers to develop firsthand contacts with the ultimate consumers as a group (Rothschild, 1973:83). The automobile customers, in fact, controlled so much of GM's return-on-investment performance that the Sloan-Brown designers were not to be satisfied with delayed feedback information regarding the buyers' past choices. Instead, they wanted anticipative information covering the consumers' future purchasing decisions as well as their decision-making processes. Sloan (1926), in a general discussion of such vital anticipative design work, emphasized:

> Planning involves looking into the future, and the future is the most uncertain element in business, as it is in life. Make the future even a very little less uncertain, look ahead in business with some certainty for even a short distance,

and the groundwork is all laid for that business to make more money that it otherwise could.

... The forward view enables the competent manager to prevent overproduction and inflation, after which evils always follow, [such as] underutilization of capacity, unemployment, and disappearing profits. (p. 993)

Business stabilization hinges on anticipation and prediction.

Sloan's vision was so fixedly focused on the future that he geared GM's accounting system not so much to measuring past results—as was then typical—but to anticipating and shaping future performances. "By means of our accounting system," he noted, "we can look forward . . . and can alter our procedure or policies to the end that a better operation may result. In one case we are in principle, looking backward—in the other case, forward" (Sloan, 1929:92, 96). Thus the designers changed emphasis from determining past or current performance with feedback information to predicting future performance with feedforward data on environmental conditions, particularly consumer decision states. In this endeavor considerable success was achieved; for instance, with respect to the short run Sloan (1929) boasted, "We are able to forecast our operations four months ahead with a certainty that would hardly seem possible" (p. 96).

Brown discussed GM's consumer-oriented synchronization program along two dimensions: the statistical and the constructive. The former was "directed toward ascertaining the statistical facts bearing upon future consumer demand" ("Forecasting Methods of General Motors Corporation," *Iron Age*, May 10, 1928:1322). Accordingly, it was necessary to identify, then monitor continuously, variables that were sources of feedforward information on potential environmental shifts. While these environmental variations remained uncontrollable directly, their impact on performance could be indirectly mitigated by an anticipatory adjustment of the appropriate controlled variables. These anticipative adjustments Brown termed the constructive planning dimension, "designed to affect favorably the consumer demand for [GM's] products" (Brown, 1929c:258).

Brown separated these constructive efforts into short-term and long-term factors. Short-term factors of influence on demand were those that could "quickly be called into play to offset unfavorable developments. They include special sales stimulus, more intensive advertising, or even temporary underpricing, whenever these seem called for by a falling off in anticipated and logical demand" ("Forecasting Methods," *Iron Age*, May 10, 1928:1322). The long-term constructive efforts include "those relating to consumer appeal in style, functioning, serviceability, etc." Here, "the engineers and salesmen work hand in hand" in order to improve the probabilities of consumer acceptance.

The statistical aspect of the GM designers' synchronization work built the incoming communication channels from the environment that provided the information GM's executives needed to coordinate their decisions with the consumers' desires. There were two thrusts to the Sloan-Brown designers' statistical effort: determining the *quantities* of autos demanded and establishing the *qualities* consumers desired. "A theoretically perfect coordination of business," Brown (1929c) asserted, required "an exact knowledge, in advance, of what the public wants," "an exact knowledge of *how much* the public wants," "and a regulation of production . . . to fit these ultimate consumer demands for both kind and quantity of goods" (p. 258).

In 1921 the GM designers faced a bewildering environmental situation, for Durant had not established any units to study the economic variables that influenced automobile demand. Consequently, the basic periodicity and growth factors operating in the finished-automobile market were unknown to GM.

To gain an understanding of consumer demand, the Sloan-Brown designers supported a series of demand and price studies, primarily under Bradley's direction. By 1926 Bradley's economic analysts had isolated four factors that determined automotive demand; three of these factors influenced the industry as a whole and the fourth affected the individual firm. According to Bradley (1926d), "for each industry taken as a whole," three major economic factors had to be appraised and dealt with: (1) seasonal variation, "i.e., differences in activity which may be attributed to the change in seasons," (2) business cycles, or the "operation of those factors which are generated within the general business situation itself, and which make for the alternating periods of depression, revival, prosperity, and crisis," and (3) "growth, which is primarily dependent upon the increase of the country in population and wealth after a condition of stabilization in the particular industry has been reached" (p. 3). "Each representative member of an industry is affected not only by the three factors enumerated, but also by competition from other members of the same industry" (p. 3).

Information on seasonal buying habits was vital for planning purposes, as these repetitive patterns accounted for most of the high-frequency demand fluctuations in the automotive industry. In spite of the growing popularity of the closed car, consumers still avoided the purchase of automobiles in the winter months and delayed until the spring of the year. "There must, therefore, be available for delivery in the peak month of April almost four times as many cars as during December or January, and about twice as many as during each of the months of July, August, September and October" (Bradley, 1926d:10-11). So, with past seasonal data in hand, GM could anticipate these repetitive changes in environmental demand. In fact, Sloan (1926) could claim: "We know from experience that if our dealers sell at retail 5,000 automobiles of one make in a given week, the next week they will probably sell 5,200 if past experience for the season indicates that degree of expansion. Thus we have a reasonably definite measure of what to expect this week, next week, and the week after" (pp. 993–994).

The GM designers did not attempt to coordinate production *exactly* with actual seasonal demand, for they recognized that "factories should operate on as level a line of production as can reasonably be attained; and that radical changes . . . should be eliminated as far as possible" (Bradley, 1926c:156). To smooth production, GM required that its dealers accumulate "finished cars during the winter period, to take care of the spring rush" (Sloan, 1926:996). "The resultant production schedule," as Bradley (1927) described, "is then one having seasonal characteristics, but without the radical changes indicated by the trend of sales to consumers" (p. 419). In this way GM not only held down its own production costs and capital investments, but in addition the dealers usually paid for the winter inventory accumulation in interest charges to the firm's GMAC subsidiary.

With respect to the somewhat longer-term demand fluctuations, it was necessary to predict how the auto market would contract or expand each year with variations in the business cycle and how it would grow in relation to the economy as a whole. The expensive consumer-durable nature of the automobile meant that the fluctuations and growth of the economy in general had considerable influence on GM's demand.

In 1923, in an attempt to develop feedforward data on growth-dictated demand variations, the Sales Section of the Advisory Staff undertook "a comprehensive study of the total automobile market" (Sloan, 1964:136). In particular, it attempted to predict "the size of the total market for the next few years, the market potential in the various price classes, the probable effect of price reductions on the size of the market, the competitive relationship of new and used cars, and when the so-called 'saturation point' would be reached in the market" (p. 137). While the Sales Section's study underestimated the long-term growth potential of the market, it provided important feedforward information breakthroughs for the GM design group, thereby helping to set "sales strategy and productive capacity."

One of the reasons the Sales Section analysts misgauged the market's potential, according to Sloan (1964), was that they failed to appreciate "the continued growth of the economy" fueled by the boom in auto sales "and the effect of general economic conditions on the sales of the industry in any specific year" (p. 137). Soon, however, Bradley and his staff of economic analysts clearly demonstrated the connection between potential automobile demand and economic activity as measured by income distribution (p. 137).

Bradley's group continued to study "the ups and downs of the automobile business in relation to the business cycle and saw that when business, and hence national income, was on an ascending trend, car sales increased at an even faster rate than income; and when business was on a declining trend, sales decreased at a faster rate than income." When additional data became available on the overall economy, they were able to "demonstrate the remarkably close correlation between car sales and personal income."

The importance of this relationship was that variations in personal income *preceded* demand fluctuations and that demand amplitudes would be amplified beyond personal-income fluxes. By understanding these relationships and by monitoring the personal-income variable, the GM decision makers had the feedforward information to adjust their controlled variables, such as inventory and capital-investment levels, before the environment shift actually occurred.

The importance of generating such feedforward information can be seen in the designers' continued effort to refine and broaden their data base. In 1926 H. G. Weaver, assistant to the director of the Sales Section, published an important article, "General Motors 'Purchasing Power' Index," describing GM's development of a county-by-county determination of personal income. By 1929, the GM designers had improved their anticipation of forthcoming business conditions by tracking "agricultural conditions, as reflected by crop reports, prices of farm products, the livestock situation; the state of general industrial employment; the trend of commodity prices; the volume of industrial production; credit conditions as indicated by bank statistics and interest rates; the probable effect of fluctuations in the securities market on business sentiment; and so forth" (Brown, 1929b:236).

Still other market variables had been isolated and were being monitored by 1931. R. H. Grant (1931) mentioned keeping a close watch on the auto inventory in the hands of consumers, "looking at an automobile not as an automobile but as six years of transportation" (p. 197). Additions to this inventory were constantly compared to the reductions from it in an attempt to provide predictive information which, in turn, could be used to set the divisions' indexes of volume. "According to our figuring," as Grant put it, "there are about 12,000,000 years less of transportation in this inventory at the moment than has been considered normal since 1925" (p. 197). Grant also looked at gasoline consumption to confirm that demand was accumulating during the early 1930s. "With 12,000,000 car years out of this inventory we used 9 per cent more gasoline ... than we did in 1930 ... which means we have some 22,000,000 people working to increase automobile demand when we get a return to economic soundness" (p. 197).

In addition to having developed numerous leading barometers to the economy's growth patterns and business cycles, the GM designers had accumulated considerable experience with past patterns and cycles that could be used to estimate future variations. By 1934, for example, Pound's (1934) officially authorized history of GM noted the periodicity of automobile demand during the 1920s.

The trade began a succession of three-year cycles, apparently as a result of its own circumstances and enthusiasms, as follows:

1921	Black year	1924	Black year	1927	Black year
1922	Good year	1925	Good year	1928	Good year
1923	Banner year	1926	Banner year	1929	Banner year

(p. 216)

Pound emphasized that these were relative descriptors, since the passing years also brought a growth in volume.

When, for 1932, GM's prognosticators still drastically underestimated the deepness of the depression's trough, the firm pursued further demand-prediction studies. In 1939 GM published a series of articles under the title *The Dynamics of Automobile Demand.* (See Chapter 11 also.) By that time, GM also had begun to provide the decision makers of other firms with important business barometers. For example, it had published the "General Motors-Cornell World Price Index, a free weekly statistical bulletin that Cornell University compute[d] from field data cabled by G.M.'s farflung overseas divisions" ("General Motors IV," *Fortune,* 1939:150). All this highly sophisticated work demonstrates how far the designers' thinking had progressed from GM's initial 1923 demand study.

Finally, to predict the demand for its products GM needed to assess its competitive position within the automotive industry. Two dimensions needed to be considered: product price and overall quality. Not only would the results of such studies be valuable for measuring current demand, but they also offered the potential for improving GM's future performance. So if GM could discover the correct pricing and if it could better the competition's product offerings, it could assure itself of an increased share of the then comparatively stable market. In turn, the increased volume would mean improved returns on invested capital.

In conjunction with Bradley's early demand-estimation work, GM's designers supported a series of price studies. Despite the Ford Motor Company's incessant price reductions, GM found that cost to the consumer was becoming a less critical performance variable. That is, with the tremendous initial demand for automobiles essentially filled by the early 1920s, additional price reductions could not attract sufficient volume to maintain the same return on invested capital (or perhaps even the requisite cash flow). Bradley (1927) concluded, for instance, that to offset only a 5 percent price drop would require a 34 percent increase in sales volume. While a small manufacturer might be able to realize such a volume increase "and thereby not suffer an actual reduction in aggregate profits" (p. 432), a larger manufacturer would need "a tremendous gain" (p. 433). For the industry as a whole, the necessary annual increase was approximately 1,400,000 cars, a figure "entirely out of the question" (p. 433). Thus, the industry had reached or was fast approaching the point where lowered prices could not "attract business in volume sufficient to compensate for the reduced profit margin per car" (p. 433). And GM's planners had anticipated analytically what Ford was learning experientially at a much greater cost.

Another investigation that bolstered the GM designers' conclusion about the diminishing importance of price centered on the growing popularity of installment selling. In 1927, a GM-sponsored study of consumer credit by Professor Seligman culminated in *The Economics of Instalment Selling,* a two-volume work which Sloan (1964) believed further accelerated the "acceptance of installment selling among bankers, businessmen, and the public" (p. 306). With installment selling growing in popularity, initial cost to the consumer meant even less than it had previously. Ford Motor Company's low prices then were attracting fewer customers.

Even during the lengthy 1930s depression, short-term, tactical price competition played only a small role in GM's adjustments to the consumer market. It was written at the time that Sloan "believes that the demand for automobiles is chiefly a factor of the national income, and . . . less responsive to price cuts in a depression than at other times. This is because an automobile is a postponable purchase; the prospect who feels unsure of his job can all too easily make the old bus do another year" ("General Motors IV," *Fortune,* 1939:145).

Just as GM avoided price reductions, it also avoided price increases. In general, the GM designers recognized that ill-advised price changes could amplify undesirable demand variations inherent in the economy. Brown and Bradley thus introduced a standard volume accounting technique that helped "lessen the deepening of troughs and the raising of peaks in the course of the business cycle." For instance, "following the 1929 crash, this pricing principle was rigidly adhered to, in spite of markedly subnormal profits. . . . During the low levels of 1931," a price increase "above the defined 'standard price,' in order to improve return on investment . . . inevitably would have [created] a further depression in consumer demand" (Brown, 1957:50–51). In sum, then, GM showed little tendency to raise or lower prices in the face of temporary demand variations.

Probably the most important reason why GM did not vary its prices in response to short-run demand fluctuations stemmed from the minor competitive role price played once Ford had filled the initial demand with his Model T. Even after the lengthy depression of the 1930s, for instance, GM's Customer Research Staff confirmed the relatively weak effect of price reductions: "Dependability, operating economy, and safety" jockeyed for the lead among desirable characteristics, while "low first cost" came in "a poor seventh or worse" ("General Motors IV," *Fortune,* 1939:142).

Thus, product characteristics presented the ultimate key to synchronizing the firm with its environment and to overcoming its competition. Accordingly,

GM's designers established positions and units to monitor the firm's competitors. Consider an example presented by Sloan (1927a): "We send representatives abroad to study foreign methods and foreign cars. We have an engineering office in London with representatives in the other countries to keep us advised at all times as to what progress there may be along European lines that General Motors can capitalize. Again, we are seeking the facts" (p. 550). To keep current with overseas styling developments, GM also commissioned European custom coachbuilders to do special design studies (Schneider, 1974:13).

Domestic surveillance was much more extensive, for the GM designers initiated "an elaborate technique for keeping abreast with its competitors. At the proving grounds near Milford, Michigan, 42 miles from Detroit, every competitive model [was] tested under the most exacting road conditions. In three weeks a car is subjected to as much testing as it gets from the average motorist in three years" (Rukeyser, 1927:376). The test engineers compared the competitive makes to the divisions' products, which were purchased—as the Director of the Proving Grounds, O. T. Kreusser (1926), noted—"in the regular way" (p. 29). "Upon receipt of the car" Kreusser's staff inspected "it thoroughly" and made "a detailed record of assembly shortcomings, weight, dimensions, quality of finish, and . . . many angles of engineering specifications." A 2,000 mile break-in and adjustment period followed, and then tests began under "controlled" conditions.

The GM Proving Ground emerged out of Sloan's frustration with testing four-wheel brakes on public roads. Instead of traffic interruptions, Sloan wanted rigidly controlled, completely comparable conditions (General Motors Corporation, 1926). Besides testing design innovations, the new GM Proving Ground "developed standardized test procedures and measuring equipment, and . . . became the corporation's center for making independent comparisons of division products and the products of competition" (Sloan, 1964:109).

As early as 1925, 250,000 miles per month were being driven on the GM Proving Ground (Hagemann, 1926:434). More than 100 distinct tests were made to each car, covering "fuel economy, horsepower, maximum and minimum speeds, acceleration and deceleration, steering, turning radius, braking, temperature rise of cooling water, transmission of power, wear of all moving parts, action of all mechanisms, condition of car at various mileages, general comfort of riding at various speeds and on different kinds of roads, shimmying, action and wear of tires, wind resistance, vibration, visibility of road, action of car in cold weather, and every conceivable factor which may have any influence on the user" (p. 433). In addition, measurements were made "of all different body designs so that the most harmonious and attractive combinations [could] be developed" (p. 433). The exhaustive data were charted for graphic presentation so that relative performance could be analyzed and "the best economical combinations of qualities for each [price] class" (p. 434) could be worked out in detail.

While GM's comparative product testing helped in assessing as well as improving the short-term demand for its products, the firm could not rely on such delayed feedback information for long-term product planning. The future product characteristics that consumers would desire had to be anticipated long in advance if they were to be made available when demanded.

Brown (1929b), in his article "Forecasting and Planning as a Factor in Stabilizing Industry," stressed that in addition to regulating the *quantity* of production to fit the demands of the ultimate consumers, "in many lines of business it is even more important to consider the elements of utility and style." Otherwise, "what might appear from statistical indications of past consumption to foretell a continuing demand for a given product . . . may be altered overnight . . . by a change . . . of public taste and preference for some substitute or improved article" (pp. 181–182). The avoidance of such dangers, as Brown (1929c) noted subsequently, necessitated an "extensive program of research . . . to increase consumer appeal of our products through improvements in design, functioning, serviceability, style and so forth. I call these 'long-term' factors because, from their nature, they require a continuing program of some duration before their influence can make itself felt on consumer demand" (p. 258).

The GM designers first attempted to provide long-run synchronization information to GM's decision makers through the establishment of Kettering's Research Laboratories. "Compared to competitors like Ford, whose engineering departments were dominated by self-taught mechanics," GM's research unit "was quite progressive. Moreover, Kettering's academic contacts" gave GM "ready access to the best professional expertise of the day" (Leslie, 1983:98). As was noted earlier by Sloan, foreign contacts were important too; and "to keep track of them, Kettering stationed a network of correspondents abroad" (p. 188).

In Sloan's (1927a) words: "Our research laboratories are no different from our field trips in principle. We are searching for the facts that we may . . . add value to the performance and effectiveness of our products, just the same as in the field work we are trying to learn more about the distribution of our products" (p. 550).

Kettering held identical views, stating to an inverviewer, "It is our organization's business to discover, analyze, and recommend improvements in products and in methods" (O'Shea, 1928:359). And he went on to explain the underlying factors with which he worked:

> The fundamental force which changes business is quite simple. Two and one-half . . . million . . . people are born into this country every year. In 10 years 25,000,000 new people enter the picture.

So long as we have younger generations, so long as we have 25,000,000 new people every 10 years, we will have changes. Their views are new, their tastes are new, their likes are new—and emphatic.

Their effect upon the market is important even beyond the proportion of their numbers. For instance, market research on automobiles has indicated that probably over 50% of automobile purchases are guided, not by the heads of families who write the checks, but by youths—people under 21. (p. 359)

Unlike their fathers, who would be satisfied with any good car, the automobile-conscious youngsters were "painfully and passionately aware of anything which is not the exactly correct thing, and equally proud of the right car or style or color." So their fathers "yield to the greater selective will of the younger generation" (p. 448). Because "the younger generation . . . wants results delivered instantaneously," Kettering saw that his job was to "prepare for the demand now. . . . For if a demand should require five years of preparation, we must start . . . now. Similarly if the trend . . . has been steadily year after year toward lower chassis, more powerful motors, faster cars, more silence, we know that . . . cannot change overnight, so we must prepare methods to meet a continuance of these demands" (O'Shea, 1928:450). "General Motors Research Laboratories Plan Improvements for Years Ahead of Production," the title of a July 27, 1929, *Automotive Industries* article, thus captured the essence of Kettering's efforts.

As an example, Kettering noted the laboratory's prediction of the demand for color in automobiles and its development of Duco lacquer. And, since the country was "becoming ear-minded as well as eye-minded," due to radio, Kettering's laboratory had a soundproof room where researchers made "experiments to distinguish between neutral and unpleasant noises and [to] locate noises which indicate waste energy as in gears" (O'Shea, 1928:451).

The General Technical Committee decided which research projects would be undertaken by Kettering's Research Laboratories "to assure . . . systematic change-making" (Kettering & Orth, 1932:57). Because the work on all projects began with a thorough literature search, GM maintained an extensive library of "the leading technical and scientific books in the automotive and related industries, and files of the leading industrial publications . . . all carefully indexed by an experienced librarian" (Hagemann, 1926:361). Specifically, during Kettering's tenure the library's collection rose "to 23,246 volumes and 1,398 scientific reports" (Leslie, 1983:188). In developing a complete bibliography on a subject, such sources as the Engineering Societies Library in New York and the patents library of the National Automobile Chamber of Commerce were also used (Hagemann, 1926:361). "No pains" were "spared to make this preliminary survey comprehensive and thorough" (p. 361). For those projects that resulted in potentially profitable developments, GM maintained a Patent Department in the Advisory Staff for inventions, infringements, and litigation (Mott, 1924:524). If a project was unsuccessful, its reports still were "not thrown away, but . . . as carefully recorded, indexed and filed as those on worthwhile ideas" (Hagemann, 1926:361). "The slight cost of filing records of rejected proposals" was warranted as it prevented the "duplication of thousands of hours of work" on future projects (p. 361).

GM added still another organizational unit to monitor the efforts of external inventors. On September 1, 1925, Sloan created the New Devices Committee "as an organized effort to encourage outsiders" with ideas to submit them to GM for "examination and appraisal" (Sloan, 1941:188–189). As Sloan wrote in a GM pamphlet describing the Committee's activities to prospective inventors: "No matter how able our own organization may be or how extensive our research and engineering activities, we realize that useful developments may emanate from the outside. We, therefore try to give proper consideration to everything submitted to us" (General Motors Corporation, 1929d:1). The pamphlet explained how the inventor should proceed in presenting his invention to GM. By 1939, the New Devices Committee had given "a serious and cordial hearing to the offerings of outside inventors, having received 75,000 ideas" ("General Motors IV," *Fortune,* 1939:150). While many proposals had no commercial value, some notable exceptions rewarded the firm amply for its openness. One early success, for instance, was the synchromesh transmission that eliminated gear clashing (Sloan, 1941:189). No one in GM "had ever seen or heard of" its inventor. Some years later, the same individual developed a workable idea for the automatic transmission (p. 190)— an expensive option that would yield the firm a handsome return on its investment.

GM's most important unit for synchronizing the firm's production offerings with the consumers' taste was the GM Customer Research Staff. The designers developed this central-office group to discover the buying public's most desired automotive improvements and to forward this information on to the engineering and styling staffs (Pound, 1935:129). "In a business characterized by huge volume and high preliminary expense, innovations cost money. They are profitable when they win acceptance, burdensome when they fail. To eliminate guesswork, General Motors has undertaken to guard against marketing blunders by studying public moods and desires in advance of production" (p. 129).

The GM Customer Research Staff went back to 1926, when H. G. Weaver (then an analyst in the Advisory Staff's Sales Section) developed his correlational formulas for calculating United States buying power by counties ("G.M. III," *Fortune,* 1939:138). Weaver, who had started his GM career in 1918 under Sloan in the United Motors Corporation, ultimately became the head of the Customer Research Staff.

As GM's internal control problems subsided and Sloan became more concerned with environmental synchronization, the importance of the Customer Research Staff increased. The depression was an added impetus. In 1933 Sloan wrote, for instance, that his firm's market-survey activity had been pursued in the previous two years more exhaustively than before, constituting a "Proving Ground of Public Opinion," devoted to "finding of facts" about "the attitudes of the practical motorist toward various aspects of merchandising and service . . . vitally important [to] customer good-will and continued patronage" (p. 92).

The Customer Research Staff operated "in close co-operation" with the Research Laboratories, the Art and Color Section, the Proving Ground, and the "various divisional engineering, sales and service organizations, supplying them with data which are constantly flowing into the central office direct from owners of cars of all makes and ages all over the country" (Sloan, 1933:93). Although Sloan acknowledged that this approach was not new, he emphasized that old techniques—"casual contacts and personal impressions"—were ineffective, for "our business is too big; our operations too far-flung" (p. 93). More importantly, he saw that the firm was caught in the midst of a "kaleidoscopic era characterized by swift movements" in social and economic conditions as well as "rapid changes in the tastes, desires and buying habits of the consuming public." Thus GM needed to keep its products and policies "sensitively attuned to these changing conditions. . . . A more intimate, detailed and systematic knowledge of the consumer's desires will afford the corporation a sound and progressive basis upon which to meet the new conditions as they unfold" (p. 93).

The Customer Research Staff generated long-term environmental feedforward data from its extensive consumer surveys (Customer Research Staff, 1937:7). The Staff sent out questionnaires, called on owners, and digested customer reactions "flowing into the corporation through miscellaneous channels" (Sloan, 1933:93). More specifically, between 1933 and 1938 Weaver reportedly "sent out 15,000,000 questionnaires, made an average of a speech a week, carried on an enormous correspondence and built up a formidable battery of charts and files. From it all he and his staff of 37 . . . winnowed . . . 185 public reactions which . . . found their way into . . . GM cars" ("Thought-starter," *Time*, November 14, 1938:68). In addition, Weaver's "special list of 100,000 motor enthusiasts" played a particularly important role in providing GM's decision makers with critical feedforward information; for, as described in 1939, "these fans think about eighteen months ahead of the general public. Then he queries on such things as whether they think running boards will disappear, whether engines should be in the rear, how they would react to a Diesel car, etc." Thus, Customer Research defined "for the engineers, in Buck Weaver's words, 'the boundary lines of public acceptance' " ("General Motors IV," *Fortune*, 1939:141).

In addition to discovering consumers' desires, Weaver spent about half his budget "to study consumer habits and psychology" (p. 141). Specifically,

Weaver and his staff tested the relative merits of different advertising techniques, how to handle customers in the showroom, and how to handle owners, that is, potential repeat customers. Along these lines, the Customer Research Staff made some extremely important discoveries. For example, they moved beyond early surveys by *Saturday Evening Post, Colliers,* and others which had concluded that three-fourths of all car purchasers decided what make to buy before visiting a dealer. Such surveys were designed to convince GM of the importance of its advertising budget, which then was the largest in the world. However, Weaver's "figures showed that while a third of the predetermination was due to advertising, the greater part was the direct result of the owner's past experience and that of his friends. This conclusion put a wholly new emphasis on the importance of adequate service," which could be supplied solely "by a reliable, well-capitalized set of dealers" ("G.M. III," *Fortune,* 1939:105).

GM's combined customer-survey activities reputedly comprised 10 percent of the total spent on market research by all of United States industry in 1938, or approximately $500,000 out of $4,500,000 ("Thought-starter," *Time,* November 14, 1938:66). In comparing the market-survey techniques of the three major auto manufacturers, *Time* writers observed that Ford depended "almost entirely on its dealers' reports on consumer tastes," Chrysler on "registration tabulations, dealer suggestions and sales records," and "occasional direct surveys of buyer opinion," while only GM carried "on constant customer research in the full sense of the word" (p. 66). No wonder GM's market share doubled between 1918 and 1938!

The Customer Research Staff not only collected state-of-the-environment information, but also helped shape it as "a highly polished sales and propaganda" unit (p. 69). That is, "keeping G.M. on good terms with its customers often" entailed "interpreting the engineers to the customers" ("General Motors IV," *Fortune,* 1939:141). Thus Weaver and his staff supplemented "his questionnaires (themselves rare specimens of commercial good nature) with a steady barrage of booklets, advice, user's and buyer's guides"—in short, "propaganda of all kinds. . . . These manifold mailings" made it sound "as though G.M. thought its customers were individually as well as collectively important."

Weaver's Customer Research Staff also provided a critical focus for the advertising campaign mounted against Chrysler's dramatically streamlined Airflow cars introduced in 1934. As Jim Ellis (1968), one of the advertising agency executives involved, put it, "If the Chrysler design really captured the popular fancy, GM could be faced with a crash job of redesigning and retooling which could cost, some people estimated, close to $100 million" (pp. 101-102). In keeping "tab on car-owner attitudes," Weaver discovered that only "10 to 20 per cent of the people liked" the revolutionary Chrysler styling while "[19] to 27 per cent rather violently disliked it, and the rest of the people were undecided" (p. 102). Given that most people wavered, GM's advertising needed "subtly to suggest that Chrysler had made a wrong guess." The resultant ad campaign expounded the theme: " 'An eye to the future, an ear to the ground.' " The statement suggested that GM also looked "ahead, not back," but, more importantly, "did not make rash changes, without consulting the people's desires."

Whether from GM's ads or just a lack of visibility stemming from low depression-era sales, Chrysler's Airflow flopped. "Chrysler in effect admitted failure by switching advertising emphasis from Airflow styling to engineering" (p. 103), touting less obvious technical advances such as aluminum pistons, hydraulic brakes, and all-steel bodies. With the Airflow's premature demise, GM became the undisputed style setter for the domestic auto industry.

GM's export-market analysis staff and its Canadian market-research group provide two more examples of the firm's considerable environmental synchronization activities. G. F. Bauer (1925b), the Secretary of the Foreign Trade Committee of the National Automobile Chamber of Commerce, cited GM's Export Company for deeming "market research highly essential, warranting the employment of an expert and a staff of special assistants" (p. 561). Hanson A. Brown, division manager of General Motors of Canada, conducted market-research programs similar to those of the Export Market Group and the domestic Customer Research Staff (Shidle, 1933:666).

GM's designers also established the General Motors Fleet Sales Corporation in 1930 (Baird, 1935:188) to coordinate the sale of all GM passenger cars and trucks to big fleet operators (Pound, 1934:377). Its head, C. E. Dawson, described the need for this specialized synchronization unit: "There was a lack of uniform procedure, coordination, systematic effort, and many other essentials" (Baird, 1935:188). To improve GM's fleet sales performance, Dawson's staff "compiled comprehensive records to show where our market was, who was getting the business, and where effort was most needed. Now we keep these records constantly up to date and they are to us what a compass is to a sailor. We know who our prospects are, how many units they operate, the make and age of each unit, and many other facts. If one of our accounts begins to slip, we know it promptly and are governed accordingly" (p. 188).

Did GM's fleet-sales performance improve? According to Baird (1935), "Sales of all makes of General Motors passenger cars and trucks to fleet owners in 1929 totaled approximately 17,200 units; in 1934 the total was approximately 31,000 units" (p. 188). A remarkable increase, since business had peaked in the boom year of 1929 and total sales dropped tremendously over the next five years.

The competition GM faced was not limited to the other automobile manu-facturers, for the firm existed in an environment that offered direct substitutes for the automobile and truck. During the 1920s and 1930s the principal nonautomotive competition came from the railroad industry, which "was mak-ing a frontal attack on motor-vehicle transportation. Organized at national and state levels, the railroads promoted legislation to restrict and harass automobiles, and especially trucks" (Douglass, 1954:166). To avert a potentially dangerous battle between these giant industries, Sloan—upon urging from A. J. Brousseau, the president of the Mack Truck Corporation—headed "a movement to develop a Highway Users Conference. . . . Through the efforts of the conference, a joint Committee of Railroads and Highway Users sought to bring about a moderation and reconciliation of the differences of view between rail and highway transportation." To Sloan, moreover, cooperation between the auto and rail industries was becoming more and more vital since GM's diesel divisions and research staff were rapidly developing diesel-electric locomotives. Once perfected, these engines would need to be marketed to the railroads as replacements for their steam powerplants.

CONNECTION OF THE ENVIRONMENTAL COMMUNICATION CHANNELS TO THE CORPORATE HEADQUARTERS

GM's planners connected the most vital environmental communications channels directly to the corporate headquarters, ultimately to the all-important Finance Committee of the Board of Directors. Environmental connections were attached to the corporate headquarters rather than to the divisions for four reasons. First, the corporate headquarters assumed a time perspective, or planning horizon, more in keeping with the frequency of the incoming synchronization information. That is, most of the incoming intelligence data dealt with longer-run policy issues, such as product-development and resource-allocation decisions. While at first nothing precluded the divisional managements from occasionally taking the longer-run view, they had a natural tendency to concentrate on the shorter-run, higher-frequency problems associated with producing and distributing current models or at most preparing for next year's line. As Sloan (1964) observed, "When you inject a piece of long-range research and development into this situation you are superimposing on an already loaded organization something to which it cannot properly give its attention" (p. 255). Over time, Sloan's centralization efforts accentuated the divisional tendency to avoid long-run planning. Thus, the corporate headquarters' decision pace became more attuned to the frequency of the environmental information flows.

Second, conducting most of the external intelligence functions at the corporate level saved money. Central staff personnel, such as Bradley's econometrics analysts, Kettering's research scientists, and Weaver's consumer researchers, cost a great deal to support. Hence divisional duplication of these efforts could not be justified, especially after Sloan had painstakingly built GM's internal communication network purposely to disseminate such performance-improvement information among the divisions. Valuable divisional market findings, of course, could not be overlooked; but, here again, Sloan's extensive committee structure assured the planners that such discoveries would be identified, verified, and promulgated.

Third, the corporate headquarters needed to be apprised directly of any production or distribution problems the divisions encountered. Had such environmental response information gone only to the divisions or had it come to the corporate headquarters via the divisions, it might have been ignored, delayed, and/or distorted—as had been the case under Durant's administration. The evaluative data embedded in this intelligence information meant that it needed to be routed directly to GM's corporate staff and officers. This need for reliable environmental feedback, then, explains why Sloan and Brown made their own personal visits to the dealers. It was, after all, during their 1924 trip west that they learned of the divisions' overproduction and that "faulty transmissions, bad paint jobs, and other technical defects were hurting sales of Chevrolet and one or two other models" (Chandler and Salsbury, 1971:552). Shortly thereafter, the extensive comparative tests conducted at GM's corporate Proving Grounds supplied the central executives with reliable data on the products of the divisions and their competitors. Without making sure that evaluative information reached the headquarters first, GM's planners could not be certain that the divisions were meeting the environment's demands.

The fourth, and most important, reason for feeding environmental intelligence back through the corporate headquarters first and then downward to the subordinate divisions was that GM needed to be synchronized with its environs as an integrated whole. Admittedly, the divisions, not the corporate headquarters, did most of the responding to the environment. But still these individual environmental reactions had to be coordinated wherever and whenever they were important enough to affect other units or GM as a coordinated entity.

Because they were so crucial in the aggregate, even some short-run, high-frequency (and, therefore, highly repetitive) decisions such as setting and revising monthly production schedules demanded central coordination. Otherwise, the divisions acting on their own initiative—again as they had under Durant—might inadvertently saddle the corporation with excessive inventories requiring many months to eliminate and interim financing to carry.

Similarly, longer-run product-improvement matters had to be appraised and decided as a coordinated package if GM was to keep in step with competitive demands and opportunities, yet not overstep its financial limitations. For example, should the innovative but still expensive synchromesh transmission discovered by the New Devices Committee be introduced by the low-priced Chevrolet to fight Ford's Model A or should it be offered first by Cadillac to compete with Packard's high-quality cars? While Chevrolet's market share gains were more important to GM than Cadillac's, it was problematic whether the Chevrolet price could cover such an initially expensive option (Sloan, 1941:190). Since matters like these had to be decided by the corporate headquarters, the environmental intelligence information needed to be directed to that unit.

In sum, then, GM's planners connected the firm's most critical environmental linkages to the central headquarters, in particular to the Advisory Staff, Financial Staff, and corporate officers. In these units and among these executives the incoming intelligence was discussed, digested, summarized, and blended with internal information flows.

To formulate the firm's coordinated response to its economic surroundings, the planners ultimately had to focus all of this information on a particular unit; otherwise, there could be no formal integration of the various possible courses of action into a unified plan. Accordingly, a condensed package of recommendations, requests, proposals, and performance reports went to GM's all-important Finance Committee for the overall review and resource allocation necessary to adjust the firm's strategy.

The outsider-dominated Finance Committee was ideally constituted to guide GM's unified reaction to its economic environment. For many years it was composed primarily of outside members who represented the large stockholders' interests. These individuals, who were associated either with the banks or the duPonts, helped secure the necessary capital for GM's product improvements and its accompanying facilities expansions (testimony of Walter Carpenter, *U.S. v. Dupont* 1956:2730). And as knowledgeable representatives of the investment communities, these men could judge whether the past performances and future prospects of GM's divisions warranted further infusions of outside capital or justified the internal retention of earnings. Hence, GM's Finance Committee held the strategic responsibility of "watching [GM's] earnings position."

Brown brought to the Finance Committee his extensive financial skills. Moreover, since he headed GM's Financial Staff, Brown provided the group with the reliable performance information it needed to evaluate the divisions' achievements and prospects. Reliable financial data were so essential to the Finance Committee's evaluative reviews that (as was first mentioned in Chapter 5) the whole Financial Staff came "under the direct control and supervision of the Finance Committee" (GTX 130, *U.S. v. DuPont*, 1956:3249).

Sloan, for his part, contributed his vast knowledge of GM's operations. His chairmanship of or membership on the Executive, Operations, and various Interdivisional Committees gave him ready access to a vast pool of internal and external data.

GM's planners created further information sources for the Finance Committee by linking it closely with the Executive Committee, which held the oversight responsibility for the divisions' performances. "Cooperation . . . between these two committees," as Brown (1927) noted, was "furthered by the . . . common membership on the part of several individuals" (p. 7). During the 1920s those individuals holding joint memberships on the Executive and Finance Committees were Pierre duPont, Raskob, Brown, and Sloan. Because of the overlapping nature of Sloan's committee structure, these executives also held positions on or were in close contact with the Operations and Interdivisional Committees (see Chapter 9). Thus, the extensive internal-communications network Sloan built to coordinate GM's internal operations helped to coordinate its external adjustments.

The Finance Committee members, then, had ample opportunity to convey their performance expectations to the divisions. In turn, they could check on whether the divisions were responding to the environment as they had been instructed. As a final motivator, the Finance Committee administered the various GM bonus plans, fixing the total bonus allotment and approving the president's recommendation for its distribution (testimony of Carpenter, *U.S. vs. DuPont*, 1956:2730).

Unfamiliar with the automobile industry, the bankers on the Finance Committee (Baker, Prosser, and Stettinius in the early days) played the role of watchdogs rather than innovators (Forbes, 1974:136), leaving the duPonts, their associates, Sloan, and Brown to make the necessary corrections. The 1924 inventory crisis, for example, upset members of the Finance Committee, especially the bankers (Sloan, 1964:131-132). They wished to know from Sloan: "Why were steps not taken earlier by the Operating Divisions to curtail their production? . . . What steps will be taken to assure effective control of production schedules in the future?" (pp. 131-132). The committee also questioned Chevrolet's decline in quality and requested a report on its causes, remedies, costs, and effects on sales (Chandler and Salsbury, 1971:553).

Even after the early crises had subsided and new controls had been implemented, the Finance Committee carefully watched short-term performance as it provided an early warning of long-run problems. The Divisional Indices became the principal vehicle for the Finance Committee's monitoring of GM's operations. As early as 1925 the firm's production schedules, purchasing commitments, employment, and even pricing were tied to the Divisional Indices prepared by GM's central headquarters (Chandler and Salsbury, 1971:553). Members of the Finance Committee found the Divisional Indices to be crucial in their resource-allocation decisions. "Actual output, sales, and profit could be continually checked against earlier estimates" (p. 554). After this analysis and explanation of the discrepancies, the Executive and

Finance Committee could determine if "general business and marketing conditions or managerial ability accounted for these differences." In turn, "they
could allocate capital funds and executive talent more rationally" (p. 554).
Moreover, the bankers and the duPonts were able to oversee effectively
"their massive investment by doing little more than a careful reading of the
statistical reports and regularly attending the meeting of the Finance Committee"
(p. 554).

Eventually, the basic approach followed by the Finance Committee as well as
GM's other corporate bodies in judging divisional performance was one of
comparison. Both internal and external comparisons were made.

GM's specialized-parts divisions were judged on their ability to meet the
competitive prices of external manufacturers. Since the auto divisions were not
compelled to buy parts from the accessory divisions, the latter units were
forced to meet the prices and quality of outside manufacturers. A manager of a
buying division who found prices charged by a selling division out of line was to
notify the Executive Committee, which "would then have the price differential
investigated" (Chandler and Salsbury, 1971:500). Supposedly, "if the selling
division was unable to bring its prices down, and if its production was not
vital for assuring an adequate supply of critical items, then the division could
be sold or put to other uses" (p. 500). This prospect, of course, served to keep
the parts and accessory divisions' prices synchronized with environmental
prices.

With respect to the internal comparisons, the finished auto divisions' costs
were compared to each other at the same point in time and to themselves at
different points of time. For such comparisons the designers had to develop
special techniques to discount for the divisions' fluctuating monthly or yearly
production levels. Otherwise, the varying output volume would have resulted
in monthly cost variances "attributable to changes in rate of plant activity.
Such cost figures are valueless for comparative purposes. . . . It is, therefore,
essential that there be a definite method established whereby the effects
of fluctuating volume upon overhead costs may be eliminated" (Bradley,
1927:419).

Brown and Bradley's standard-volume technique filtered seasonal changes
and business-cycle effects from the divisional performance variations so that
internal comparisons could be made directly. The GM standard-volume or
base-pricing method made "it possible to determine at one glance whether a
certain division—or a department within a division—[was] producing with
greater or lesser efficiency than the norm, and why" (Drucker, 1946:66). And
since GM used uniform accounting practices throughout all divisions, the costs
of divisions making similar products or employing similar methods could be
compared quickly (p. 67). More importantly, with the standard-volume method
a divisional manager would be held "accountable for a deterioration of produc-

tive efficiency even when it [was] concealed by an increase in total profits" and would be credited "for any strengthening of managerial efficiency, even when as the result of bad business conditions, his division operate[d] at a loss" (p. 66). Thus, with the standard-volume method, the GM designers had an objective standard. In more detail, Sloan (1964) himself explained, "Changes in these unit costs reflected changes only in wage rates, material costs, and operating efficiencies and were not affected by year-to-year changes in volume." They thus helped gauge "the efficiency of our performance from one month to the next as well as from one year to the next" (p. 146).

Internal cost comparisons, of course, represented only part of the finished-auto divisions' evaluations. The corporate headquarters and the Finance Committee also had to judge the car divisions' performances against external standards: their competitors' achievements. In addition to the product comparisons conducted at GM's Proving Grounds, the auto divisions were evaluated in terms of their market shares in the appropriate price classes. The R. L. Polk Company automobile-registration statistics "supplied General Motors with a clear picture of the share of the market each of its products enjoyed and how that share was changing" (Chandler and Salsbury, 1971:553). In turn, "the percentage of each price group that General Motors might reasonably be expected to obtain" (Sloan, 1964:134) was kept in mind when reviewing the divisional sales projections and setting the Divisional Indices.

By focusing on market share, as opposed to sales volume, GM again could factor seasonal and business-cycle influences out of the evaluation, thereby developing the true competitive picture of each division's product acceptance and distribution efficiency within its price class. Hence, a division losing sales volume, as Cadillac did during the 1930s depression, would be considered to be doing well if it increased its market share, since its management exerted no control over the "shrinkage of the market for higher priced cars" (Drucker, 1946:68).

Even though a division's management would not be held responsible for a well-run but poorly performing unit, the division could be eliminated if its prospects in the environment were not good. In fact, the Cadillac was almost scrapped in favor of its cheaper stablemate the La Salle when the demand for expensive cars temporarily declined during the depression (Cray, 1980:278). The Finance Committee, however, concentrated on the divisions' sustained economic potential, that is, "an average rate of return to be realized over a period including both good and poor years" (Brown, 1924c:420). And only if "a previously expected average rate of return upon capital employed in a given operation is no longer obtainable, the result may be a deliberate restriction upon further expansion, or even a curtailment of volume with release of capital for employment in more profitable channels" (pp. 417–418).

Besides establishing the communication channels to evaluate the divisions during the synchronization stage of the design process, GM's planners ultimately had to provide similar information linkages to evaluate the firm as a whole. There, the emphasis swung from the short-run, microscopic evaluation of the divisions within the firm's immediate economic environment to the long-run, macroscopic evaluation of the whole firm—an integrated unit within the much larger social and political milieu.

11

Evaluating the Performance

Evaluating the firm's overall performance completes the system-design-for-performance-control model. At GM, the two most immediate groups that performed the final evaluation function were the full Board of Directors and the actual—as well as potential—stockholders. GM's long-run policy changes in the areas of goal definition, strategy formulation, structural organization, decision-maker training, internal coordination, and environmental synchronization (chronicled throughout earlier chapters) received their ultimate approval as part of the Board of Directors' performance evaluations. For their part, the stockholders reviewed GM's performance, not to improve it directly, but rather to judge whether the corporation's stock offered a good investment opportunity in comparison with the stock of other firms.

To perform their evaluative work, these groups needed reliable information. The Executive and Finance Committees plus the Financial Staff provided this information to the Board of Directors. And for the stockholders' data Sloan engaged an independent auditing firm. He also supplemented this verified financial information with numerous descriptions of GM's managerial techniques, etc.

Along with these reviews, the evaluation by nonstockholding groups—labor, dealers, and consumers, for example—must be considered. As was more and more the case in the middle and late 1930s, these secondary clienteles were less than fully satisfied with how GM served their interests, especially in light of how well it served the stockholders. Considerable friction developed, particularly with labor. Since the parties could not arrive at what both considered an equitable split of the benefits generated by GM, the government intervened as the final arbiter of the various disputes. Moreover, GM's considerable return-rate and market-share success thrust the firm into a position of prominence in American industry. This visibility, coupled with the dramatically altered social, political, and economic environment of the 1930s, focused the government's attention on how well GM served nonstockholding groups and what impact its decisions had on the economy in general.

In response to this new scrutiny from its secondary clienteles and, more importantly, from the government, GM began to assign executives and create units to monitor the long-term social, political, and economic environment. Such monitoring was to ensure that GM was prepared for the long-run changes that would affect its performance in subsequent planning cycles. The environmental surveillance addressed here differs from GM's synchronization efforts (discussed in Chapter 10) only in its longer-run orientation. The present chapter covers GM's attempts to sense, understand, and cope with major societal shifts (e.g., the changing relationships among business, labor, and government) as opposed to the shorter-run emphasis assumed in the synchronization phase.

EVALUATION BY THE BOARD OF DIRECTORS

"In the broad sense the interests of capital," as Brown (1924a) put it, "are represented by a board of directors and such subcommittees as may be constituted having jurisdiction over policies; while authority and responsibility to operating results are lodged with executive officers and subordinate departments" (p. 195). So GM's proprietor-dominated Board of Directors, as the pinnacle unit in the corporate hierarchy, played the ultimate role in evaluating the planners' design work and operating achievements. For example, in addition to approving financial policies and major capital investments on the Finance Committee, Pierre duPont, as Board Chairman of General Motors, considered his second task reviewing top-management performance (Chandler and Salsbury, 1971:580). To this end he met with GM's top executives frequently and stayed close to Sloan (p. 580)—who, he felt, had brought the state of GM's organization and management "to a point of great perfection" (p. 575).

Sloan too held that the GM Board of Directors served an extremely important audit function for the stockholders, albeit at a more aggregated level than the Finance Committees' divisional-performance reviews. In describing the Board's functions Sloan (1964:187) noted that it met monthly and sometimes more often; it elected GM's officers and, more importantly, the Executive and

Finance Committee members; and it acted on legal and general corporate matters, such as declaring dividends or issuing additional securities. Furthermore, as Sloan explained: "The General Motors board of directors has still another . . . unique function of great significance . . . an 'audit' function," that is, "a continuous review and appraisal of what is going on throughout the enterprise" (pp. 187–188). Sloan saw GM as both large and highly technical with problems "too many, too diversified, and too complex" for the Board to be able to tackle the technical operating problems, especially since its many outside members had numerous external responsibilities. Nevertheless, he believed the Board could and "should be responsible for the end result" (p. 188) through its review of past and projected performances.

Accordingly, Sloan kept the Board thoroughly apprised of GM's activities. He mentioned that the Executive and Finance Committees reported monthly and other standing committees periodically. "Staff vice presidents and top executives" discussed "developments in their fields." A visual display presented GM's "financial, statistical, and competitive" positions along with "a forecast of the immediate future" and "a summary of the general business outlook." Board members then asked questions and sought explanations. From Sloan's vantage point, "this audit function" was "of the highest value to the enterprise and its shareholders." Indeed, he could not "conceive of any board of directors being better informed" (p. 188).

The Board could well be satisfied with GM's progress. "With the big depression—from 1930 to 1934—there was contraction in General Motors. But this time, unlike 1920–21, and despite its greater severity, the contraction was orderly" (Sloan, 1964:199). *Barron's* in an August 18, 1930, article, "General Motors Stronger in This Depression," observed that "as early as Sept. 30, 1929, inventories had declined more than $25,000,000. . . . Cash increased more than $30,000,000" (p. 11). So even before the stock market crash began in October 1929, GM was ready.

In explaining how GM increased its market share, earned profits and paid dividends while many durable-good producers failed or nearly went bankrupt, Sloan (1964) said: "We had simply learned how to react quickly. This was perhaps the greatest payoff of our system of financial and operating controls" (p. 199). In fact, GM had been charting the decline in car sales throughout the late spring and summer of 1929. Keeping inventories in proper trim was vital, as they absorbed huge amounts of capital and required much time during slow sales periods to reconvert to a life-sustaining cash flow. Thus "dependable forecasting and planning were of outstanding importance during those difficult years. Production by all divisions was held in reasonable bounds" (Brown, 1957:71).

Not only did GM not have to borrow money as it did in the 1921 recession, but it was able to come to the aid of several endangered Detroit banks. Brown, most notably, worked hard to avert a banking crisis; to this end, he used GM's capital structure to support the weakened financial community. More importantly to the firm itself, GM's strong liquidity position allowed time to shift its pricing strategy and to adjust its organizational relationships.

EVALUATION BY THE STOCKHOLDERS

Besides communicating extensively with stockholder representatives on the Finance Committee and its parent Board of Directors, Sloan built strong communication links directly to the stockholders themselves. As early as 1919, Sloan (1964) had suggested to Durant that, "in view of the large public interest in the corporation's shares, we should have an independent audit by a certified public accountant" (p. 25). Durant agreed and as a result, Sloan retained Haskins & Sells, the firm that had audited United Motors' accounts. Indeed, he felt so strongly about providing GM's primary clients with reliable and detailed information on their investment that the audit was simply a first step in a prodigious effort to inform the stockholders. "We take the position that our stockholders are entitled to know the status of the business which in the aggregate they own," Sloan (1929) wrote. "No management can have a proper appreciation of its responsibilities as trustees should it approach the question from any other viewpoint" (pp. 96–97).

For Sloan himself, providing the stockholders with information had two thrusts: group publications and individual correspondence. He took great care in writing such group publications as GM's annual report, which was quite comprehensive for the day. "In our annual report we try to state all the facts the stockholders should know," Sloan (1927c) emphasized; "We, knowingly, hold back no information they are entitled to as partners in the business" (p. 13). These reports contained "not only statistics revealing the Corporation's financial position, but also explanations of procedure and policy" (Pound, 1934:418). As the reports went to "banks and trust companies" also, "the stability of the Corporation's security structure [was] in part due to the frankness and completeness of its published statements." Not surprisingly, GM was one of the first American corporations to issue complete quarterly reports (p. 417) containing "detailed statements" of its financial and operating positions (Sloan, 1927c:13). Sloan also sent special messages to the stockholders several times a year; as he put it, "telling them of our hopes and ambitions and explaining developments

which should give them a better insight into our position" (Sloan, 1929:97).
Sloan's special messages took the form of pamphlets on such topics as:

- How Members of the General Motors Family Are Made Partners in General Motors
- Plants and Products of General Motors
- Installment Selling: A Study in Consumer's Credit with Special Reference to the Automobile
- Development of Installment Selling
- General Motors Acceptance Corporation: The Sales Financing Organization of General Motors
- The Export Organizations of General Motors
- Motorizing the World
- General Motors Institute of Technology: The Training School for Employees of General Motors
- Financial Control Policies of General Motors [by Bradley]
- Decentralized Operations and Responsibilities with Coordinated Control in General Motors [by Brown]
- The Principles and Policies behind General Motors [by Sloan].

Many of these pamphlets were also advertised in such magazines as the *National Geographic* with an invitation to the public at large to write for them. These advertisements undoubtedly lured new stockholders to the GM fold as well as influenced potential customers for GM products.

Sloan's correspondence with individual stockholders was also copious. He wrote to each new stockholder (Pound, 1934:419) and "every stockholder" who sold out received "a letter of regretful query from Mr. Sloan, asking if there was anything wrong" and how GM could retain his or her "friendship" ("General Motors IV," *Fortune*, 1939:148). The many responses from the recipients of Sloan's letters, in turn, were carefully answered (Pound, 1934:419).

As the number of GM shareholders increased along with the firm's prominence in financial and national affairs, the task of directing public relations became so time-consuming that Sloan found it necessary to put the work in the hands of a corporate staff group. In 1931, when corporate public relations was still an infant field (Golden, 1966:78), Sloan chose Paul Garrett to head GM's new Public Relations Department. Given that GM's primary client was the stockholder, Garrett's appointment was quite appropriate, for he had been the "financial editor of the then financially minded New York *Evening Post*" ("General Motors IV," *Fortune*, 1939:148). "Through Garrett's leadership public relations in General Motors became a functional activity at the top corporate-policy level. Dividing the nation into eleven regions, Garrett maintained the quality and tone of Sloan's pattern and implemented it with techniques and instruments of contemporary scientific communications" (Douglass, 1954:187).

By 1939 the Public Relations Department was reported to have a staff of more than 50 and a $2 million budget for institutional advertising, stockholders' reports, and other items such as Sloan's speaking trips ("General Motors IV," *Fortune*, 1939:148). It also monitored the publication of 36 plant and divisional house organs, and published "the biggest one itself—*G.M. Folks*, a *Life*-like monthly that cost $7000 an issue, shows its 250,000 G.M. readers pictures of each other at work and play, and explains the latest G.M. financial report or wage plan in simple photographic form." In addition, it distributed "thousands of handsome educational booklets" on topics such as "automotive metallurgy," "how a Diesel works," and "how to drive safely"; produced short films for the public and a ten-minute newsreel each month for GM's field organization; ran "the Parade of Progress, G.M.'s industrial medicine show"; and helped develop the firm's New York World's Fair exhibit.

What were the results of Sloan's long effort to nurture the well-being of GM's stockholders? A once highly skeptical investment community was impressed. In 1933, the investor-oriented *Magazine of Wall Street* reported:

> In a nutshell—and something which many stockholders gloomily realize cannot be said of all large corporations—it seems fair to make the blunt assertion that GM is managed in the interest of stockholders. It is in a strong liquid position; has consolidated its activities and is correlating future plans; is strengthening its competitive position; is efficiently managed; is laying the foundation for a resumption of large future earnings; and will continue to show earnings, pay dividends, and maintain assets at most conservative valuations. Its foreign holdings are well protected and should be convertible through foreign banks in the event war endangers ownership. In spite of its towering capital structure, GM may be expected to show good earnings at the end of the current year if business continues to climb upward at its present rate.
>
> Professional opinion has it that the stock at its recent quoted value, is both a good short-time speculation and a long-pull investment. (McClary, 1933:300)

Correspondingly, a 1933 survey found GM shares in more investment-trust portfolios than any other stock ("General Motors Leads," *Barron's*, February 19, 1934:13). And the number of GM shareholders had risen from 4,739 in 1918 to 342,384 in 1936 (GM *Annual Report*, 1936:62).

EVALUATION BY NONSTOCKHOLDING GROUPS

Beyond the narrow confines of the financial and business community, however, GM's success during the late 1930s was becoming all too problematic for Sloan and his colleagues. The social upheavals stemming from the great depression meant that maximizing the well-being of GM's primary clients, the stockholders, was no longer sufficient justification for the firm's continued existence in the larger community. Long-stable equity relationships (i.e., societal parameters) among capital, consumers, labor, small business, and the general public were being questioned and realigned. "There ensued," as Brown (1957) summarized the era, "years of depression, of radical political innovations, of marked stresses and strains in many relationships affecting business operations" (p. 72).

Under these circumstances, GM's comparatively high return rates and market shares started to draw considerable attention from nonstockholding groups also dependent on the firm for their well-being. Labor and the dealers, particularly, were becoming restive about their respective payoffs relative to GM's stockholders. More importantly, under the Roosevelt Administration's leadership various federal agencies—e.g., the National Recovery Administration (NRA), the National Labor Relations Board (NLRB), the Federal Trade Commission (FTC), and the Antitrust Division of the Justice Department—began to look carefully at how GM and its competitors conducted themselves in relation to these nonstockholding groups. In addition, the federal government was becoming concerned about the impact GM and the automobile industry exerted on the economy in general.

Indeed, by the end of the 1930s Sloan appears to have become quite cautious about GM's capturing too large a share of the market and thereby inviting the federal government to bring antitrust actions against it. Thus, he appears to have "placed a brake on General Motors' continued growth" (Cray, 1980:311). For 1938 GM's overall share of the passenger-car market was about 44.8 percent, which was almost right on Sloan's target. "Our bogie is 45 per cent of each price class," says Alfred P. Sloan; "we don't want any more than that" ("General Motors," *Fortune*, 1938:158).

But, in reality, it was far too late for Sloan to downplay GM's prominence. Just as Chevrolet had grown in importance within GM and as a consequence attracted considerable corporate scrutiny, GM as a whole had become too important to be considered just another business by the nation. GM, along with its two major competitors, Ford and Chrysler, stood out for several reasons. First, they operated in what had become America's bellwether industry: as the auto industry went, so went the remainder of the economy. The NRA approved a total of 557 industry codes, for example, "but because of the vital relationship of automobile manufacturing to the process of economic recovery, no other code was so favored by President Roosevelt or Administrator for Industrial

Recovery General Hugh S. Johnson or received so much of their attention" (Fine, 1963:vii). Second, the big three automakers had gained an oligopolistic hold on their market, sharing among themselves 90 percent of the sales. And after being severely weakened by the depression, the few remaining independents were dropping out of the competition with alarming rapidity: "Cord and Auburn ended production in 1937, while venerable Reo, Hupp, and Graham were near collapse" (Cray, 1980:311). Third, each of the big three producers had in some way opposed the economic reconstruction policies of the Roosevelt Administration (pp. 311–312).

While Henry Ford was by far the most outspoken critic and vehement opponent of Roosevelt's New Deal, Sloan and his colleagues also were extremely recalcitrant in accepting the new business relationships being proffered by the government (pp. 275, 281–284). GM's designers particularly resented and resisted the government's incursions into what had long been their exclusive decision-making domain. Nevertheless, Sloan and Brown were simply getting a taste of the strong medicine they themselves applied to GM—"Centralized Control with Decentralized Responsibilities" (Brown, 1927) they had somewhat euphemistically called it.

Having intervened when GM's divisions were once badly managed without an eye to the overall corporate good, undoubtedly they felt they had done an exemplary job correcting GM's course and advancing everyone's well-being. For instance, GM's officers no longer participated in stock-market manipulations nor made speculative inventory purchases—two practices that dangerously amplified the business cycle's natural fluctuations. They found it hard to accept then the fact that in a broad context GM's actions could be construed by many as motivated by narrow self-interest and conducted at others' expense. Hence, they believed, government intervention in their business was unwarranted and, more importantly, downright dangerous. Had not their success proven beyond all doubt that they had earned the right to manage GM's affairs as they saw fit? Had not their performance-control system made GM the nation's leading business enterprise? How could any knowledgeable and reasonable observer of the economy question their abilities or motives? And, for that matter, what did the government really know about managing a business, meeting a payroll, or creating jobs?

Sloan (1936) held: "Government, as such, creates nothing. It provides a structure within which industry may create" (p. 352). Too much structure emanating from the central government, of course, meant reduced industry initiative. So with the "shifting of power between politics and economics," "government has everywhere come more and more into fields before dominated by private enterprise" (p. 361). Sloan thus concluded that increasing "political management, irrespective of degree, can have no other result than lowering the ceiling of industry's ability to contribute toward human progress" (p. 363). Five years later Sloan (1941) still questioned, "Shall we continue the system of free enterprise, or shall we accept the only alternative—regimentation of industry by a political bureaucracy, and we have been drifting very rapidly in that general direction in recent years" (p. 131).

Unfortunately for GM, many influential observers—especially those in the Roosevelt Administration—were unwilling to restore the firm's freedom of action in the economic sphere. These critics attributed GM's success not to superior management but simply to oligopsonistic and oligopolistic prowess. In response, Sloan (1936) asked: "Does honesty of purpose reside only in smallness? Does the mere process of pooling common interests and talents of an enlarged group . . . in an enlarged radius of activity constitute in itself an offense against honesty, social justice or security?" (p. 364). And he warned that "it is only by concentration in large units, making possible the use of the most efficient machinery, . . . by extensive advertising, by distribution on a national [or] world basis . . . that our unsurpassed standard of living has come about" (p. 369). Later, Sloan (1941) would explain that GM became "large through a process of evolution, but only because it was rendering a service to the community. As its volume of business expanded it became able to do more for workers, stockholders and customers" (p. 144). For example, "Large-scale operations can support costly research, can carry new developments through the initial stages where losses are inevitable" (p. 145).

Critics who might concede that GM's growth resulted from expert management and promoted industrial development still called for a more equitable distribution of the benefits than had been obtained. After all, just because the Sloan-Brown design team were reputed to be one of the country's best management groups they were not to be granted free rein to advance the stockholders' aims at the expense of other interested parties.

The organizers of the labor movement, for instance, had become particularly adamant on the latter point. Furthermore, these views were gaining widespread worker and public support and receiving the government's enthusiastic sanction. At long last, the labor unions were winning the right to improve the well-being of their primary clientele, the workers, just as management had long held the right to serve the stockholders. And like it or not, a government had arrived on the new social, political, and economic scene that felt it had the right and obligation to balance the long-term equity positions of the various interdependent interest groups as well as to protect the well-being of the country as a whole. The laissez-faire business environment was gone; GM would have to operate under a markedly altered set of long-term relationships.

More than anything else, GM's immense economic success caused its greatest problems in the middle and late 1930s (Fine, 1969:178). Had it been less successful GM could have avoided the spotlight of public attention. The firm's large annual market shares and the fact that between 1933 and 1938 it made 82 percent of the auto industry's profits had a "social significance that the antimonopolists of the New Deal" ("General Motors," *Fortune*, 1938:158) constantly pointed toward. It made no difference that twenty years previously GM teetered on the brink of bankruptcy, nor that the Sloan-Brown performance-control system might have propelled GM to the forefront of American business, nor that they had published numerous articles and made many speeches so all would know, competitors too, how they conducted their business. GM had grown big and successful, so it was easily tainted with a monopolistic image.

GM's well-publicized success over the years also engrossed groups such as organized labor. First, GM could easily afford to contribute to the improved well-being of workers in a time when most firms contemplated bankruptcy (Gates, 1936:498). Second, if an extremely prominent employer like GM could be organized, the rest of the nation's business could be expected to follow with little resistance. Here, then, it did not matter that GM had pursued relatively enlightened labor policies, especially compared to Ford. Nor did it matter that the success of GM's management methods also contributed much to the prosperity of the firm's workers.

Sloan, after years of inspired and painstaking work, deeply resented both the monopoly and antilabor accusations and the many investigations into GM's activities that resulted from them. But the assertions got more and more press in the middle and late 1930s. At first Sloan spoke out against them and the government's as well as labor's mounting intrusions into GM's affairs. (See, for example, Sloan, 1934.) To his chagrin, however, the more he and his colleagues defended their policies, the stronger the spotlight on GM became (Cray, 1980:312–313). Seeing this, Sloan began to accommodate reluctantly to the new social and political environment.

While Sloan adjusted to the new milieu more gracefully than did Ford, he did not bring to this new task the inspired leadership he had shown until the early 1930s. He would follow, even if somewhat reluctantly, and he would not reverse a change once it was made, but *he would not lead GM into the era of broadened corporate responsibilities.*

Labor

The first major adjustment GM needed to make in its relationships with the secondary clienteles came in the field of labor relations. Long blessed with a largely unorganized and cooperative work force, GM's designers did little to establish communication links and bargaining committees with the workers as an organized group. "Before 1933," Sloan (1964) himself admitted, "General Motors had no dealings with labor unions, except for a few craft organizations in the construction field. For this and perhaps other reasons we were largely

unprepared for the change in political climate and the growth of unionism that
began in 1933" (p. 405).

With the extensive worker movement and the accompanying labor legislation of the 1930s, then, the situation on the labor front shifted substantially.
Even though GM's workers were better off than, say, those in Ford's factories
and its average hourly pay rates were considerably higher than most manufacturing firms' (Fine, 1969:22), workers began to realize how poorly GM served
their interests. Most workers complained, for instance, of the "speed-up" in the
pace of work during the 1930s (pp. 55–58). Others expressed "the resentment
of men who had become depersonalized, who were badge numbers in a great
and impersonal corporation, cogs in a vast industrial machine" (p. 59). And
although GM paid high hourly wage rates, the workers' annual income was
often comparatively low because of seasonal variations and economic fluctuations.
In other words, "The irregularity of employment in the automobile industry
meant that the well-publicized high hourly wages of the auto workers did not
necessarily become translated into equally high annual earnings" (p. 61). The
gyrations in GM's production schedules also demanded excessively long hours
during busy periods. Under the strain, many workers suffered from severe
physical and nervous exhaustion and "even had encountered domestic difficulties because of their tiredness after work" (p. 62). Moreover, had the workers'
representatives gained the opportunity to examine GM's books in detail, they
undoubtedly would have been even more disgruntled upon learning how well
the firm's stockholders and executives had benefitted from its operations.

The first change in the labor-relations field came with the passage and
signing (on June 16, 1933) of the National Industrial Recovery Act (NIRA),
which established the NRA. No longer could the GM designers ignore the labor
movement.

The NRA was designed to achieve two goals: (1) "to remove some of the legal
restrictions of the anti-trust laws, permitting certain areas of collaboration
among industrial units engaged in specific fields" and (2) "to promote the
aspirations of organized labor" (Brown, 1957:88). Donaldson Brown was appointed
"Chairman of an Industry Committee" (p. 89) to write the code for the auto
manufacturers. The auto manufacturers' representatives knew that the administration was "particularly anxious" for the automobile industry, "the pacesetter
of the economy," to be "among the early code adopters" (Fine, 1963:48). "The
committee unanimously rejected the opportunity offered by the law to seek any
relief from existing anti-trust statutes" (Brown, 1957:89). Such a dispensation
was unnecessary in the auto industry because its oligopolistic structure prevented
the ruinous price competition that plagued other industrial sectors. Moving
on, the industry committee quickly turned its attention to the second NIRA
objective, the famous Section 7(a) of the law, "which extended the right to
employees to organize and bargain collectively through representatives of their

own choosing" (p. 88). The committee, "after prolonged negotiations," drafted a statement "which incorporated the fundamental position of the industry. [NRA head] General Johnson approved it and then obtained its acceptance" by the American Federation of Labor. Those involved next "went with the General to present the case to the President" (p. 91). President Roosevelt, after making a few minor revisions, issued a statement on March 25, 1934, describing the settlement as "a framework for a new structure of industrial relations—a new basis of understanding between employers and employees" (p. 92).

"General Motors" however "felt it desirable to draw up a statement of its own policies concerning labor relations" (Brown, 1957:94). Accordingly, the firm issued a document—drafted by Brown, approved by GM's Executive Committee, and signed by Sloan—on August 15, 1934, explaining "to personnel throughout the organization, including factory employees" (p. 94), the firm's attitude toward employee-employer negotiations. It read in part:

> Collective bargaining is to be understood as a method of intercommunication and negotiation between employees and management whose objective is the maintenance of harmonious and co-operative relations through mutual understanding with respect to terms and conditions of employment.
>
> . . . It must be made clear that collective bargaining does not imply the assumption by the employee of a voice in those affairs of management which management, by its very nature, must ultimately decide upon its own responsibility. It does not mean collective employer-employee management and must be limited to employer-employee relationships.
>
> Management is charged with the responsibility for promoting and maintaining the best long-term interests of the business as a continuing institution. Therefore, while management should exhaust every means . . . to settle all problems of employer-employee relationship which may arise, it cannot agree to submit to arbitration (which is a surrender by both sides to the authority of an outside agency) any point at issue where compromise might injure the long-term interests of the business. ("General Motors Sets up Basic Policies," *Automotive Industries*, September 15, 1934:322–323)

One of GM's primary concerns with the institution of collective bargaining, then, was to preserve the boundaries of its controlled domain, i.e., its decision-making authority. To Sloan and his colleagues, "what made the prospect seem especially grim . . . was the persistent union attempt to invade basic management prerogatives" such as setting production schedules, establishing work standards, and disciplining workers. "Add to this the recurrent tendency of the union to inject itself into pricing policy, and it is easy to understand why it seemed, to some corporate officials, as though the union might one day be virtually in control" (Sloan, 1964:406). As Sloan saw it, then, the stakes were extremely high. GM had to keep up its opposition to the union's organizing efforts.

The key ploy the Sloan-Brown group used to thwart unionization was to reject majority rule in selecting employee representatives for collective-bargaining purposes and to give "unswerving support" to "collective-bargaining pluralism" (Fine, 1969:30). Both of these provisions were contained in the NIRA automobile code and GM's own (subsequent) statement on labor relations issued to its workers. In regard to the latter, Brown (1957) stated, "Certainly the principles embodied in this document were far removed from any concept of industry-wide bargaining or national unionization" (p. 94). In effect, these provisions "made it unlikely" that a "united trade-union front" could be presented to GM for collective bargaining "even if the majority in a particular plant or unit were able to agree on a bargaining representative" (Fine, 1969:30). To insure further that it would not be facing bona fide union representatives, GM began to support the creation of " 'employee representation plans' or perhaps more commonly 'company unions' " (Sloan, 1934:523).

Meanwhile, as a consequence of its efforts to frustrate meaningful unionization in its plants, GM encountered numerous local strikes over the next several years. Of these work stoppages, the April 23, 1935, Toledo Chevrolet strike was the most important in the entire industry during the NIRA years (Fine, 1963:387). Because the workers had struck the plant which produced all of Chevrolet's transmissions, they halted Chevrolet production all over the country, forced GM "to retreat" from its public refusal to negotiate, and achieved what was hailed by a future union president as "our greatest single step forward" (p. 387). The strike had caught GM's designers "napping" (p. 402).

Determined not to be placed in such a vulnerable position again, GM moved half of the Toledo production facility to Saginaw, Michigan, and Muncie, Indiana. More generally, the firm's planners had learned that GM's hand would be strengthened in the future if it carried larger inventories of semi-finished and finished products and if it adopted a policy of "diversification of plants where local union strength is dangerous" (Fine, 1969:49). "General Motors" as Sloan (1941) put it, "is striving for decentralization [i.e., dispersion] of its plants and has been working in that direction for many years. As a matter of policy, in the last four or five years we have established units in fifteen or twenty smaller places" (p. 197). But given the highly integrated nature of automobile production, the dispersion of its capital facilities could not buy GM much performance protection. Hence, GM was "willing to make substantial concessions to union demands" (Fine, 1969:53).

Still, it was too early for GM's management to capitulate. Labor, racked with much internal dissension and uncertainty, did not present a unified front. Furthermore, it did not always receive unanimous public support. Hence, other approaches could be tried to limit union power.

First, as the worker's unions grew in their power to damage GM's performance via strikes and slowdowns, the management created more communication lines to monitor the workers and their organizations. In this way, unfortunately, GM became involved in a major espionage effort. Indeed, "the best customer of the labor spy agencies before 1936 was GM" (Fine, 1969:37).

Second, in addition to attempting to secure information on union activities, GM spent considerable time and effort to create "a favorable public opinion" on this matter. GM's Executive Committee viewed "the development of good public relations in plant cities" to be "just as much a responsibility of the executives as the conduct of ordinary business" (pp. 52–53). GM's intensive effort to cultivate the press and influential citizens of Dayton, Ohio, for example, had resulted in favorable articles and "'cooperation' in suppressing labor union 'propaganda'" (p. 53). As a GM source put it, "If . . . we could have our own employees and the public of our plant cities think and say 'WHAT HAPPENS TO GENERAL MOTORS HAPPENS TO ME' this would be the most effective protection against efforts to undermine our corporate goodwill" (p. 53).

GM's labor espionage and propaganda activities only stalled off the advent of company-wide collective bargaining. Ultimately, such negotiations would be necessary to bring management and labor together in a cooperative dialogue aimed at creating long-term industrial peace.

In spite of several successes achieved under the NIRA, labor leaders began to back away from the auto industry code. Among other things, "they held that collective bargaining on a local basis, with representatives of employees chosen from among themselves, was the equivalent of 'company unions,' subject to employer domination" (Brown, 1957:93). Then, on May 27, 1935, the United States Supreme Court ruled the NIRA unconstitutional. The first round in redefining the relationship between capital and labor was over.

Although organized labor minimized the results achieved by the NIRA, "the automobile code had important long-range consequences for the automobile manufacturing industry" (Fine, 1963:428). It raised wages and reduced hours permanently, led to greater industry concern over irregular employment, and sped up the introduction of new car models in the fall rather than at the beginning of the year (p. 429). This move leveled employment somewhat by increasing low fall sales and decreasing high spring sales. "In 1935 General Motors," for its part, "had already set up a revolving fund of sixty million dollars to build up an inventory in periods of low seasonal consumer demand to maintain employment" (Douglass, 1954:179). Most importantly, the automobile code made management aware of the need for improved labor relations and for the first time legitimized the position of organized labor (Fine, 1963:429). Future efforts by the unions to organize workers would be easier. The ground had been broken.

Much to the consternation of GM's officials, the NIRA's demise did not halt the federal government's attempt "to prescribe rules" for conducting labor-relations activities (Fine, 1969:50). On July 5, 1935, the Wagner National Labor Relations Bill became law, establishing a three-man non-partisan National Labor Relations Board. The NLRB could require employers to cease unfair labor practices and could use the federal circuit courts to enforce its orders (Fine, 1969:50). Undoubtedly, to the Sloan-Brown group the most undesirable aspect of the new bill was its stipulation for the election of the employees' bargaining representatives by majority rule. In effect, this proviso authorized the closed shop and marked the end of the company union. Convinced that this latest piece of New Deal legislation would also be declared unconstitutional, Sloan and GM refused "to obey the Wagner Act's provisions permitting labor to organize freely and to bargain collectively" (Cray, 1980:283).

Another major confrontation with labor was in the making for Sloan and his colleagues at GM. On their side, the workers were ready. Times were relatively good compared with the early 1930s. In addition, the NRA and then the Wagner Act "gave them heart. The overwhelming reelection of President Roosevelt in 1936 had reinforced their courage" (p. 288). By late 1936 and early 1937 strikes were erupting throughout the GM system. Two of the more important of these disputes were the sitdown strikes begun on December 30, 1936, at two critical Fisher Body Plants in Flint, Michigan. Sloan was incensed. Not only had the workers gone on strike, but they had occupied company property illegally. Brown (1957), acting as the corporation's chief labor negotiator, stated, "Our position was that when the struck plants were restored to the management, we would be willing to enter into negotiation" (p. 95). The workers, however, would not vacate the plants. Even more galling, Michigan's newly elected Democratic governor, Frank Murphy, refused to order the National Guard to evict the strikers. "To make matters worse," Sloan (1964) later complained, "it appeared that the UAW [United Auto Workers] was able to enlist the support of the government in any great crisis. . . . President Franklin D. Roosevelt, Secretary of Labor Frances Perkins, and Governor Frank Murphy of Michigan exerted steady pressure upon the corporation, and upon me personally, to negotiate with the strikers who had seized our property, until finally we felt obliged to do so" (p. 393). Brown, representing GM for Sloan, entered into discussions with John L. Lewis, representing the UAW-CIO (United Auto Workers-Congress of Industrial Organizations). "The negotiations with Mr. Lewis extended over some days, resulting in an agreement as to the general basis on which further negotiations would be conducted. . . . Thereafter, the more detailed problems were taken over by the central labor relations staff, under the chief guidance of C. E. Wilson, then Executive Vice President of the Corporation" (Brown, 1957:96-97). When all was over, "GM had been compelled to sign its first agreement with a union, and it had for the first time agreed to recognize an international union as a party to the collective-bargaining process" (Fine, 1969:309).

Thus, it took the major sitdown strike of 1937 to convince GM *finally* that it had to negotiate long-term contracts with labor if its production lines were to run again without interruption. To keep the antagonists separated over the years, these agreements became "rigid packages in which the only variations came through formalized grievance procedures and new contract negotiations" (Yates, 1983:247). Accordingly, after 1937 Sloan elevated the Personnel Staff to a corporate-level activity that "served the corporation in two ways: as a specialized staff of experts on which the corporation can rely for advice and consultation; and as a group of executives entrusted with line responsibilities in union negotiations and in administering the provisions of the contract" (Sloan, 1964:392). Less important labor-relations disputes still came under the jurisdiction of divisional managers. In other words, some fine-tuning, high-frequency, communication channels continued to operate at GM's lower levels to handle localized disputes and grievances.

"The GM sit-down strike of 1936–37 was, all in all, the most significant American labor conflict in the twentieth century" (Fine, 1969:341). A new era of labor-management relations opened with new ideas and approaches. The *Fortune* editors in viewing the GM sitdown "in terms of the 1933–37 strike wave" felt it was "a landmark, measuring how far labor had traveled in less than three years and through some 4,000 strikes" ("The Industrial War," *Fortune*, 1937:166).

So in the end, then, GM had to shift abruptly and dramatically its relationship with labor. While Sloan and his colleagues retained control of their most valued managerial prerogatives, e.g., pricing and scheduling (Sloan, 1964:406), GM now had to bargain collectively with labor's representatives over wage rates, working conditions, grievance procedures, etc. By steadfastly insisting on conducting business as usual for too long, the Sloan-Brown planners had unwisely allowed the workers' long-smoldering frustrations to exceed their threshold of restraint (Fine, 1969:63). Once the flash point was reached and the favorable governmental winds blew, a series of short, but catastrophic strikes erupted. Despite their brevity, these stoppages changed how GM (as well as America) would conduct business. Almost overnight the firm's workers with the government's support had elevated their client status in the GM system. In the future, there would be no customers' cars, executive bonuses, or even stockholder dividends unless the workers were sufficiently satisfied with benefits that their resources, i.e. labor, earned.

Dealers

Labor was not the only group dissatisfied with how it was treated by the auto manufacturers. Dealers too had voiced concerns about their well-being. Primarily, the dealers complained that the auto companies overloaded them with cars, failed to protect their territories when granting new franchises, and compensated them inadequately for inventory losses when models were changed or franchises cancelled. Another serious concern dealt with the ease and short notice (usually 30 days) with which a dealer's franchise agreement could be terminated by the company. Less serious problems also arose from the dictatorial tendencies exhibited by some factory representatives. All of these abuses had long existed in the industry because of the manufacturer's economic power.

Ford, by far, had earned the worst reputation in the industry, and Sloan, for his part, had worked hard to prevent shortsighted tactics from being foisted on GM's dealers (Nevins and Hill, 1963:62). "Markets must be defined and protected," Sloan (1941) cautioned. "Enough dealers of the right size to deliver the potential of those markets and no more" (p. 201). Sloan undoubtedly understood the dealers' positions, for he had started as a small businessman himself, tirelessly visited GM's dealers since the early 1920s (see Chapter 10), and really considered the dealers as GM's partners. So in contrast to his insensitive and laggard response to labor's demands, Sloan showed more empathy and initiative in addressing the dealers' severe problems during the 1930s ("G.M. III," *Fortune,* 1939:106). In fact, he actually considered "the dealer problem as the last big job of his career" (p. 106).

With the depression the dealers' viability became extremely problematic, and their sensitivity increased sharply as the manufacturers' representatives pushed to maintain sales volume. Serious misunderstandings became commonplace. Hence, the dealers appealed to the federal government for relief and protection. Throughout the automotive industry, "many dealers . . . would have liked to see the N.R.A. use its power and influence to strengthen their collective-bargaining power as compared to the power of the manufacturers just as organized labor looked to Washington to equalize the bargaining power of employer and employeé" (Fine, 1963:134). But since the automobile companies did not want to limit their freedom in purchasing supplies and retailing cars, they astutely avoided invoking the antitrust provisions provided in the NIRA. All the dealers could do then was to write their own code to protect themselves from ruinous competition within their own ranks. "The result was that the motor vehicle retailing code, which was approved on October 3, 1933, left the status quo with regard to factory-dealer relationships undisturbed" (Fine, 1963:133). Even so, the NIRA helped to strengthen the National Automobile Dealers' Association (NADA) and served to prod Sloan into making further improvements in the relationship GM had with its dealers (p. 463).

In late 1934, Sloan devised GM's Dealer Council to provide systematic communications between the top executives and selected dealer representatives. From his extensive travels, Sloan had seen that the dealers appreciated direct contact with corporate (as well as divisional) executives and believed something more formal and regular was needed than occasional visits to different dealers: "Out of these early field trips, therefore, grew a related idea, that of bringing representative" dealers into GM's "conference rooms" for "a continuing series of round-table discussions on distribution policies" (Sloan, 1964:291).

The Dealer Council included "forty-eight dealers, divided into four panels of twelve each." As president of the corporation Sloan chaired the council and each year "chose a different panel of dealers, representing all car-manufacturing divisions, all sections of the country, and all types of territory and capital commitment. . . . The vice president in charge of the Distribution Staff and other top officials of General Motors were also members" (p. 291).

The Dealer Council's specific charge was to help forge "policies on which an equitable dealer selling agreement could be based" (Sloan, 1964:291). With this purpose in mind, the dealers came, in groups of twelve, three times a year, "to chew the rag with the Distribution Group" ("G.M. III," *Fortune*, 1939:105). Both headquarters staff and dealers who attended reported the atmosphere was "candid," with the corporation "freely" discussing its policies and "cordially" receiving suggestions. These meetings, like Sloan's field visits, helped eliminate "the hazard of misinterpretation of both G.M. policies and dealer beefs by the army of 'over-zealous' subordinates" (p. 105).

Interestingly, in 1936 GM made substantial alterations in its contractual relationships with dealers. As was written then: "Heretofore the trade generally regarded dealer contracts primarily as an expression of manufacturers' rights but in the new agreements dealers' rights are more expressly declared. . . . The most important changes incorporated in the revised agreement affect clauses relating to cancellation of franchises and to allowances in connection with clean-up of outgoing models" ("GM Makes Dealer Contract Changes," January 11, *Automotive Industries*, 1936:37). Most notably, GM had inserted a 90-day rather than a 30-day cancellation clause in its contracts. Purportedly, the NADA supported Sloan's new "policies with enthusiasm" ("G.M. III," *Fortune*, 1939:106)

By 1938, "Chrysler and Ford . . . followed G.M.'s lead with more liberal contracts and policies of their own" (p. 106). But by then "the less patient dealers had begun to take matters into their own hands by the legislative route. As a result of their agitation, four states [passed] dealer-licensing laws". More significantly, in early 1938 Congress, after hearing lengthy testimony from the NADA and the manufacturers, directed "the Federal Trade Commission to investigate the policies employed by manufacturers in distributing motor vehicles, accessories, and parts, and the policies of dealers in selling motor vehicles at retail as these policies affect the public interest" (USFTC, 1939:1).

Under some pressure from this Federal Trade Commission investigation into manufacturer-dealer relations, the GM designers created in January 1938 still another contact point between the top corporate executives and the dealers: the Dealer Relations Board. "It acted as a review body," and as Sloan (1964) explained, "enabled the dealer who had a complaint to appeal directly to the top executives of the corporation." Given that Sloan himself chaired this board, the "divisions made very sure they had a sound case and were observing all the equities" (p. 294). From its external perspective *Fortune* reported that the power of "over-zealous" divisional executives "to make trouble" had been "further limited by the creation of a Dealer Relations Board," and, in particular, that the divisional sales managers had "become more scrupulous than before on such matters as cancellation" ("G.M. III," *Fortune*, 1939:105). Similarly, the impartial Federal Trade Commission investigators concluded that GM's Dealer Relations Board "as a final review agency in case of aggrieved dealers appears to be a move to safeguard in some measure against arbitrary treatment by factory field representatives pressing for volume, or otherwise interfering with or directing the conduct of the dealer's business" (USFTC, 1939:176).

Sloan's efforts to improve GM's long-term relationship with its dealers, however, did not blunt the FTC's final criticism. It wrote: "The Commission finds that motor-vehicle manufacturers, and, by reason of their great power, especially General Motors Corporation, Chrysler Corporation, and Ford Motor Co., have been, and still are, imposing on their respective dealers unfair and inequitable conditions of trade" (USFTC, 1939:1075–1076). A lengthy list of specific problem areas then followed.

Sloan and GM faced even more embarrassment in another dealer-related matter. Shortly after President Roosevelt signed into law the resolution directing the Federal Trade Commission's investigation, the Justice Department had a federal grand jury in South Bend, Indiana, return antitrust indictments against GM, Ford, and Chrysler and their auto-financing subsidiaries ("G.M. III," *Fortune*, 1939:109). "Both Ford and Chrysler signed consent orders agreeing not to force their dealers to use the corporations' captive finance companies. General Motors refused. . . . Sloan, Knudsen, and fifteen other uncomfortable General Motors executives found themselves in the criminal dock, facing a jury of their peers" (Cray, 1980:312). After hearing six weeks of testimony, the jury acquitted the executives but "found General Motors and three of its finance subsidiaries guilty of conspiracy to restrain interstate commerce" (p. 312). The Sloan-Brown designers had been dealt a hard blow. When GM was much smaller, they had formed GMAC simply to provide financing for their dealers and consumers who could not find it elsewhere. But now that GM was much bigger, some of their long-practiced strategies were being declared illegal.

Consumers

Surprisingly in this era of criticism, there was one secondary clientele that appeared fairly well satisfied with GM's efforts. Consumers, for their part, voiced no significant criticism of GM, and they continued to reward the corporation with an increasing share of the market. Since there were no organized consumer groups to publicize GM's questionable policies toward automotive safety and planned obsolescence, the consumer received little or no critical information on which to judge GM's decisions. The firm's products appeared as good or in many cases better than the competitors'. Even the USFTC, one of the few governmental agencies then concerned with the consumers' interests, concluded in its 1939 *Report on Motor Vehicle Industry:* "Active competition among automobile manufacturers, although some of them have made very large profits, gave to the public improved products, often at substantially reduced prices. . . . This has been especially true of those manufacturers who . . . obtain large volume . . . through competitive improvement in motor-vehicle construction, style, performance, and safety" (p. 1074).

Another area that gained the attention of the federal government during the late 1930s was the auto manufacturers' new-car pricing policies. The problem was not that consumers were being overcharged, but that the manufacturers refused to lower prices during periods of reduced demand. Instead, they held prices constant, cut production, and furloughed workers. Given the national economy's sensitivity to the auto industry's performance, such cutbacks soon reverberated and amplified the general economic downturn. Here was "a prime example of the social spectacle that President Roosevelt called 'rigid prices and fluctuating payrolls'" ("General Motors IV," *Fortune,* 1939:145). Naturally, Roosevelt disliked the manufacturers' environmental response pattern because it saddled the government with a difficult unemployment problem. After hearing Knudsen's disquieting testimony on this matter, the Senate established a Temporary National Economic Committee to conduct further investigations into what appeared to be monopolistic pricing practices (p. 142). Again GM's size and prominence had thrust it into the limelight. The government unleashed still another potential threat to the Sloan-Brown group's managerial prerogatives.

To quell such criticisms of its pricing policies, GM conducted a detailed study "of the effect of price and price changes upon the demand for automobiles. The study developed into a broad analysis of the many factors affecting automobile demand." It was presented at a joint session of the American Statistical Association and the Econometric Society in 1938, under the title of "The Dynamics of Automobile Demand" (General Motors Corporation, 1939:iii). Not surprisingly, the study concluded that a manufacturer using "even the most daring financial policy" (p. 134) could not use the price variable to increase sales volume significantly. Thus, there was little GM could do to help the national economy when it turned down.

MONITORING THE LONG-TERM ENVIRONMENT

Throughout the 1930s Sloan and Brown held the primary responsibility for dealing with the major, long-term shifts in GM's social and political environment. Brown (1957) described how GM's monitoring of long-run environmental shifts related to his own advancement within the corporation:

> My election as Chairman of the Finance Committee of General Motors in 1929 marked a definite shift of emphasis in my business activities. During the period 1921 through the summer of 1929, I had been preoccupied largely by problems of internal management control. . . .
> During the 1930s I found myself increasingly concerned with events transpiring outside of the corporation; events not directly related solely to GM operations, but with important impact upon its business affairs. . . . Free from the detailed responsibilities in connection with internal coordinated control, I gave more and more of my time to problems arising out of the interdependence of the corporation and the national welfare, and to broad questions of economic, social and political trends. I continued, of course, with primary concern over matters of coordinated financial control, and remained on most of the internal Policy Groups. Later I was designated by Sloan as his representative on matters related to labor policies. (p. 72)

Because of his experiences during the 1930s, especially in the labor arena, Brown suggested to Sloan that GM create a policy group to monitor the firm's social and political environment. "Rapid changes," as Brown (1957) argued, "were occurring in social and economic conditions, requiring a continual adjustment of corporation policies, organization attitudes and management technique" (p. 102). Brown was concerned that these changes, which "resulted from the pressures of public opinion," not only should be "understood" but "their effects . . . carefully projected." Like them or not, such trends "were destined" to affect "the conduct of business, and hence had to be realistically appraised. . . . The proposed policy group would . . . develop such information on a reliable basis for the general guidance of the corporation management."

Although the "Policy Group—Social and Economic Trends" was created and Brown was named its chairman, it operated for only a few years (p. 102). With the depression receding and war approaching there seemed to have been less and less need for a social, political, and economic policy group even though major long-term environmental changes would continue to buffet the corporation.

In retrospect the demise of this committee appears as part of a—then new—trend toward isolation at GM. Rather than having learned from its protracted battles in the political and legal arenas during the 1930s that its prominence now made continued success more than ever dependent on long-term cooperation among government, investors, management, labor, dealers, and consumers, GM looked more and more inward. As Sloan (1941) himself put it, "For me the essential ingredient—the heart, if you please—of our organization is a group of not more than 10,000 workers whose skill in management, in engineering and in science as well in the special crafts makes possible the work in which all the others are engaged." GM's "220,000 workers" were peripheral; even its "hundreds of thousands of stockholders" (p. 193) were becoming less important in light of GM's financial strength.

The formal beginnings of this movement can be traced to Sloan's major reorganization of top management in May 1937. Having just lost the major (sitdown) battle with labor and using as his take-off Lammot duPont's resignation as Chairman of the GM Board, Sloan moved "to reduce the influence of Wall Street and other outside directors" (Cray, 1980:310). Financially, GM no longer depended on the capital markets, as the necessary funds for expansion and improvements could be generated solely from profits (p. 310). This independence allowed Sloan to reorganize the Board's structure. Though he left the Board's overall composition essentially unaltered, Sloan had himself elected as Chairman—as well as made Chief Executive Officer—and Brown elected as Vice Chairman.

More significantly, Sloan with Brown's (1957:68) support combined the Board's Finance and Executive Committees into a single group: the Policy Committee. "All duties and responsibilities previously held by the two governing groups were consolidated in the new nine-member Policy Committee" (p. 68). However, the fourteen-member Finance Committee in May 1937 included ten outside directors (from the duPont and banking interests), while the new Policy Committee had only three outside members (p. 69). And the "six active officers—Messrs. Bradley, Brown, Knudsen, Sloan, Smith, and Wilson— ... dominate the all-important policy committee, whose other three members are two duPont men and a Morgan partner and which has never been turned down by the Board of Directors in any recommendation it has made" ("General Motors," *Fortune*, 1938:44). In effect, the Sloan-Brown consolidation of the Board's top committees served to "diminish the active participation of duPont-Morgan interest in the executive affairs of the Number One automobile manufacturing organization of the world" ("General Motors Enters New Phase," *Automotive Industries*, May 8, 1937:679).

Now Sloan no longer had to contend with outsider dominance on GM's top oversight committee and the significant tensions that had long existed on this front. There had been, for instance, the lengthy dispute between Sloan and Lammot duPont over safety glass, and that was but a single battle of a protracted war over the use of DuPont products in GM cars. Then, too, Pierre duPont had strongly opposed Sloan's insistence that Raskob resign his Finance Committee Chairmanship because of his involvement as Chairman of the Democratic National Committee in the 1928 political campaign of Al Smith (Brown, 1957:70). Only reluctantly did the duPont interests agree to the reorganization proposed by Sloan and "wholeheartedly and enthusiastically" (p. 68) endorsed by Brown. Even the duPonts and their associates could not counter the combined prestige of Sloan and Brown—the creators of GM's performance-control system which had demonstrated its financial worth beyond all doubt during the deep depression of the 1930s. (See, for example, GTX 194 and 196, *U.S.* v. *DuPont*, 1956:3428-3433 and 3435-3439.)

Accordingly, the increased insider orientation and the contraction of the Board's top committee further diminished an already far too limited source of outside opinion. This heightened isolation occurred, incongruously, at a time when the corporation most needed a broad input of views.

Ironically, Sloan (1941) wrote of the need to broaden his perspective: "My responsibilities had expanded enormously. At Hyatt, big as it was, I had been obliged to consider the interests of only a few stockholders, a few customers and three or four thousand workers" (pp. 143-144). There, "the absorbing problems of industrial management were largely limited to the fields of engineering, production and distribution" (p. 144). "But as president of General Motors, I realized our thinking affected the lives of hundreds of thousands . . . as we expanded, the economic welfare of millions was becoming linked with the welfare of General Motors" (p. 144). Supposedly to Sloan this increase in size meant

> that industrial management must expand its horizon of responsibility. It must recognize that it can no longer confine its activities to the mere production of goods and services. It must consider the impact of its operations on the economy as a whole in relation to the social and economic welfare of the entire community. For years I have preached this philosophy. Those charged with great industrial responsibility must become industrial statesmen. (p. 145)

Moreover, even before GM's sitdown strike Sloan (1936) had written: "To my thinking, 'business bigness' must have an even greater respect for the equities of others—a greater recognition of the broader responsibilities of industry. This is because of the greater power and influence, for good or bad, that flows from bigness itself" (pp. 369-370).

There can be no doubt whatsoever, then, that Sloan knew the proper role to play after the middle 1920s when GM's growth accelerated: the broad-minded industrial statesman, i.e., the planner who exercised social, political, and economic leadership in the general interest and without narrow partisanship. Sloan's record, however, fell short of the standard. Back in the late 1920s he slowed the introduction of safety glass; throughout the middle 1930s he frustrated labor's aspirations and fought the government's economic reforms; and in the late 1930s he turned GM inward.

Sloan's personal isolation was further accentuated by his growing deafness, an ailment that caused him considerable embarrassment (Cray, 1980:309) and undoubtedly made him increasingly dependent on his long-time GM associates whom he knew and trusted. (It should be noted here that prominent gerontologists view hearing loss as even more emotionally upsetting and personally isolating than vision loss. See, for example, Butler and Lewis, 1982). Moreover, Sloan invariably traveled in a chauffeured limousine and a private railway car; contact with a broad spectrum of people who could keep him in touch with the country's changing social and political milieu was unlikely.

In conclusion, Sloan's *managerial* achievements from 1918 to 1938 were truly monumental. He, with the help of colleagues like Brown, had organized and rationalized the tremendous productive forces unleashed by Henry Ford twenty years earlier. And as Ford's initiative had done for Ford and FM, Sloan's very success helped change the world in which he and his GM operated. Using GM's performance-control system, the modern industrial enterprise could grow much larger than even Ford had envisioned.

Ford did not like the new world of the 1920s he had helped to create because it required the large enterprise to design performance-control systems for survival. Similarly, Sloan disliked the new business environment of the 1930s because it required the large corporation to operate its control system within the parameters set by the larger society. GM's performance-control system was not enough for government, labor, and eventually consumers. Probably more than anything else, these groups could not accept Sloan's narrow view of GM's clientele now that the corporation had such a broad impact on the general well-being. However, as had earlier been the case with Henry Ford, Alfred Sloan's accomplishments led him to turn inward rather than outward. After 1937, GM's success allowed Sloan to narrow rather than expand the corporation's view of its responsibilities to the country as a whole. The most damaging result "of a system that discourages attention to matters far outside the purview of [management's immediate] jobs" (Cordtz, 1966:118) was GM's mishandling of the Nader/Corvair affair 30 years later. "G.M.'s remarkable commercial expertise," helped little "with a challenge" that "was political and sociological rather than economic" (p. 117). After this notorious episode of corporate irresponsibility, even the firm's insulated management questioned "whether G.M. was sufficiently in touch with large areas of social and political reality." But few improvements were made, for Sloan's effort to "get the facts" usually stopped at the economy/society boundary.

But unlike Ford's shortcomings, many years were needed for Sloan's long-range failings to become unmistakable. Ford's were already at hand.

PART III

The Ford Motor Contrast

12

Repudiating Planning

The Ford Motor Company situation between 1918 and 1938 provides a dramatic contrast to the GM case history. While the Sloan-Brown performance-control system propelled GM from failure to (economic) success, Ford's antagonism toward such systems drove FM from success to failure.

By the early 1920s the country's automobile plants had expanded productive capacity to several million units. Ominously, as most American families completed the purchase of their first car, capacity had begun to equal, then exceed demand. Next came the sharp, unexpected, and severe postwar recession; with it automobile demand plummeted. The auto industry thus had entered a period of tough competition rather than of easy growth (Chandler, 1964:95).

In 1921 GM's prospects for a major success in the new automobile market were weak indeed, for FM completely dominated the business with its inexpensive Model T. FM exhibited its strength by a stunning performance during the economic crisis of 1920–1921 when industry sales sagged by over 25 percent (General Motors Corporation, 1968:2): FM had a strong hold on 60 percent of the market while GM had a weak grip on but 15 percent. GM, subsequently, admitted FM's "factories were efficient, its distribution network well organized . . . its employees the envy of all workers, and its consistently high profits . . . the envy of most businessmen" (p. 2).

FM's enviable position at the beginning of the 1920s was by no means a short-run historical aberration. Fabulous success had become the standard of performance at FM. "In the adventurous history of the industry," as Nevins and Hill (1954) extolled, "no other concern had so spectacular a history as the Ford Company, and none had gained so solid a position" (p. 490). With an initial investment of only $28,000, FM's founders had built a business whose dividends by March 1913 had exceeded $15 million and whose assets were more than $22 million (p. 490). More specifically, "for the fiscal year ending September 30, 1914, sales were 12.74 times the plant inventory—as against a ratio of only 3.67 for General Motors" (p. 491). As the 1920s boom period unfurled, FM's dominance seemed secure and unchallengeable (Rae, 1965, p. 96).

Much of FM's early success was due to Henry Ford himself. Ford had been responsible for a large part of the Model T's design, its extreme standardization, and its endless series of price cuts which in the highly elastic market had swelled FM's profits phenomenally. Henry Ford's vision appeared to assume profound proportions, and within his company he ascended—as Sloan at GM did in the 1920s—to the dominant position. Unfortunately, "on so rich a diet of publicity and social acclaim, any ego might well have gorged itself; Ford's, in the year 1914 and after, was no exception. . . . Now as never before the Yankee mechanic who had 'arrived' by rule-of-thumb became the man of destiny; he began to feel that the 'inner guide' to which he alluded so often could never do him wrong" (Sward, 1948:63).

Had Ford practiced more humility, he probably would have encountered considerably less humiliation later. More importantly, the historical facts called for humility, for Ford lagged in producing a cheap car for the multitudes. As early as 1896 Charles Duryea wrote of the need for such a vehicle (Wik, 1972:235). Later, "Ransom E. Olds insisted that a simple, inexpensive, utilitarian auto should be produced." So in 1902 Olds built 425 cars that sold for under $400. "In 1903 he sold 2,500" and proclaimed "that the era of the fad had terminated and the era of utility begun" (p. 235). Numerous other makes followed. "After a decade of experience, Ford finally introduced his Model T in 1908 and the assembly line in 1913" (p. 237). Ford's timing was perfect, so "Henry Ford deserves credit for being eminently more successful than his competitors" (p. 237). Yet, Ford definitely was not the first to conceive of making low-price cars for the masses.

Along with Ford's growing self-assurance came an abdication of power by men with the personalities and knowledge to influence Henry to modify his often unfounded opinions. The first design-oriented executive to depart was James Couzens. Couzens, a minority stockholder in FM, was as "supreme in the business field—marketing, advertising, bookkeeping, finance—as Henry Ford was supreme in production" (Nevins and Hill, 1954:570). Accordingly, "until 1915 the Ford Motor Company, as a business, was run by Couzens. He set up the dealer organization, managed sales, bought materials and parts, approved capital outlays, enforced cost discipline, kept the books, watched the earnings, and held and paid out the money. He was a superb organizer, and he had a brilliant sense for both large issues and small" (Galbraith, 1960:158). Thus "the more prosaic aspects of the company's life, the tiresome details of administration, control, finance, purchasing, sales, and distribution were dealt with by James Couzens and dealt with so effectively that Ford was consistently to underrate their importance in the future" (Jardim, 1970:200–201). And, "everyone [except Ford] knew that much of the extraordinary success of the company was attributable to the organizing capacity and commercial acumen of [Couzens]. . . . Down to his resignation in 1915 the public properly gave him almost equal honors with Ford for the company's achievements" (Nevins and Hill, 1954:570–571).

Ford and Couzens both possessed strong, dictatorial personalities that kept each other in check and the company running efficiently. Yet, as many could see, their cooperative truce was precarious and at very best temporary. In the long run, the two strong personalities could not tolerate each other. Having only a minority interest in FM, Couzens resigned. The inevitable break, as Nevins and Hill (1954) describe it, "came when Couzens objected to a pacifist statement by Ford as injurious to company interests" (p. 571). While the resignation supposedly resulted from this minor issue of FM's public image, the two had a deeper disagreement over the Model T's future (Sward, 1948:64). Ford thought the market was endless and consequently favored unbridled expansion to match it; Couzens, on the other hand, believed it finite and wanted to expand cautiously and conservatively. After Couzens' departure Ford remarked to John F. Dodge "that Couzens' leaving had been 'a very good thing for the company.' Henceforth, he said, the brakes were off; he would be free to expand at will" (Sward, 1948:64).

Couzens eventually departed as a minority FM stockholder as well, and was joined by the other minority stockholders, who as a group held 41.5 percent of the FM stock. From 1915 on, Ford became increasingly disgruntled with having to pay huge dividends in return for the minority stockholders' minuscule original investments, believing that their absentee ownership was an antisocial practice (Nevins and Hill, 1957:89). Besides, Ford had definite plans for the profits which were being drained from the firm: the massive FM expansion.

John F. Dodge and his brother, Horace E., made the first formal move in the battle which was to end with Henry and the Ford family in complete control of FM. The Dodges brought suit against Ford to compel him to pay the dividends retained to finance FM's expansion. When the courts decided for the minority stockholders, Henry Ford threatened to withdraw from FM and found a competing company. Ford not only induced the minority stockholders to sell, but at prices favorable to the Ford family, who secretly bought the stock through several agents. The entire minority interest cost $70,000,000, a sum borrowed from various financial interests. Given FM's healthy cash flows, paying this debt appeared routine. Ford had won his freedom. But "with no restraint upon him whatever, would Ford with his unorthodox plans for expansion, manufacturing, and marketing be able to steer a successful course through the still uncharted shoals, reefs, and narrows of the automotive world?" (Nevins and Hill, 1957:113).

Unexpectedly another hurdle appeared in Henry Ford's path toward complete power within FM: the 1920 recession. The economic downturn had done severe damage to automobile demand and helped topple GM's William Durant. "Ford detailed his financial worries, pointing out that there would be coming due on the first of the year [1921], or soon thereafter, $18,000,000 in income taxes and $7,000,000 in bonuses to Ford employees ... and, in addition, there remained $33,000,000 to pay of the $70,000,000 that had been borrowed to buy out the minority stockholders" (Beasley, 1947:103). In spite of his pressing financial demands, Ford, like Durant, had to shut down operations. Soon after Ford's financial predicament became public knowledge, a New York banking firm offered him a large loan but insisted on selecting a treasurer who would have complete control over the FM finances (p. 106). To Henry Ford, this solution was intolerable, so he hurriedly built "$88,000,000 worth of parts and supplies" into "93,000 surplus cars," which had been neither sold nor ordered. "In the midst of a national depression in January 1921, Model T's began to rain on every Ford dealer in the United States" (Sward, 1948:76). Thus, "the banks, friends, relatives and in-laws of the dealers paid off Ford's creditors in New York and Boston. They did it reluctantly and sometimes with fury in their hearts" (Galbraith, 1960:157). Dealers who rebelled lost their franchises.

Henry Ford had strengthened his stranglehold on the firm. Since Ford obtained control of the company without financial backing, no one could force reorganization on him, as had the duPonts upon Durant at GM.

Over the next two decades, "the immense Ford organization ... cried for a unifying administrative control. This would require dozens of gifted executives and a sound plan for relating the many separate activities to a central directing office" (Nevins and Hill, 1957:269). But instead Ford dismissed his best executives and dismantled the central-office units that could have designed a performance-control system for FM.

Ford's brutal elimination of his central office force illustrates quite memorably his antagonism toward organizational design, as well as his almost sadistic (Sward, 1948:183) treatment of designers' administrative creations. When one group "of eighty office workers" came to work one morning, "their office was bare. It had been stripped of every last piece of furniture. ... Another group of salaried workers ... arrived ... only to find that their desks had been chopped to pieces with an ax." Sorensen (1956), one of Ford's favorite antidesign lieutenants, proudly related another such foray against a FM cost-accounting office: "One Sunday morning Ford and I went into the record room Hawkins [a soon-to-be-dismissed designer] had set up. We found drawer after drawer of cards and tickets. Mr. Ford took one drawer, held it bottom up, and its contents spilled on the floor. We did the same with all other cards until the entire record system was thoroughly fouled up" (p. 40).

So as GM and FM entered the competitive environment of the 1920s that resulted from stabilized auto demand, GM bolstered its planning staff and performance-control system, while Ford discarded his designers and disbanded the departments that could have formed the basis for an effective control system similar to that developed at GM. Sloan, the trained engineer and experienced business man, searched endlessly for reliable data on which to base his decisions; "Henry Ford's philosophy," on the other hand, "was 'We must go ahead without the facts; we will learn them as we go along.' It was his working principle while designing his cars" (Sorensen, 1956:38).

By the 1920s FM had become the giant of American industry. Still Ford felt that "organization in the company offices . . . must be kept in hand, lest the tail wag the dog" (Nevins and Hill, 1957:271). And "through the twenties, thirties and into the forties, Henry Ford, aging and autocratic, became increasingly resentful of the organization without which his company could not be run" (Galbraith, 1967:90). Ford's strong antiplanning attitude toward systems of performance control made FM a giant ship without a helm. The rudderless firm's performance soon reflected this lack of corporate direction.

13

Adopting Antiplanning

Henry Ford's antiplanning work at FM will be traced by considering his negative actions with respect to the seven design phases of the system-design-for-performance-control model. Examining some subsequent policy changes of the FM designers who corrected Henry Ford's many performance-control-design errors after his death yields additional insights. Although this latter developmental work occurred after the period of analysis of this book, it is presented here to help elucidate Henry Ford's blunders during the 1918–1938 era.

DEFINING THE GOAL

After Henry Ford gained complete authority over FM, he diluted the influence of FM executives who wanted to design a corporate headquarters for FM. Not only did the elder Ford remove executives with design ambitions, but he also eliminated the performance-control units already present in FM's minuscule central office. Individual dismissals and resignations eliminated selected managers, and massive purges removed whole departments.

Ford quickly dismissed or forced the resignation of notable executives like Norval Hawkins, C. Harold Wills, John R. Lee, and Frank L. Klingensmith. Norval Hawkins left FM only to join the newly developing GM headquarters group. In staffing GM's corporate offices, Sloan's "greatest find," according to

Chandler and Salsbury (1971), "was Norval H. Hawkins, Henry Ford's brilliant sales manager; Hawkins . . . became head of General Motors' new Sales Analysis and Development Section" (p. 497). Besides having further developed the strong distribution and dealer network begun by Couzens, Hawkins had introduced FM's first comprehensive accounting system. Among other things, Ford "did not like Hawkins' systemization of company forms" (Nevins and Hill, 1957:146).

C. Harold Wills, though not directly interested in the firm's central-office business functions, played a key role in designing FM's car offerings. Wills was FM's most important engineer; but since he was always in conflict with Ford, he had to go. Much of the conflict stemmed from Wills' constant efforts to improve the Model T in the face of Ford's desire to freeze the design and boost production. Since Ford wanted design fixed, he had little need for an engineer of Wills' caliber. Moreover, Ford was irritated that Wills was to receive FM dividends, as part of a special agreement. "Just as he was irked by minority stockholders who drew large returns on an original investment absurdly small, Ford was perhaps annoyed at the idea of paying Wills special sums in addition to a very high salary" (Nevins and Hill, 1957:146) for work he definitely did not want performed.

In contrast to Wills, John R. Lee left FM on his own accord with little pressure from Ford, who may have wanted him to stay. "Lee was in all probability aware of Ford's hardening temper, and Wills's departure undoubtedly influenced him. But it is significant that the popular executive who, with Dean [Samuel S.] Marquis, had been the symbol of liberal Ford welfare personnel policies, should leave the organization at this time" (p. 147). Besides conducting the firm's personnel operations, Lee had built on Hawkins' accounting system (Jardim, 1970:85) and maintained important external linkages for FM, having served as Ford's political agent in Washington during World War I (Sward, 1948:190).

Frank L. Klingensmith was probably the most important of the central-headquarters executives to depart shortly after Ford acquired all the FM stock. Before his rather turbulent dismissal, Klingensmith was FM's treasurer and vice president in charge of many central-office operations. According to Sward (1948), "It was this man's province, as Couzens' successor, to select and train the company's office personnel. He was responsible for most of [Henry Ford's son] Edsel's business training" (p. 192). Klingensmith was discharged because he questioned the way Ford survived the financial crisis of the 1920–1921 recession. "To Ford, accountants were nonproductive and liable to become 'experts.' Klingensmith was a 'good bookkeeper' who became too much of 'a banker's man'" (Jardim, 1970:227). Instead of agreeing with Ford's flooding his dealers with unsaleable cars, Klingensmith suggested more usual credit avenues through the banks. Ford thought that Klingensmith had aligned with the bankers, so he was dismissed (p. 209).

The executives who left FM at this time—and only a handful have been cited—could have formed the nucleus for a strong central-design group. Under Couzens, Hawkins had been in charge of sales, Lee responsible for personnel, and Klingensmith in charge of accounting. Coupled with Wills in engineering design, the Couzens group might very well have provided a system of performance-control for FM as the Sloan-Brown group did for GM. The talented executives who left FM, as Galbraith (1960) observed, "were not superannuated bureaucrats. Most of them were in their prime, and with rare exceptions they were grabbed up by General Motors, by Chrysler when it came along, or by one of the smaller rivals. Those who went to Chrysler and General Motors had the pleasure of helping their new employers end Ford's leadership of the industry" (p. 159).

Along with expelling individual decision makers and designers, Ford ruth-lessly pared FM's entire central-office staff. Employees in the FM offices and administration buildings could either go to the factory and become productive in Ford's eyes or leave the company. In short, Ford planned a thorough housecleaning of the company's headquarters, "determined to throw out 'everything that did not contribute to the production of cars'" (Nevins and Hill, 1957:156).

Two very important designers remained and tried in vain to give the elder Ford's slashings some semblance of planning and order: Edsel Ford and Ernest Kanzler, Edsel's able assistant. Although "an orderly reorganization such as Kanzler and Edsel could have conducted would have benefited the company immensely" (p. 167), Henry Ford's "handling of the office problem was ... both cruel and inept ... smashing and hacking, and not in accordance with a constructive plan. Here he completely missed the opportunity which General Motors seized and developed under Alfred P. Sloan, Jr." (p. 167).

However, the 1919–1921 dismissals and purges did not finish off the FM headquarters. Ample talent remained, "and for a time it seemed as if a coherent administrative plan might be evolved" (Nevins and Hill, 1957:269). Edsel and Kanzler saw an opportunity for building: "'Mr. Kanzler was trying to set up a corporation structure which would have a central administrative system simi-lar to any corporation,' said Herman L. Moekle, then in the Treasurer's Office. 'Particularly he wanted control to be in the office of the president'" (p. 269), the position occupied by Edsel Ford. Throughout the next five years Edsel and Kanzler worked energetically, and despite minor flaws, made what they man-aged well-integrated and smoothly running (p. 272). "While Henry Ford was issuing spectacular statements that kept him in the public eye, Edsel was shouldering the day-in-and-out responsibilities, making decisions on adminis-tration, manufacturing, taxation, and sales" (p. 272). Yet there was no question that the ultimate authority was Henry himself (p. 272).

Edsel's power was further diminished when Ford finally dismissed Kanzler, over Kanzler's attempt to force a change in the Model T. Kanzler, probably more than any other man at FM, had the system designer's touch. He "could lay out a complicated production schedule, appraise sales possibilities, or write a letter of clarity and literary distinction with equal facility, he seemed early destined for leadership" (Nevins and Hill, 1957:271).

Kanzler first exhibited his considerable organization-design talents when he was the Fordson tractor plant manager's assistant. There the production/shipment relationship intrigued him and, perceiving "that failure to synchronize the two built up 'inventory' " (p. 155), Kanzler established new routines for receiving, producing, and shipping. "So exact were his schedules that supplies arrived practically as needed, and freight cars bringing in wheels, radiators, castings, etc., were utilized a few hours after their arrival to dispatch completed tractors" (p. 155). Here then was a designer in the classic GM tradition of Sloan and Brown.

Even the elder Ford was impressed, and he shifted Kanzler to FM's Highland Park Plant to synchronize freight movements between this central production unit and FM's many outlying assembly plants. Kanzler soon inaugurated ten-day reports from dealers to help him regulate the production process. "The ten day reports . . . were important in maintaining a rapid flow *from* the factory as well as to and through it" (p. 265). Steady maintenance of the cycle would save millions in inventory holding costs. When Kanzler arrived at Highland Park, FM's 36 branch assembly plants "were operating amid considerable confusion, receiving their shipments irregularly and in jumbled form" (p. 266). Kanzler's subsequent reorganization improved service, increased production, "and cut inventory float some $40,000,000. This great saving was of crucial importance in cutting the total costs of manufacture and enabling the company to lower prices" (p. 266).

If Kanzler had restricted himself just to production improvement, it might not have been necessary for Ford to remove him. But Kanzler did not. Although the chief reason for this dismissal was Kanzler's desire to abandon the Model T in favor of an improved car, his "departure in 1926 was not unrelated to his efforts to systematize a firm that its head wanted to keep loosely organized" (Nevins and Hill, 1963:313).

Edsel Ford remained as the lone design-oriented executive of FM. While in basic agreement with his father's production policies, Edsel differed from him markedly in the field of administration. No matter the question, Edsel's work was "constructive, judicious" (Nevins and Hill, 1957:270). Before making a judgment, he examined "all available facts" and consulted "with those who had assembled them." Instinctively he drew upon others' abilities. "As a business man he could intelligently build an organization, encourage the originality of others, and work successfully as a part of the whole."

In contrast to his logical, deliberate, and open-minded son Edsel, "Henry Ford had a glacial indifference even to the most modest requirements of institutional organization and procedures" (Greenleaf, 1964:25). Henry Ford's "was the greater talent, but . . . he should have perceived . . . that the company, with its increasing complexity, was now at a stage where Edsel's mind and methods were more likely to serve it well" (Nevins and Hill, 1957: 271). Instead, the elder Ford pitted executives like Charles Sorensen and Harry Bennett against Edsel to diminish further his remaining influence.

So rather than become a modern corporate hierarchy, "the Ford Motor Company remained a dictatorship" (Hounshell, 1984:293). Henry Ford himself dictated broad company policy and specified the car designs. Charles Sorensen, until the "snowballing power of Harry Bennett" loosened his grip, "dictated all aspects of production." Bennett eventually became "a virtual dictator" within FM.

By World War II, after many years of catastrophic misrule under Henry Ford and his henchmen, FM began to falter badly. In 1941 Edsel Ford, "then profoundly discouraged" (Nevins and Hill, 1963:252), felt the firm's disastrous course could not be corrected. By 1943, Senator Truman's Special Committee Investigating the National Defense Program issued a harsh criticism of FM while it congratulated GM for its role in the war effort (Burlingame, 1957:157). In fact, FM's war-related "performance was so deficient that its seizure by the government was discussed," as was "the uniquely insulting proposal that it be managed by the Studebaker Company" (Galbraith, 1967:90–91).

Rather than seize the firm and be forced to manage it, government officials discharged Henry Ford's grandson, Henry Ford II, from military service in August 1943. "One reason why Secretary Knox released him from the Navy," according to Nevins and Hill (1963), "was that high government officials hoped that he might put an end to the growing chaos in management. . . . Another reason lay in the intercession of Ernest Kanzler and other anxious Detroit observers" (p. 254).

Given the endless intrigues and massive confusion within FM, however, few inside or outside observers expected much from the twenty-five-year-old heir to the Ford legacy. Yet Henry Ford II, by combining his grandfather's tenacity of purpose with his father's goodwill and judgment, managed with much help to revive FM's corporate brain.

The designers primarily responsible for the creation of FM's corporate intelligence after World War II were Henry Ford II, Ernest R. Breech, Lewis D. Crusoe, and the Thornton group—ten executives, including Charles ("Tex") Thornton, eventual head of Litton Industries, Robert S. McNamara, and Arjay Miller, later Dean of Stanford's Graduate School of Business Administration. Ernest Kanzler too played an important role even though he had left FM twenty years earlier. Kanzler had maintained his close friendship with Edsel as the brother-in-law of Mrs. Edsel Ford and had stayed in touch with the FM business operations as a director of the company financing FM's cars, Universal Credit Corporation (Nevins and Hill, 1963:313). After the war he became an important advisor to Henry Ford II. Though unable to return to Ford himself, Kanzler helped the young Ford in selecting his designers for the new FM.

Kanzler's most important find for the reorganizing company was Ernest R. Breech, then the president of Bendix Aviation, on whose board Kanzler served. Breech was well-suited to the design task, for not only had he acquired broad experience in accounting and finance, but he had considerable experience in GM's financial operations.

Breech, an extremely ambitious man, started his career by winning the gold medal on the Illinois CPA exam. Shortly thereafter he organized a branch of the Illinois Manufacturers Association, which pioneered "in the early development of comprehensive financial control systems that would in time be adopted as standard by all well-managed business firms" (Hickerson, 1968:52). In January 1923, Breech took the auditor's job with the Yellow Cab Company and soon became controller.

Breech joined GM when the General Motors Truck Division at Pontiac, Michigan, purchased the Yellow Cab Company (Hickerson, 1968:55). He served for a time as the controller of the consolidated operation, Yellow Truck and Coach (p. 57), and then as general assistant treasurer in GM's critically important New York headquarters, where he handled accounting and finances (p. 61). In this job he designed the GM Management Corporation Plan, the successor to GM's original Managers Securities Company motivational scheme.

Breech's later experience at GM broadened his perspective considerably, for he came to head GM's relatively autonomous aviation subsidiaries: North American and Bendix Aviation (Sloan, 1964:367–368). "Though particularly strong in finance, he had the general reputation of being a 'management trouble shooter.' It was even intimated that he was in the line of succession to head General Motors" (Nevins and Hill, 1963:313).

With a satisfying past and a promising future at GM, Breech obviously was reluctant to leave. Although he viewed FM employment with considerable foreboding, he was enticed by Henry Ford II to make an inspection trip of FM. Breech found the company's balance sheet to be "about as good as a small tool shop" and that the controller did not know what "standard volume" meant (p. 315). Seeing that FM was losing about $10 million a month, with virtually every aspect in need of attention, Breech declared, "I was the cleanup man for GM, but this one was *really* a mess." The only bright spot was that the unusual postwar demand for cars would allow FM "to make money even while it was getting its house in order."

So Henry Ford II managed to hire Breech as FM's number two man: executive vice president. To Breech, the main enticement was the challenge

(Hickerson, 1968:133). After a month at FM, Breech found the challenge adequate: "For the first time in my life I was overwhelmed. Not afraid, but badly disturbed. Our problems seemed almost insuperable. Things were in a mess. It would take years to get them under control" (p. 130).

Breech quickly recruited Lewis D. Crusoe, another GM executive. Crusoe started his GM career in the Fisher Body Division, where he rose to divisional controller and after 1930 acted as GM's assistant treasurer. With his "immense knowledge of labor, overhead, cost accounting, tooling, sales, and management in general . . . he was precision in the flesh and demanded accuracy of those who worked with him" (Nevins and Hill, 1963:317). Having worked for Breech at GM, Crusoe became Breech's executive assistant at FM. Shortly he was installed as head of FM's newly created Division of Planning and Control, with Thornton under him directing Planning, and in November 1946 he became company controller. In April 1947, Crusoe was vice president in charge of finance (p. 327).

The "Thornton group" over which Crusoe had charge was, after gaining experience in the automobile industry, to supply FM with six vice presidents and two presidents. Originally, the group grew out of the Air Forces' Office of Statistical Control. Thornton and his fellow officers "were sure that the techniques they had developed in the Air Forces would be valuable in business" (p. 308). After consulting Kanzler, who as a former official of the War Production Board knew of Thornton's work (p. 310), Henry Ford II agreed. Not surprisingly, to all the members of the Thornton group, FM "was a shocking revelation of inefficiency. It was 'completely different from what we ever imagined.' Financial controls 'in any terms in which we were thinking' simply did not exist" (p. 311). In response to an early question to an official in FM's Controller's Office about expected financial results six months hence, they got: " 'What do you want them to be?' It seems financial records were so kept that they could be juggled to produce whatever result was desired! [Similarly,] the organization of practically every part of the company was hopelessly confused" (p. 311).

To sum up, Henry Ford removed designers from FM. When designers were reintroduced by Henry Ford II, they found the firm in shambles.

Along with removing FM's designers, Henry Ford rather capriciously removed various client groups from the organization's concern. A group might receive much attention for a while, only to be rejected completely shortly afterwards. Ford's treatment of six potential client groups will be considered: (1) stockholders, (2) management, (3) workers, (4) dealers, (5) customers, and (6) Ford and his family.

Henry Ford viewed the people who had supplied the capital for the original Ford Motor Company as parasites who served no useful social function. Moreover, he came to equate "systems of management" as mechanisms to serve such absentee owners. He wrote in the 1930 *Saturday Evening Post:*

> Business has been regarded as a pump which can be profitably used to supply the needs of the owners. . . . If the pump could be made to operate automatically and thus relieve its owners from the labor of operating the pump handle, so much the better; it would be regarded as a great advance in business method. In reality it would be a great reverse. Systems of management are designed to relieve owners from running their own pumps. . . . They do not see that their pumps are necessary to social salvation. (Ford and Crowther, 1930:24)

Clearly, Ford was not interested in serving any stockholder who was not active in management. But neither did he seem interested in those stockholders who were in management: Couzens was forced out, and Edsel, the other major stockholder, was treated poorly.

Nor was Ford interested in serving his management. His best managers were forced out of the company, usually under the most humiliating circumstances, and the remaining less competent or more submissive lieutenants worked in an atmosphere of utter fear. In short, "everybody was on edge" (Nevins and Hill, 1957:296).

For a while it seemed that FM workers were one of FM's primary client groups, for John R. Lee and Dean Samuel S. Marquis, as has been mentioned, had fashioned exemplary personnel policies. To promote the workers' welfare, a Sociological Department was established in 1913. After the five-dollar minimum wage was adopted in 1914, this unit "increased immensely in importance, for it investigated every Ford worker and his family in relation to the wage, and organized a growing number of services for them" (Nevins and Hill, 1957:13). However, the FM experiment with improving industrial relations "proved short-lived" (Meyer, 1981:6). By 1921 the Sociological Department "had no economic reason for being, so waiving its other functions, Ford sheared it off, and with it his man of good will, Dean Marquis" (Sward, 1948:79). Although Ford had promised that "once he had cleansed his house of [stockholding] 'parasites' " (p. 80) he would share profits with his workers, that promise was empty. For FM's 1921-1922 clear net profit was $200 million, but "if he so much as recalled the idea of 'cutting melons' with his 'partner-workmen,' he kept it to himself" (p. 80). And with Lee's resignation and Marquis' dismissal, conditions for FM workers deteriorated rapidly. Ford's River Rouge Plant "was a machine-age nightmare" (Galbraith, 1960:154) by the mid-1920s. As one of FM's executives put it, "Ford was one of the worst shops for driving the men." Similarly, as Forbes (1927b) quoted Ford, "We make no attempt to coddle the people who work with us" (p. 16). By the 1930s FM was the worst place to work in the automobile industry, without a doubt. Long after GM had reached an accord with labor, Ford still refused to settle. It was not until 1941 that "the United Automobile Workers, CIO, finally organized Ford, the last of the auto makers to yield" (Galbraith, 1960:154). Even after that FM's union relationships were long to remain the worst in the industry.

Ford dealers were treated as poorly as stockholders, managers, and workers (Forbes, 1927c:17-19; Sprague, 1927:26-35). Ford considered FM dealers not partners—as Sloan did at GM—but parasites who contributed nothing to the company's success yet earned substantial profits because of the Model T's early popularity. So Henry Ford "consistently refused to cultivate them, to cater to their wishes, to ask their advice. He considered them in the same category as his factory organization. He retained absolute control, his word was law and he, and he alone, could decide what to do, when and how to do it" ("Ford Dealers Rebel," *Business Week*, April 2, 1930:9).

Thus FM dealers were spied on (as were FM executives and workers), forced to buy expensive FM tools, left for years without the aid of a FM national advertising campaign, and required to exist on the lowest profit margins in the industry. "Unbridled factory forcing," however, presented the most difficult problem for the dealers. "In good times and bad, and all too often in complete disregard of local needs and conditions," FM imposed excessive quotas on automobile and parts sales. "For countless numbers of his men, the rule of factory forcing spelled ruin. Even his more successful representatives groaned under the load" (Sward, 1948:208).

Along with FM's unbridled forcing went the extended shutdowns of 1927-1928 and 1932-1933. Since Ford failed to adjust to consumer demands, FM's cars became hopelessly out-of-date. FM production facilities, moreover, were not geared to rapid alterations. Thus when Ford finally revised his models, extensive design changes were necessary and production had to be halted completely for many months. During these shutdowns, dealers were left without automobiles to sell (just as workers were left without jobs). Many were in dire straits before FM resumed production (Lewis, 1976:200).

Seesawing between factory forcing, on one hand, and delivery starvation, on the other, as well as being hit by many other abuses, Ford dealers took a terrible beating. With the onset of the depression in 1929, conditions deteriorated still further. While "General Motors did the most to protect its dealers," Ford "cut the markup or commission given the dealers from 20 to 17.5 per cent of the list price. Then he increased the number of dealers so that there would be a Ford dealer at all 'cross-roads'" (Macaulay, 1966:14). FM's "crossroads policy," in fact, had been tried with disastrous results several times previously when sales lagged or Ford just suspected his dealers of indolence (Bennett, 1951:43-44). During the depression, the practice was particularly damaging to the established, well-capitalized franchise since it put a new, low-overhead dealer on its doorstep. GM instead offered many of its best dealers "territorial protection" via an "infringement charge ($25 and up) paid by G.M. to one dealer and levied against the other" ("G.M. III," *Fortune*, 1939:105). By the end of the 1930s, then, Ford had succeeded in decimating the ranks of the powerful dealership force that had been assembled earlier under Couzens' and Hawkins' guidance.

Henry Ford often talked of serving the customer, and for an extended period of time he made significant contributions to the consumers of his cars. He greatly improved and changed their life styles. Yet after their life styles shifted, he refused to provide them with more suitable transportation. He wanted to build forever a car, the Model T, that he more than any other man had made obsolete. The improved roads resulting from the automobile revolution fostered by the Model T meant that ruggedness and simplicity could be traded (literally) for smoothness and style. Yet Ford refused to adapt at all until demand had dwindled to almost nothing. Customer be damned.

Ford resisted not only superficial styling changes and unnecessary accessory additions that benefited consumers little, but also safety features. Just as Sloan opposed GM's introduction of high-cost safety glass (and only used it after FM introduced Triplex glass on the Model A), Ford retarded the introduction of important safety advancements. His fight against quick and positive hydraulic brakes bordered on mania. "For a time the mere mention of hydraulic brakes in his presence was highly dangerous" (Nevins and Hill, 1957:253). Still earlier, a collapsed front suspension member on a Model T resulted in Ford himself being thrown into a ditch and shaken; Joe Galamb, an FM experimental engineer who saw the accident, worked feverishly over the next months to develop an improved front suspension. Rugged tests showed he succeeded admirably, yet, as Galamb reported: "All Mr. Ford would let me do was increase the wall of the tubing . . . so that . . . people wouldn't see that there was any change. We sold twelve million cars after that with the old radius rod" (Stern, 1955:139).

When Churchman's (1971) terms are applied, it is doubtful that FM under Henry Ford would qualify as a system. It was an organization without a *permanent* client. Workers might be served for a while, only to be discarded ruthlessly; a similar fate awaited FM customers, and so on. Perhaps the only client throughout the period of 1918 to 1938 might be Henry Ford himself. Yet even here it is extremely difficult to discern the elder Ford's value system.

Ford, himself, certainly benefited handsomely from FM's operations. While averse to dispersing profits to his former stockholders, "once in possession of the entire holdings of the company, the Ford family—Henry Ford and his wife and son—were to withdraw cash dividends for their private use for the next seventeen years at the rate of more than $25,000 a day" (Sward, 1948:70). In 1925 alone, according to the *New York Times,* the Fords' personal dividends, apart from salaries, exceeded $14 million (p. 201). Beyond Ford's overt capitalistic behavior, he also expressed strong beliefs about profits as a measure of a business firm's service; he claimed: "An industrial organization has to be formed for a purpose, and the quality of its performance does not have to be guessed at or argued about. The balance sheet gives the results. No corporation can continue over a period of years to have income exceed outgo otherwise than by performing a service" (Ford and Crowther, 1930:25).

But if Ford and—perhaps—his family were the firm's clients, it is still impossible to determine any systematic design on Ford's part to insure this group was well served. Indeed, over the next two decades, there were relatively few years where income exceeded outgo.

Ford's failure to pick a client for his firm to serve can be attributed to his own variability. "In Henry Ford, except for a few beliefs like that in the cheap car, there was a singular lack of consistency and dependability . . . no certainties, only an 'unlimited uncertainty'" (Nevins and Hill, 1954:581). Thus, Ford observers saw in him discordant elements that were "never properly fused, because his mental processes lacked training, and his character wanted discipline" (p. 587). So side-by-side in Ford were the "untutored farm boy, the engineer, the industrial manager, the social planner, the entrepreneur, and the hard-driving plant executive; the apostle of rural virtues and the prophet of mass production; the isolationist and the internationalist; the plodder and the seer" (p. 582).

Only after Henry Ford's death was a definite FM client group identified and a performance-control system established to serve it. The "former autocratic control of the company [was] now replaced by trained executives under shareholders' control" (Nevins and Hill, 1963:xv).

FORMULATING THE STRATEGY

In formulating FM's strategy Ford virtually ignored product diversification and obsessively concentrated on channel integration. Besides resulting in a poor match with the consumers' changing tastes, Ford's approach extended FM into many areas that overtaxed the organization's decision-making power. In short, FM's high internal variety helped little in overcoming environmental fluctuations because it was not matched to the environment.

Channel Integration

There can be little doubt that Henry Ford was preoccupied with the idea of channel integration. After an extensive study of business consolidations and mergers in both America and Europe, Farnham (1929) concluded: "At present the Ford Motor Company probably represents one of the most complete examples of the vertical [channel-integrated] combination in this country" (p. 11). The fabulous profits earned by the Model T were poured into the acquisition of many FM subsidiaries in the early 1920s. For example, in December of 1922,

> the Michigan Land & Lumber Company, a subsidiary, purchased 30,000 acres of timber property, including sawmill and logging railroad; in February, 1923, the Company acquired the Allegheny Plate Glass Company . . . ; in the same month, the Fordson Coal Company was incorporated . . . to take over coal properties in West Virginia and Kentucky; in June, the Company obtained permission from the Federal Power Commission to develop electric power . . . for a new . . . manufacturing unit; in September, 400,000 acres were added to the holdings of its timber-owning subsidiary; in October, the factories of the Johansson Gauge Company . . . were purchased; etc. (Seltzer, 1928:119–120)

Between 1924 and 1926, large additions were made to the River Rouge Plant's blast furnaces, rolling mills, and other units; new assembling factories were opened throughout the country and in Denmark, Germany, and Japan as well; in addition, "tire, battery, and textile manufacture were introduced into the Company's Highland Park Plant" (p. 120). As if that diversity was not enough for FM and Henry Ford, he also acquired iron mines, coke ovens, artificial-leather works, motion-picture laboratories, 1000 Ford-owned coal cars, paint and varnish plants, a wood-distillation plant, a cement plant, a paper mill, the Ford fleet in lake and ocean service, soybean farms, and a commercial air fleet (Ford Motor Company, 1924). Moreover, he "entered the banking, newspaper and grocery field," and was "reported as entering the rubber and textile field" (Farnham, 1924:261).

Although the listing provided of FM's holdings is by no means exhaustive, these holdings were undoubtedly exhausting to manage properly. By opening large, cut-rate grocery stores at FM's Highland Park Plant, River Rouge facility, and Lincoln works, Ford incited a boycott of his products in the Detroit area (Forbes, 1927:9). Even so, FM's rubber plantations presented some of the most extreme difficulties (Nevins and Hill, 1957:237).

When the technology involved in an enterprise was more closely related to that used in the automobile, FM fared better. Having long had its own cast iron foundries, for instance, steel making was not a totally novel addition. "Considering their limited background for steel work, and the relatively few expert employees they imported from outside, the performance of the Ford officials was impressive" (p. 292). FM, in fact, did much to introduce high-strength vanadium steels to the American steel industry. One executive even claimed that FM saved money by making its own steel (p. 292). It is doubtful, however, that this executive considered either the diversion of scarce managerial talent and capital resources from more profitable opportunities or the lack of any measurement system upon which to gauge cost comparisons.

Ford's extensive capital investments also reduced his firm's strategic flexibility. In the mid-1920s, for instance, less wood was being used in automobiles, and FM was stuck holding excess lumber capacity. More seriously, FM's extensive capital investments in raw-material supplies altered its cost structure. "By 1926, nearly 33 cents in such assets backed each dollar of sales, up from 20 cents just four years earlier, thereby increasing fixed costs and raising the break-even point" (Abernathy and Wayne, 1974:113).

Besides reducing FM's ability to absorb a loss in volume and still maintain its profits, the integrated channels failed at crucial times, as in 1927 when the Model A finally replaced the Model T. "The newly introduced laminated safety glass was purchased for the Model A," for example, "as were upholstery fabrics, even though Ford had been extensively integrated into materials for both components in manufacturing the prior model" (Abernathy, 1978:142). Only after considerable time passed could FM match the prevailing industry practices for producing these raw materials and reestablish "the previous degree of backward integration."

While Ford extended himself needlessly into the production of basic commodities, he neglected the much more important parts and accessories businesses (Chandler, 1966:462). "Ford, with much the same resources as General Motors, could have as easily developed these businesses" (p. 462), but, in this area where FM had considerable management expertise, Ford failed to utilize it. Such an oversight was very dangerous since FM production could be halted through a wholly uncontrollable mishap with an outside supplier. Moreover, without a multiple-source rule in effect the probability of a production stoppage was high. Ford himself recounted one such incident when the plant of the Diamond Manufacturing Company burned down: "They were making radiator parts for us and the brass parts—tubings and castings. . . . We had enough stock on hand to carry us over, say, for seven or eight days, but that fire prevented us shipping cars for ten or fifteen days. Except for our having stock ahead it would have held us up for twenty days—and our expenses would have gone right on" (Ford, 1922b:167–168).

FM even purchased from outside sources many of its finished car bodies—an integral component in the automobile's manufacture ("Mr. Ford Doesn't Care," *Fortune*, 1933:128). So whereas GM, to improve control and coordination, had internalized its body-building operations by acquiring Fisher Body, Ford often allowed this vital function to remain under external direction. As with raw materials, this reliance on outside suppliers typically occurred when FM's own units could not handle the technological innovations demanded in the market. A "more complex" Model A body, for instance, could not be manufactured in sufficient volume; hence more than half of the new model's requirements had to be "purchased from the Briggs Manufacturing Company" (Abernathy, 1978:142). Five years later (1932) FM was still making sizable purchases from Briggs as well as from Murray Body and Budd Manufacturing ("Mr. Ford Doesn't Care," *Fortune*, 1933:128). Ford thus exposed his firm to the possibility of coordination delays since the design functions were spread among several administrative structures.

While failing to integrate FM's body operations completely, Ford carried his questionable channel-integration concept so far that he was buying up junked cars and running them through an automobile disassembly line. The salvaged materials, in turn, were recycled (see "Ford's 'Demolition Line,'" *Business Week*, July 2, 1930:13).

So Ford channel-integrated before the consumers, through the extensive ownership of basic industries, and after them, with the salvage of their used cars. Yet he delayed closing his production-consumption system: FM's entry into consumer and dealer financing was quite belated. As he was strongly opposed to credit purchasing, the most he would allow along these lines was the creation of the Ford Weekly Purchasing Plan in 1923 (Nevins and Hill, 1957:267-268). The results of this program were disappointing, for it was nothing more than a savings plan. "Had Ford liberalized the plan" and "delivered the car when a considerable part of the total had been paid, he might have tapped the lower stratum of the population where price was still more important than style or comfort. But this would have amounted to a form of time-payment, which he was not yet ready to accept" (p. 269). It was not until 1928 that Ford finally relented and created his own finance company, the Universal Credit Corporation (UCC), run by the able Kanzler. Ford's response was a full ten years after GM created GMAC. (As discussed in Chapter 5, GMAC was formed to lower the initial out-of-pocket costs of GM cars, which were then experiencing extreme difficulty in competing with FM's cheap Model T.) Never happy with the UCC investment, Ford decided to shut it down in 1933 even though it was in sound financial condition (Nevins and Hill, 1963:62). In response, Kanzler proposed that UCC be sold rather than closed completely. Ford accepted this proposal and so ended FM's short history of providing a credit source for its products. "It was a triumph of Ford's prejudice over his business interest" (p. 62). Thus, while Ford extensively integrated his firm, he kept the vital activity of credit provisioning outside its controlled sphere. He refused then to admit that the mass production and mass distribution of consumer durables—no matter how cheap—demanded mass finance to clear the market pipeline.

What was Ford's rationale for his extensive program of production-oriented channel integration? Most likely he wanted to control all the variables involved in the manufacture of the Model T—to make FM a closed or self-contained organization. While he disclaimed this motive, his disclaimer made little sense:

"We have been gradually going back to the sources of raw material. We have our own mines and timber tracts and the like, not because we have any particular desire to be self-contained, but because our manufacturing program is not safe as long as we are threatened with market fluctuations in the price of raw material" (Ford, 1922a:262).

And Ford was aware of the tremendous amount of (useless) variety to which he was subjecting himself and his decision makers: "If the concerns which supply us were organized on the same fundamental basis as we are, then we should not have to bother controlling our own sources. We would be better off not controlling them because a company making iron and steel as a sole business ought to develop more economies than a company making them as a side issue" (p. 262). Nevertheless, Ford plunged in anyway and overloaded FM's decision capacity, for "no firm or management team has the expertise to do *everything*. Any wholly self-contained empire would surely collapse of its own weight (as Ford's nearly did)" (White, 1971:80).

Many years later, "under the presidency of Henry Ford II," FM was "getting out from under the expensive sideline luxuries that [had] been a drain on the primary business of selling cars at a profit" ("Ford Divests," *Business Week,* March 16, 1946:28). The *Business Week* editors concluded that "Ford has always earned a far lower net return . . . than any other important automobile company. [One] reason generally ascribed was . . . that it engaged in too many sidelines" (p. 28). For instance, "the Brazilian rubber plantation, sold late in 1945 to the government of Brazil for $250,000, had represented a constant drain, losing more than $20,000,000 from 1927 to 1945" (Nevins and Hill, 1963:323). Other FM divestitures included soybean processing factories (for oils), farm and mineral lands, and iron-ore freighters.

Ford compounded his excessive channel integration by providing no means for measuring the benefits contributed and costs incurred by the various units within his integrated channel. Consequently, it was impossible to isolate inefficient components. "Imagine a company that boasted of a higher order of integration than any similar organization in the country, with ownership of a steel mill, glass plant, timber stands, maritime fleet, etc., and no accurate knowledge of which individual operation was paying its way, or which was padding the cost of the company's end product by open-market standards" ("The Rebirth of Ford," *Fortune,* 1947:84).

FM was unable to evaluate the cost/benefit contributions of its various subsidiaries simply because it aggregated all operations into a massive Gordian knot. Ford, who "disliked and distrusted accountants all his life" (Bennett, 1951:39), always "wondered why his accounting department did not consist of one man with a large burlap bag. Incoming money, Ford felt, could go into the bag and bills and wages could be paid out of it. What was left could be considered profit" (Ford Motor Company, 1953:21).

Not surprisingly, then, materials were just transferred between the channel-integrated units without reference to price or costs. Along with the lack of transfer prices went an admittedly inexpensive aggregated accounting system: the old merchandising type. "You took the purchases" at the input interfaces, "took the sales" at the output interfaces, "and the difference between the two was profit" (Nevins and Hill, 1963:328). So here was "a company with assets of $900 million, and an annual gross volume in the neighborhood of $1.5 billion, operating on a set of books that would put a country storekeeper to shame—no budget, no controls, just a how-much-comes-in, how-much-goes-out kind of calculation that would do for tax purposes" ("The Rebirth of Ford," *Fortune*, 1947:84). Thus all particular information about any unit was lost, and FM decision makers "if seeking responsibility for loss could see only the total result of a melange of activities. . . . The share of each [component] in an overall loss was impossible to determine" (Nevins and Hill, 1963:326).

FM's avoidance of segregated performance-measurement schemes as well as its neglect of market-referenced—or even negotiated—transfer prices can be traced to Henry Ford's dislike for accounting and finance executives (like Couzens and Klingensmith) and the extreme cuts he made in their ranks. Along these lines, Jardim (1970) quotes Charles Martindale, the company's overworked auditor after 1920: "The accounting department eternally found themselves coming up short . . . when we took physical inventory, running into millions of dollars. . . . You wouldn't believe some of the techniques I used. The one that the accountants got the biggest laugh out of was the method of pricing material requisitions" (pp. 204–205). Martindale was so short-handed that eventually he just weighed the requisitions after having arrived at an average price for a given weight—certainly a brilliant adaptation to pressure, but hardly sufficient to control FM's performance.

Thus, the new Ford management after World War II had to rebuild the FM accounting staff. Crusoe built a strong group mostly around recent college graduates, as the experienced FM personnel simply did not have the appropriate training and skills. Eventually, the new central Controller's Office set "basic financial and accounting practices and [issued] statements of uniform accounting to be observed by all divisions. Obviously, this is essential if uniformity and consistency are to prevail" (Rickard, 1950: 573).

The accounting procedures of the 1920s, 1930s, and early 1940s had to be deaggregated under Henry Ford II. In 1950, E. B. Rickard, the controller of FM's Ford Division, mentioned that the accounting data "provided no key as to the ability of any specific component" to generate a profit satisfactory to the capital invested. With a single profit figure, "it was not possible to determine whether the return on investment made in the steel mill, forge shop, or motor plant was sufficient to warrant its continued operation." Although special studies could shed light on specific operations, "the accuracy of such statistical excursions into the accounting records" could not be assured. As a result, "only a few top executives could be held responsible for profit performance" (p. 569). To correct this situation, the new management segregated components with clear organizational and operating responsibilities. Then Crusoe and his new financial staff installed "individual accounting systems for individual operations." Once the steel mill had its own books, for example, its inefficiencies were no longer "hung on the cost of the car" and steps were taken to drive its operating costs down to levels "in line with open-market quotations" ("The Rebirth of Ford," *Fortune,* 1947:88, 204).

Along with the new segregated accounting system went the use of transfer prices among the channel-integrated units. Now, for example, the Parts and Equipment Manufacturing Division produced to compete with outside vendors and "market prices . . . provided a ready yardstick for measuring the effectiveness of its operations" (Rickard, 1950:571). Thus the profits and losses shown in its financial statement reliably indicated the division's contribution to FM profits. So the use of competitive transfer prices became "the essential core of Ford's decentralized [divisionalized] system," putting "pressure on each division to operate . . . as if it were in business for itself" (Stryker, 1952:158). Breech added that revenue control of intracompany sales "gives invaluable guidance in make-or-buy decisions[s], provides a check on supplier prices, and is a useful test of performance" (Hickerson, 1968:142). In Breech's judgment, requiring FM's parts divisions to sell competitively to the end-product divisions had "contributed enormously to the improved profit position" of FM.

Where external markets did not exist, negotiated transfer prices were used. For example, the Rouge Division made items such as the V-8 engine and body quarter panels for which there were no directly competitive prices. So FM had "the manufacturing divisions and the end product divisions negotiate prices that are mutually satisfactory in the expectation that the selfish interests of each divisional management will produce realistic transfer prices" (Rickard, 1950:571). Since these prices were not "determined on a scientific basis," Rickard noted that FM was "not yet satisfied as to the significance of the resulting profit figures reported by these respective divisions."

Crusoe's new segregated accounting system, market-referenced transfer prices, and expanded accounting staff showed many components in FM's integrated operation to be notoriously inefficient compared with external sources. Consequently, many FM properties were sold as the reorganized company came to concentrate more on automobiles. Since most of these activities "were clearly not vital to auto manufacturing . . . the decision to sell them was an easy one for Henry Ford II and Ernest Breech to make in 1946" (White, 1971:83).

Product Diversification

Ford's product-diversification attempts were as puritanically sparse as his channel-integration efforts were expansive. Standardization throughout the company was epitomized by the "any color so long as it's black" policy. Prior to that decision Ford had standardized the chassis under the various Model T bodies.

Ford even kept the Model T chassis essentially unchanged from year to year. When Ford permitted a few needed changes, "he persisted in his policy that every improvement in the Model T 'should be interchangeable with the old model, so that a car should never get out-of-date'" (Beasley, 1947:94). Thus virtually no significant design improvements were made between 1908 and 1927; "the fifteen-millionth Model T Ford chassis, produced in 1927, was identically the same type of chassis as that which rolled out of the Piquette Street plant in 1908 when this model was first announced" (Epstein, 1928:283). Design variables became *unchanging parameters* and thereby useless to FM for countering environmental variety and controlling performance.

Similarly, FM's production facilities had not been designed with environmental adaptation in mind. With the Model T chassis design frozen, plants were constructed solely for its volume production. In particular, FM's giant River Rouge Plant was a brilliant technological achievement of integrated production (producing nearly all the Model T's parts) but far less impressive as a business venture due to its extreme inflexibility. Moreover, most machinery within FM plants was "based upon the idea of one-purpose machines" (Faurote, 1928:769) and could not be adjusted to new models. Instead they had to be scrapped and replaced or, at best, extensively rebuilt.

In contrast, Knudsen, Ford's former production expert who began running GM's competing Chevrolet Division in the early 1920s, advocated universal—high variety—machine tools that could be adjusted rapidly to design changes. In fact, on assuming the Chevy command, Knudsen gradually discarded all machines, installing in their stead "new heavy type standard machines (not single purpose)" (Knudsen, 1927:66).

While Ford apparently was unconcerned with adjusting FM's design and production variables to match the environment as time passed, he did attempt to counter the environment's new variety with one variable: price. Ford invariably set that at the lowest possible figure.

Over time, however, Ford's growing list of unprofitable investments and expensive hobbies absorbed a considerable cash flow. And since there were limits to how much his managers could squeeze from FM's suppliers, workers, and dealers, even Henry Ford had to observe some restraint in pricing.

In the market place, furthermore, low price exerted less influence on sales performance. First, other manufacturers quickly adopted Ford's original low-cost production techniques. And as these competitors, especially GM's Chevrolet, captured some of FM's volume, they could afford to compete in the price arena, for increasing volume allowed fixed costs to be spread over more and more units. Thus firms like GM were able to maintain profits while lowering their prices; when the better equipment found on a Chevrolet was taken into consideration, GM's car became very competitive with the Model T. Conversely, Ford's "precious volume, which was the foundation of his position, was fast disappearing. He could not continue losing sales and maintain his profits" (Sloan, 1964:162). So for the first time Ford found himself facing significant competition in the price class where he long held exclusive sway. The once impenetrable price barriers to entry that FM had erected around the low-price field started to crumble. Chevrolet invaded Ford's market space from the top.

Ford's second problem with his low-price strategy was that the lowest segment of this market niche began to shrink during the prosperous 1920s. For FM especially, the competitive advantage of its low prices was eroded "by the great development of the installment-payment practice. The initial cash payment required for the purchase of other cars became relatively little more than that required for the Ford [which] further redounded to the advantage of other producers" (Seltzer, 1928:122–123). Accordingly, "from 1919 to 1926, the percentage of all GM cars sold on credit rose from 32% to 56%. It was clear that installment buying was enabling the mass of consumers to upgrade their car-buying habits, to indulge their taste for more expensive cars than they could afford for cash on the barrelhead" ("Selling to an Age of Plenty," *Business Week*, May 5, 1956:126). Ford, as was noted previously, accentuated FM's financing-related difficulties by leaving his dealers to cope with their own and their customers' financing needs as small independent businessmen.

The used car presented a third problem for Ford's marketing strategy. As Sloan (1964) analyzed his competitor's predicament: "Mr. Ford failed to realize that it was not necessary [to have] new cars to meet the need for basic transportation. On this basis alone Mr. Ford's concept of the American market did not adequately fit the realities after 1923" (p. 163). The growing used-car market had begun to provide reliable substitutes for the Model T that looked better to the consumer yet were priced about the same. "Instead of buying [new] Fords, many persons with $350 to $600 to spend . . . purchased . . . other types of used cars which were available in large numbers, and in serviceable condition, at these low prices" (Epstein, 1928:283). Hence, "a man in the market could pick up a usable car for less than a Model T—and his used car would have a gear shift, shock absorbers, and other refinements he liked" ("Selling to an Age of Plenty," *Business Week*, May 5, 1956:124).

Probably the most important factors eroding the power of FM's pricing strategy, then, were the rapid changes occurring in the technological and marketing environment. By the mid-1920s the domestic automobile market was

quickly approaching maturity. It had reached the point where the new sales—
that the Model T had long thrived on—became more and more difficult and
expensive to make (Griffin, 1926:409; Shidle, 1927a:145). "And in a market
dominated by . . . replacement demand, a standardized vehicle, however attractive,
which is produced in enormous quantities, must encounter great marketing
difficulties" (Seltzer, 1928:123). Most replacement buyers wanted more than
just a new car; they desired something different and novel—just as their first
Model T had been. The extremely dynamic technological development of the
early automobile industry made Ford's rigid standardization particularly damaging.
Automobile design underwent many improvements during this period: smooth
high-compression engines of six and eight cylinders, electric self-starters,
battery ignitions and lighting, water pumps, three-speed gearshift transmissions,
longitudinal (conventional) springs, shock absorbers, four-wheel hydraulic
brakes, balloon tires, demountable rims, numerous gauges and instruments,
colored paints, closed bodies, and streamlined styling were all being made
available to the customer. Moreover, the basically expansive economic conditions of the period meant that consumers could afford such added luxuries.
"Middle-income buyers, assisted by the trade-in and installment financing,
created the demand, not for basic transportation, but for progress in new cars,
for comfort, convenience, power, and style" (Sloan, 1964:163).

Ironically, the demand for improved automobiles stemmed from Ford's own
Model T, for its widespread use helped launch a monumental highway-building
program. "The total of concrete roads . . . increased from 7,000 miles at the end
of 1918 to 50,000 in 1927. . . . Within a decade after the war the United States
was transformed from a land of poor roads into the nation with the world's
finest system of highways" (Faulkner, 1950:109–110). But each additional mile
of improved highway diminished the Model T's demand. The characteristics—
light weight, short wheelbase, high ground clearance, and solid front axle—
that made the Model T perfectly adapted to the country's deeply rutted,
mud-filled byways gave it a choppy, swaying ride on the new high-speed
thoroughfares. Henry Ford's aversion to shock absorbers and longitudinal
springs, of course, only served to accentuate the car's ride and handling
problems. And more and more, GM's Harley Earl had the country's pulse; his
longer, lower, and heavier (closed-body) cars not only looked better, but they
rode better on the improved highways. Moreover, with an additional 10,000
miles of concrete highways being laid each year in the late 1920s and early
1930s, GM's researchers worked hard to refine shock absorbers and to improve
suspension components. That Americans might develop a taste for comfort as
well as style, however, seems to have been overlooked by Ford.

As customers came to demand the new design innovations, the Model T,
though cheap, looked hopelessly outmoded: a utility vehicle only for the most
rural of farm communities. Thus, "potential buyers of new Fords decided to

spend an additional $150 or $200 and purchase the products of Chevrolet, Essex, Star, and other companies instead" (Epstein, 1928:283). Although FM preserved its position from 1920 to 1924 by its price cuts "the Chevrolet, by style and engineering changes, could *raise* its price by $15 and still increase its sales by a percentage greater than the Ford's" (Nevins and Hill, 1957:265). And, as was reported at the time of the 1926 crisis: "The effects of the present cut on the Ford sales curve cannot be determined accurately for several months, but there is nothing to indicate that price reductions alone will . . . stem the tide . . . running against the industry's largest manufacturer. . . . Price is less effective in stimulating sales today than it ever has been" (Shidle, 1926:10).

Eventually, then, even the great FM and Henry Ford had to capitulate to the shifting environment. While Ford kept the Model T chassis essentially unchanged, he approved a few internal improvements, some new accessories, and limited styling advances. In 1923, for instance, FM offered a "line of lowered and 'streamlined' bodies" (Hounshell, 1984:274). In 1925, Ford dropped the outmoded wood-framed body for the metal body (p. 274). And in a final attempt "to reverse continued and severe slippage in the market," Ford authorized additional "style and equipment changes in 1926" (p. 275). The styling breakthrough was "the reintroduction of color choices after twelve years of black" (p. 275). "These small innovations, tardily incorporating only a few of the refinements previously introduced by most other manufacturers, were made only with great difficulty, expense, and delay; and they did not suffice to regain for the Company its lost momentum" (Seltzer, 1928:120–121).

FM's attempts to update the Model T, in turn, "halted the steady reduction of costs" and "created a severe margin squeeze" (Abernathy and Wayne, 1974:114). Prices could not be cut further and still cover costs. So ended Henry Ford's "single variable strategy" (Thomas, 1973:122). With the Model T's demise approaching, Ford too would soon have to abandon the lowest price class forever (Shidle, 1931:940).

One noteworthy competitor reviewed the situation as follows: "Ford's sales held about even in 1925. . . . But since the market . . . rose substantially over 1924, Ford's share declined relatively from 54 to 45 per cent, a sign of danger, if Mr. Ford had chosen to read it" (Sloan, 1964:154). He did not. "The old master had failed to master change" and his Model T "no longer offered the best buy, even as raw, basic transportation" (p. 162).

As Henry Ford's cherished Model T became increasingly out-of-phase with current public taste, it actually became "embarrassing" to be seen in one. "In a characteristic Ford joke of the period, one person supposedly asked another, 'What shock absorbers do you use on your Ford?' His friend replied, 'The passengers'" (Sward, 1948:196). The Ford product truly had become an inferior good. "The end came for the 'Model T' in 1927. It was a pitiful situation. Henry Ford blamed . . . the Ford dealers, saying they had grown rich and lazy.

He castigated General Motors for its 'new model every year' policy" (Crabb, 1969:398). But even Ford himself could not help but see the problem, if not the cause, for "in the early weeks of 1927, 'Model-T' production was reduced but, by May, the yards surrounding the branch assembly plants, the grounds of the Highland Park factory and at the River Rouge were filled with unsold Ford cars" (p. 400).

Two years previously, Ford had unknowingly described his own demise from the industry's leadership position when he wrote:

> Business men go down with their businesses because they like the old way so well they cannot bring themselves to change. One sees them all about—men who do not know that yesterday is past and who woke up this morning with their last year's ideas. It could almost be written down as a formula that when a man begins to think that he has at last found his method he had better begin a most searching examination of himself to see whether some part of his brain has not gone to sleep. (Ford, 1922b:43)

On finally awakening himself and seeing FM's storage yards glutted with outmoded Model T's, Ford reluctantly accepted the need to make major product improvements. The first was the Model A, "a monument to the tenacity and deviousness by which Ford subordinates managed to get his agreement to a selective slide transmission, hydraulic (or even adequate) brakes, and better tires" (Galbraith, 1960:161).

Given that small design changes involved great expense and considerable delay, "staggering" changes such as the Model A entailed horrendous losses and required extended shutdowns. "Nearly every piece of the company's monolithic equipment, laid out on the assumption that the Model T would linger on forever, had to be torn down and rebuilt" (Sward, 1948:199). "Some 15,000 machine tools" had to be replaced, "another 25,000" rebuilt, and "$5,000,000 worth of dies, and fixtures" redesigned and rearranged. Ford only made matters worse when he insisted on minimizing the space between the new machinery; moving one machine often meant moving four or five others to make room, and, worse yet, as Model A production rose, whole "departments had to be rearranged with each incremental increase" (Hounshell, 1984:287).

Ford's early insistence on the extensive use of forgings and castings in the Model A, rather than lighter and cheaper stampings, wasted more time. Cost considerations, however, soon ended this "costly, short-lived exercise" (p. 287). FM encountered additional delays when it found that machinery was not available to mass produce the Model A gas tank and rear axle designs chosen by Henry Ford himself (pp. 284-286). The Model A body design caused even more delays as it "required deep or heavy drawing which was generally unproven at Ford" (p. 285); indeed, the simple curves of the Model T body could have been fabricated in the most primitive of sheet-metal shops. Moreover, because of poor planning, numerous mechanical drawings remained dimensionless and thus worthless (p. 286). Other parts designs went directly from the drawing boards to the tool rooms without the benefit of any testing whatsoever; not surprisingly, "this procedure often failed" (p. 286).

Rather than learn the lesson of the annual model change from the expensive 1927–1928 drubbing, Ford repeated his Model T adaptive errors with the Model A—just as Sloan (1964:167) thought he would. He ignored the widely held belief "that flexibility of mind and physical production facilities" would determine "Ford's future" (Shidle, 1927b:256). So "when, after much trouble and uncertainty, Model A was finally in production, Ford resisted improvements in that" (Galbraith, 1960:161). Thus the Model A "remained substantially unchanged for the next five years" (Sward, 1948:206).

Although the Model A was immensely popular at first, in 1929 Knudsen began to close the gap again. He had converted the Chevrolet into a "six" in a brilliantly planned changeover during which the production was hardly even halted (Beasley, 1947:139–140). "Just as quickly, General Motors started advertising the Knudsen product as a 'six for the price of a four.' Because of its better appearance and smoother engine performance, Chevrolet began to eat into the business of the Ford Motor Co." (Sward, 1948:206), just as it had prior to the Model A's introduction. So "on December 29, 1929, Edsel Ford announced that the Model A had been 're-designed'" (Hounshell, 1984:295). In reality, "the body had been lowered some and lengthened by six inches," the radiator and fenders changed slightly, and several new colors offered. Again it was too little and too late.

Ford, however, persisted with the Model A until 1932. Then, lagging far behind in styling and engineering, FM was compelled to modernize "for the second time within five years" (Sward, 1948:206). Another monumental shift ensued. Reluctantly discarding the Model A in 1932, Ford experienced a "gestation of his next offering—the Ford V-8— . . . almost as labored as that of the Model A" because of the scope of retooling and "because of inexperience in the dexterous art of changing models" (p. 206).

Given the crash programs under which Ford developed his new models, it was not surprising that they were plagued with defects. For instance, the Ford V-8—though the engine block's size was a major technical advancement—had numerous problems that were not soon forgotten by consumers (Nevins and Hill, 1963:63). Because of the primitive foundry technology then available, "some of the immense castings necessary were faulty and had to be replaced." Problems also developed with the piston rings and high oil consumption, the ignition, and the water pump.

Another problem with the Ford V-8 was that it was somewhat better suited to the prosperous 1920s than to the depressed 1930s, when economical and reliable operation became paramount. As was then written: "It is characteristic of Mr. Ford's sales technique that he could not sell the Four (in 1931) against the Six, whereas his competition has had no trouble selling the Six against the Eight. Still, by this time he should have made some progress in convincing people that eight cylinders do not necessarily use more gas than six cylinders and that the Ford engine is not going to shake the Ford to pieces" ("Mr. Ford Doesn't Care," *Fortune,* 1933:133). Hence, the V-8's inadequacies—such as excessive engine vibration caused by poor engine-mount design—made depression-minded consumers cautious about what would have otherwise been an extremely popular product innovation.

The problems associated with the Ford V-8 Model were not limited to its engine. "In addition, though Edsel by a hard battle for pleasing design and smooth riding had won his father's grudging cooperation, the Ford cars were not quite as attractive in appearance, or as pleasant for passengers, as their rivals" (Nevins and Hill, 1963:63).

FM's commercial vehicles maintained their position against Chevrolet longer than did the passenger car. The Ford truck outsold Chevrolet in 1930, 1931, 1932, 1935, and 1937 (Nevins and Hill, 1963:70). But it was "on a downward curve, and after 1937, . . . never again attained that distinction." Some of this early strength must be attributed to Henry Ford's philosophy of providing power, utility, and economy, an approach more in keeping with the demands of the commercial market. More of the success of FM's trucks probably was due, however, to the continuous series of changes that kept them reasonably current. "A relatively easy assent to changes" in commercial vehicles could be won "that the industrialist would never have permitted in a passenger vehicle. Frequently, Edsel approved them, assuming that his father would ignore the commercial branch of the company's business" (p. 70).

In sum, then, Ford committed three strategic blunders with respect to the Ford nameplate. He priced it to appeal to a shrinking market segment. He refused to offer a variety of colors and accessories. He repeatedly froze its design too long, without developing the managerial and production techniques needed to introduce new models frequently and smoothly. Only in 1933 did Ford finally accede to the annual model change (Abernathy, 1978:177). As a consequence, the Ford made after the early 1920s could not match the product diversity offered by competitors or the product variety demanded by consumers.

Henry Ford also failed to offer an array of more expensive makes that could attract the volume business he had geared his firm to produce. Thus, during the prosperous 1920s as customers moved away from the low-price class, FM was hit particularly hard: its next product offering, the Lincoln, was not in the lower-middle class to attract customers moving up from the Fords or Chevys, but in the very high-price class, with its considerably smaller volume. "This otherwise fine automobile was simply too expensive for general sale" (Sward, 1948:210).

Having bought the Lincoln Company in receivership from Henry Leland, the inventor of the Cadillac, in 1922, Ford managed to remove the able Leland and his son from its management. And in spite of Edsel's efforts to maintain the car's quality and improve its styling, Lincoln remained a losing proposition. "On the average, 7,000 cars were retailed annually between 1922 and 1930. Packard and Cadillac, competing at the same price level, outsold Lincoln more than three to one, while the Pierce-Arrow, selling at a considerably higher price, marketed almost as many units as Ford's big-car entry" (Lewis, 1976:185). Throughout the early 1930s, the Lincoln did little to improve FM's market position, for its sales then typically amounted to only a few thousand units. By the late 1930s Lincoln sales had "dwindled to hundreds" (Lewis, 1976:276). "The Federal Trade Commission, in the only estimate with even a tinge of reliability, set the total net loss at $18 million between 1929 and 1937" ("Lincoln-Mercury Moves Up," *Fortune*, 1952:97). In acquiring Lincoln, Ford seemed not to be pursuing a sophisticated diversification strategy. Rather, *Fortune*, in attempting to explain Ford's Lincoln acquisition, jokingly hypothesized: "The lack of other visible motive lends credibility to the folk myth that Henry felt compelled to drive a car of his own make, but wanted something better than a Model T" (p. 97).

Much of the Lincoln's difficulty after 1935 stemmed from a strategy error in expanding FM's product line. In 1935 FM expanded its offerings via the Lincoln Zephyr. But the Zephyr had not been sufficiently *differentiated* from the Lincoln. Thus, the Zephyr's initial success simply came at the expense of Lincoln's sales (Lewis, 1976:276). Worse yet, once the Zephyr's initial demand subsided, it contributed to "the impairment of Lincoln's reputation" ("Lincoln-Mercury Moves Up," *Fortune*, 1952:98). And potential customers saw it as a cheap imitation of the Lincoln. The Zephyr, then, contributed little useful variety to the FM product line. During the 1930s depression FM needed more low- and middle-price offerings, not low high-price automobiles. While GM and the other automobile makers were adjusting their models downward to match the depression environment, FM covered the deserted territory with the Zephyr. Unfortunately for FM, few consumers frequented this abandoned territory.

Not until 1938 did FM develop a car for the low middle-price class: the Mercury. Yet even here FM failed. "If the Lincoln was a liability, the Mercury-... must be called a nonentity." It "was only a trifle longer, heavier, and more powerful than the Ford; and its reputation for being 'nothing but an overgrown Ford' became ... general" ("Lincoln-Mercury Moves Up," *Fortune*, 1952:98). Since FM had not *differentiated* the Mercury from the Ford in the eyes of the consumer, no effective variety was added to the FM controlled set. And it remained unable to match the environmental variety it faced.

In the realm of nonautomotive diversification, Ford launched two noteworthy undertakings: tractors and airplanes. FM began making tractors in 1917. By 1920 its famous "Fordson" model accounted for 35 percent of the American output, despite a poor product-safety record. "International Harvester, for all its tractor experience, was unable to stave off the Ford challenge, and General Motors, which had launched the Samson tractor in 1918" under Durant's insistence, "closed out its entry in October 1922 with a $33,000,000 loss" (Lewis, 1976:180). In 1927, however, FM lost the sales leadership to International Harvester's more powerful machine. Moreover, "other tractor manufacturers also proved increasingly competitive" (pp. 182–183). So in early 1928 FM abandoned domestic tractor production.

Ford's airplane diversification had a similar history. In 1923 Henry and Edsel, along with many others, invested several thousand dollars in the Stout Metal Airplane Company. By 1925 this firm was incorporated into FM (Lewis, 1976:168, 176). The firm's most famous product, the Ford Trimotor, was quite successful, being purchased by the military services, large corporations, and airline carriers. Before the decade's end, "Ford was probably the largest manufacturer of commercial planes in the world" (p. 177). As sales declined to almost nothing in the early 1930s, Ford again closed shop, sustaining losses of more than $10 million (p. 178).

So as he had done with the Model T, Ford allowed his firm to lose a once-commanding position in tractors and aircraft. He abandoned both ventures, heedless of their near-term synergy with auto manufacturing and their long-term sales and profit potential. Useful variety was again lost.

ORGANIZING THE STRUCTURE

Just as Ford made no attempt to match FM's internal variety to its environment's variety, he made no attempt to match FM's internal variety to the capacity of its decision makers. That is, Ford saw no point in structuring the FM decision problem by assigning manageable subsets of it to specific decision-making individuals and groups. "Ford had always been reluctant to define spheres of authority" and "had often delighted in issuing contradictory orders"; eventually such "capriciousness made the managerial process a turmoil of uncertainties, fears, and cross-purposes" (Nevins and Hill, 1963:243). Even "the chaos of the changeover" from the Model T to the Model A "failed to arouse Henry Ford to the point that he established any consistent and clearly understood system of managerial hierarchy" (Hounshell, 1984:293).

Ford recognized only one internal boundary. "Our methods of management," as he put it, "have very little of what might be called method in them—in fact, when the method part seems to be getting ahead of the management part, we begin to look around a little. We have only two divisions—office and shop" (Ford, 1926a:104). Ford, of course, did everything possible to eliminate the office. With respect to FM's manufacturing operations, Ford recognized no boundaries:

> Now it might seem that since our operations are divided into many parts, it would be necessary formally to departmentalize. The work is departmentalized but the management is not. . . . The management everywhere is interlocking. It is a little hard at first for an outsider, who may be accustomed to strict definitions of duties, to grasp how a number of men may work together without exclusive responsibilities, yet it is merely that every man is responsible for getting the work out, and instead of divided responsibilities we have a united responsibility. (p. 106)

Essentially, then, FM was without an internal decomposition scheme.

Although Ford was correct in the system sense, for a manager must be concerned with the organization's entire decision set, from a practical design sense his assertion about united responsibility is meaningless. In any large firm like FM, joint responsibility is impossible; designers must establish a decomposition scheme which, at best, approximates the ideal. Ford did not attempt to develop an organizational structure that would have promoted decision making from a firm-wide perspective without overloading the decision makers of FM.

There were probably two reasons for Ford's unwillingness to factor FM's manufacturing activities into manageable subsets. First, he distrusted all organizational structures, and, second, he wanted to control FM's whole decision set himself. Chaos resulted.

With respect to Ford's distrust of a FM organizational scheme, "he felt that danger lurked in it, while there was a creative fluidity in loose procedures which he could alter at will" (Nevins and Hill, 1957:271). Partly correct in that rigid organizational structures can inhibit adaptation, Ford carried the argument to extreme.

> "To my mind there is no bent of mind more dangerous than . . . the 'genius for organization.' This usually results in the birth of a great big chart showing, after the fashion of a family tree, how authority ramifies. The tree is heavy with nice round berries, each of which bears the name of a man or of an office. Every man has a title and certain duties which are strictly limited by the circumference of his berry. . . . The buck is passed to and fro and all responsibility is dodged by

individuals—following the lazy notion that two heads are better than one. . . . And so the Ford factories and enterprises have no organization, no specific duties attaching to any position, no line of succession or of authority, very few titles, and no conferences." (Gillette, 1928:138)

In spite of Ford's protestations about "buck passing," his failure to decompose FM created "an amorphous administrative mess" ("The Rebirth of Ford," *Fortune*, 1947:84) where it was virtually impossible to determine who decided what.

Regarding Ford's desire to control everything himself, Harry Bennett (1951) explained: "It was a one-man show, and Mr. Ford was jealous of authority" (p. 15). Another executive mentioned: "He didn't believe in a big organization. He wanted to be the whole cheese and everything had to go through his hands first. Mr. Ford objected greatly if he wasn't informed on something" (Jardim, 1970:204). As Ford himself put it, "That's the only way to get anywhere—one man rule" (Nevins and Hill, 1957:270). Unwilling to control only FM's most important decision variables, Ford attempted to decide all matters in spite of his limited abilities and many diversions.

His method in attempting to maintain complete control was to encourage his subordinates to fight over their jurisdictional limits; Ford always created factions about himself and kept these factions at each others' throats. Thus "authority, never clearly defined at Ford, had been further obscured in the [various] struggles for power, and in numerous areas could not be identified; minor officials and workers would obey any superior person" (Nevins and Hill, 1963:294). Here, then, was "an embittered, mutually distrustful group of executives—most of them without titles—with no clean lines of authority or responsibility anywhere delineated" ("The Rebirth of Ford," *Fortune*, 1947:84). With fear and uncertainty paralyzing most of the firm (Nevins and Hill, 1963:294), Ford became the arbiter of all disputes and thus maintained his illusion of complete control.

Ford, in reality, controlled little regarding FM because of his poor administrative techniques and many outside interests. Since he rarely used his power, it "was bound to be exerted most of the time by somebody else" (Nevins and Hill, 1957:271). Unfortunately, his subordinates could not be expected to take a cooperative, firm-wide perspective in their joint-decision making with each other. "The strain of competition permeated the Ford organization" (p. 275). More importantly, "the device of competition was never really successful; energy that might have been merged into constructive accomplishment was siphoned off in useless friction. . . . The rivals often slowed down tasks that might have been smoothly cooperative" (p. 275).

So instead of decomposing the FM decision set clearly, Ford muddled his structure by permitting aggressive subordinates to compose their own (generally overlapping) decision spheres. "Power was not delegated but appropriated. In the late thirties and early forties, Harry Bennett carried this technique to the logical conclusion by basing his authority on armed force" (Galbraith, 1960:155). The resultant chaos was almost unbelievable. Charles Martindale related that in 1945 "a young man was set to drafting an organization chart for the company. He found the task impossible. 'He'd bring it to me with tears in his eyes,' states Martindale, 'and gave up because there was no way of knowing who reported to whom'" (Nevins and Hill, 1963:243).

It was not until well into the reign of Henry Ford II that anyone within FM began working on the organizational morass left by the elder Ford. Thornton's planning group began the design work in 1946 with a 56-page analysis, aptly titled, "Organizational Problems of the Ford Motor Company." In this report, eventually the basis of the FM internal organizational design, the group proposed first to make the company's present structure clear; second to issue an Organizational Manual that the components could use to develop charts of their own activities; and third to construct an "ultimate plan of organization" (Nevins and Hill, 1963:329).

The Organizational Manual, also prepared by the Thornton planning group, followed shortly after the original report. Besides providing basic instruction to the FM decision makers, the planners used the Organizational Manual to convey fundamental terms of organizational design language, such as division, department, section, unit, project control, and profit center (p. 330). In designating the parts of the automobile, the manual adopted the GM standardized parts classifications, known to Breech, Crusoe, et al. to be consistent and workable (p. 330). To provide for continuing hierarchical review and control of FM, "a proposed program of operational planning and financial control" (p. 330) was presented by Crusoe and approved for implementation by January 1, 1947.

Finally, in 1949 FM was ready to release to the industry and the public "the first top-level organization chart in its history" (Allen, 1949:79). In this chart, "flow of authority descends from stockholders (the Ford family) to board of directors, to executive committee, to president, Henry Ford II, to executive vice president, E. R. Breech. At this point the chart expands to the staff vice presidents, six in number, and in charge of industrial relations, engineering, manufacturing (with which is consolidated purchasing . . .), general counsel, sales and advertising, and finance" (p. 79). Below this strong, GM-like corporate headquarters ("decentralized operation and centralized control" [Nevins and Hill, 1963:329]), six divisions were configured: namely, the Ford Division, Lincoln-Mercury, Ford International, General Production, Parts and Equipment Manufacturing, and Machining & Forging (Allen, 1949:79). By September 1950 the decomposition had been carried still further, as the FM designers "broke up its single manufacturing division into six separate divisions. Each one would be responsible for its own profit-and-loss show" ("Ford Splits Production," *Business Week*, September 30, 1950:25).

Thus each FM division became a "profit center," a term coined by the new FM designers (Nevins and Hill, 1963:328). The newly defined organization structure provided a basis for separating out the performances of the various FM components. As Breech explained the new horizontal decomposition design work, "One of our first steps in 1946 was ... breaking up our business into many smaller profit centers, each run like an independent business and held accountable for its profit performance ... but with policy control and direction from top staff executives" (Hickerson, 1968:139).

TRAINING THE DECISION MAKERS

Since Henry Ford dismissed the need to organize and systematize FM, he also failed to recognize the firm's need for well-trained managers who pursued the goals of FM for him. Along with ignoring the value orientation and factual education of FM's managers, he selected managers for their inability rather than their ability to decide FM's variable settings. Forbes (1927d) warned:

> Constituted as he is, Mr. Ford cannot deputize as deputizing is generally understood. His deputies never know when their orders will be countermanded, their powers abridged, their position abrogated.
>
> And, perhaps most regrettable of all, so gigantic has the Ford Motor Company become, so weirdly many-sided its operations, so far-flung its activities, so numerous those exercising some measure of authority, that Henry Ford can no longer supervise fully and efficiently. (p. 44)

Value Orientation

Without a well-defined client in mind, it would have been impossible for Ford to develop a definitive value-orientation scheme even had he been so inclined. "His executives," not surprisingly, "were never as generously paid as those at General Motors, although occasionally acting as master, or feudal lord, Ford would secretly reward their efforts with cash payments" (Jardim, 1970:237). Ford's small cash payments did little to motivate his managers toward FM's long-run interests, and his secretive procedures yielded nothing from the vast numbers of nonrecipient decision makers.

Furthermore, Ford's leadership style fostered in his executives a self-centered, noncooperative orientation rather than a firm-wide perspective. Ford "expected each rival for his favor to act as a counterweight and an informer against every other member of the entourage. When an executive . . . reported happily to Ford that he had just [resolved] a misunderstanding between . . . two other Ford administrators, he was told to let well enough alone the next time he discovered 'fleas on a dog's back' " (Sward, 1948:185). Thus a factor in FM's weak position in the 1930s as one manager recalled, was "the accumulation of all this friction and confusion," "executives working against each other," considerable "thieving" and appropriating "Company property for personal purposes by the executives," and "a general lack of interest in the success of the company due to the personal ambition of the executives" (Jardim, 1970:205).

According to Sward (1948), Ford once articulated his theory of motivation in a conversation with Samuel S. Marquis: "Men work for two reasons only, he said—for their wages and for fear of losing their jobs" (p. 182). Charles Sorensen, Ford's most powerful assistant, concurred fully, and "distinguished himself in the 20's as a Ford specialist in the psychology of job insecurity, if not as a dispenser of high wages" (p. 182). Coupled with Ford's efforts to inhibit cooperation, Sorensen's anxiety-provoking efforts created an extremely corrosive atmosphere within FM.

Breech, years later, in reversing Ford's negative motivational approach, first attempted to remove the atmosphere of fear. Brutal methods used under the elder were forbidden, and Ford's more savage henchmen, notably Sorensen and Harry Bennett, were dismissed by Henry Ford II himself.

In a more positive vein, FM instituted a profit-sharing plan like GM's in July of 1947. The bonus was "tied to performance for men earning roughly $8,500 or more. In computing bonuses," FM placed "emphasis on the assets at each executive's disposal as compared with the profits" generated. Accordingly, FM tried "to place as many of its activities as possible directly under men sensitive to profits, and to reduce to a minimum activities, such as staff functions, that cannot be so placed" ("Ford's New Managers," *Fortune,* 1953:168). Breech told FM's executives that the profit-sharing plan bonuses "would be based on an executive's performance compared to standards. Money spent above standard would in effect come out of executive pockets" (Harris, 1954:198). As Breech himself put it, FM's financial control system highlighted profitability by combining "the profit center system" with "a supplemental compensation plan" involving "more than forty-three hundred members of management." More specifically:

> "In the fall of 1948 we called together several hundred of our top management men. We analyzed and compared the profit performance against the standards of

each key operation and showed how performance was reflected in the supplemental compensation fund. . . . Each man in turn set up similar meetings of his own supervisors, and the process continued on down the line. These meetings were subsequently held at regular intervals. The results were almost unbelievable." (Hickerson, 1968:142)

So while Ford neglected the development of performance measurement and motivation schemes, his successors at FM established such programs to insure that FM managers pursued FM goals. Breech, in particular, "made it clear that rewards would be commensurate with performance" (p. 139).

Factual Education

Ford avoided developing factual education programs for his managers because he felt that professionals and so-called experts were nonproductive as well as negative in attitude toward new ideas (Greenleaf, 1964:41). Therefore, he wanted no part of educational schemes that produced expert managers. While there was a Ford factory school for teaching shop practice, there were no programs for teaching managerial practice. Similarly, Ford shunned "employees of specialized technical knowledge—for many years college graduates were not only not sought but not hired at River Rouge" (Galbraith, 1967:90).

Much educational work thus had to be conducted after Ford's death. For example: "Breech and his executives, sometimes not too patiently, taught modern business practice to old Ford plant managers and executives who had been innocent of such learning. They explained that 10 per cent return on a new investment might look substantial, but how much did General Motors earn?" (Harris, 1954:196). At a more detailed level, "Crusoe and the Thornton group went to officials in various departments saying: 'We're going to tell you about costs. This is a tool you should have. It is not your fault that at present you do not have it, for it was not given to you.' Crusoe felt that the rank and file of officials welcomed these ministrations. 'They were like shipwrecked sailors who haven't eaten for days and see a biscuit' " (Nevins and Hill, 1963:331).

FM's new central training department established six types of training programs ("Ford Students," *Business Week*, April 13, 1946:75). Most noteworthy of these offerings was a GMI-style program to provide future executives. This five-year work-study program was established in conjunction with the School of Business Administration of Detroit's Wayne University and led to the degree of Bachelor of Science in Industrial Management ("Factory College," *Business Week*, September 6, 1947:94).

Selection

Consistent with his approach to the motivation and education of his managers, Ford often selected individuals who possessed little knowledge about the work they were to do. For example, one man "who declined an offer to work for Ford Motor Company because he knew nothing about automotive manufacture" was told by Ford himself "the less you know, the more quickly you can learn" (Greenleaf, 1964:41). Similarly, while "Ford was exacting in his search for superior materials" (Nevins and Hill, 1957:252), when "his managers spoke of their need for a highly trained metallurgist Ford pointed impatiently at a man who was sweeping the floor nearby. 'There's your metallurgist,' he said. 'Teach him the job' " (Ford Motor Company, 1953:21). These examples reveal Ford's general policy on executive personnel selection. Ford, as one executive put it, "often said . . . that if he wanted a real job done right that he would always pick the man that didn't know anything about it. The reason was that if he picked a man who didn't know anything about the job, the fellow never got far away from Mr. Ford. Mr. Ford in that way [at least thought he] was able to control what was being done" (Jardim, 1970:227–228).

Ford's subordinates carried this practice forward. "Sorensen, with Harry Bennett's imposition," for instance, placed the inexperienced Harry Mack in charge of the new assembly line in the River Rouge plant; at the time, Mack headed "the Ford box factory, a job in no way related to automobile assembly operations" (Hounshell, 1984:290).

So what did Ford end up with in the way of decision-making talent? "The worst of his selection, of course, was Harry Bennett," who, "along with his satellite prizefighters, punks, ex-football players (and Coach Harry Kipke after he had been relieved at Ann Arbor) and assorted baccalaureates of the Michigan penal institutions, eventually made the Ford Motor Company into an industrial charnel house" (Galbraith, 1960:159).

Ford's penchant for firing good men was even more noteworthy than his selection of the wrong ones. Certainly, Ford's classic dismissal of a competent decision maker was William Knudsen (Beasley, 1947:109). Knudsen's removal was but a single instance of a general Ford policy. "It had been the old man's practice . . . to demote first-line supervisors regularly every few years or so lest they 'become uppity.' . . . Just as, early in his career, he decided not to share ownership with anybody, he apparently decided not to share management" (Drucker, 1954:114).

But Ford, as Jardim (1970) aptly points out, did not really control FM as he thought he did; his abilities—like those of all human beings—were strictly limited and more and more diverted toward his many hobbies. (See, for example, "Mr. Ford Doesn't Care," *Fortune*, 1933.) "A man cannot well be a museum director, an antique collector, a researcher in history, an educator, a promotor of aviation, a farmer, a health authority, a developer of hydroelectric power, a master of square dancing... and a political dabbler... and keep his full concentration on the manufacture of automobiles" (Burlingame, 1957:122-123). Furthermore, he failed to use competent executives who could free him to control the most important variables influencing FM's success and survival. Since Ford had to be consulted about more and more, he had less and less time available for the truly important decision variables. (In startling contrast, Sloan—who concentrated his energies solely on GM affairs—hired many of the best financial and automotive executives to amplify his power of control.)

Ford's illusion that he had the firm under his personal control was, as Jardim (1970) put it, "the central and paradoxical problem of the Ford narcissistic style.... Since men are not omnipotent, ... those areas with which it is impossible for them to cope become... the fiefs of underlings who justify themselves to the leader by face-to-face submission, maintaining in this way his own delusion of omnipotence" (p. 229).

In turn, many of the men under Ford sought to eliminate competent executives who could threaten their positions. Thus Ford established a selection process that removed capable decision makers throughout the firm. At the very time he placed the inexperienced Harry Mack in charge of the River Rouge assembly line, for instance, Sorensen was dismissing the "seasoned production men who had made the assembly line the very symbol of mass production" (Hounshell, 1984:290). This purge of the managers Sorensen disparaged as the "Model T sons-of-bitches" (p. 279) resulted "in most of the production problems with the Model A" (p. 292). With the salaried employees "needed to initiate and carry out innovation" gone, it was "not surprising that the company took nearly a year to change over to the Model A" (Abernathy and Wayne, 1974:116).

It is questionable whether Ford could ever have used a rational personnel-selection policy even if he had wanted to, for with his dislike of organization structures he refused to define roles for his decision makers. Instead, he permitted them to define their own positions. In Ford's (1926b) own words: "One man is in charge of the factory and has been for years. He has two men with him, who, without in any way having their duties defined, have taken particular sections of the work to themselves. With them are about a half a dozen other men in the nature of assistants, but without specific duties. They have all made jobs for themselves—but there are no limits to their jobs. They just work in where they best fit. One man chases stock and shortages. Another has grabbed inspection, and so on" (pp. 92-93). With "no specific duties attaching to any position, no line of succession or of authority" (p. 92), how could there be a rational selection process to fill vacancies? Furthermore, every new appointment meant an organizational upheaval while the new executive defined his own special position.

In summary, FM was left without a body of decision makers capable of providing amplified control over FM's performance. Since Henry Ford could not provide the needed decision capacity either, the firm floundered, and performance began its downward spiral.

COORDINATING THE DECISION MAKERS

Ford's notions regarding the FM structural arrangement were matched exactly by his ideas about its coordination mechanisms. Ford, in short, saw no need to provide an internal communication network for the company. "He had no patience with committees or with extended discussion" (Nevins and Hill, 1957:271). Accordingly, he let lapse into extinction the firm's important executive and operating committees. These groups had provided a forum for the principal executives, as Ford himself testified, to "confer on everything that was done" (Nevins and Hill, 1954:572).

In general, then, Ford thought of a business firm as "a collection of people who are brought together to do work and not to write letters to one another. It is not necessary for any one department to know what any other department is doing" (Ford, 1922b:92). Thus there was a minimum of clerical help and "no elaborate records of any kind" (p. 92). Ford could go still further in inhibiting internal communication:

> We do not have conferences, we do not have committees, we have no formal procedures of any kind. There is no formal method for interdepartment communication because we do not have departments—if one man wants to say something to another man, he says it over the telephone. It is rather difficult for him to say it any other way because the only men who have offices or stenographic facilities are those who must communicate with people outside of the organization.
>
> The private office with all of its paraphernalia is a good deal of a time waster, and especially so in a factory. The managers of a factory ought to know what is going on and not·sit around in the office waiting for some one to send a written communication about what is happening. The best place for a factory man's office is under his hat. (Ford, 1926a:104,106)

The clerical personnel purged during the early 1920s had conducted many of the FM internal communication tasks. According to Ford (1922a) again:

We cut our office forces in halves and offered the office workers better jobs in the shops. . . . We abolished every order blank and every form of statistics that did not directly aid in the production of a car. We had been collecting tons of statistics because they were interesting. But statistics will not constrnct automobiles—so out they went.

We took out 60% of our telephone extensions. Only a comparatively few men in any organization need telephones. (p. 261)

Along with dismantling the FM formal communication network, the elder Ford took great pains to inhibit the development of an informal communication network. The managers of the FM branch plants, for instance, "were categorically forbidden to exchange information with each other, or to discuss common problems—unless specifically told to do so. Consultations between branch heads, declared the company, represented 'neither the proper way nor a healthy condition.' Similarly, dealers were enjoined from discussing their problems with each other" (Nevins and Hill, 1957:259). Even within a single plant communications remained inhibited. In a noisy factory: "Officials had no offices—only desks in the open factory at which they stood. They couldn't keep records, and lower officials could not discuss their problems together" (p. 296). For example, Harold Hicks, a FM engine designer, could not confer with his coworkers; he testified: "Although we were in one big room, things were held quite confidential. Due to strict deportment that everyone pursued in the company, you didn't wander around out of your place of business too much to find out what the other fellow was doing. You paid strict attention to your own job" (Stern, 1955:146). The Ford-created atmosphere of fear stifled informal communications more than anything else, and it "even extended beyond working hours" (Nevins and Hill, 1957:296), so that FM executives came to avoid each other at all social gatherings.

When the new FM designers assumed command in 1945, then, much work had to be done to improve internal communications. A major effort was directed toward reducing the severe morale problems within FM (Nevins and Hill, 1963:321–322); the reduction in turn produced a more cooperative atmosphere where much-needed information communication networks could sprout and grow.

With respect to formal communications, the vertical linkages between the corporate headquarters and the divisions were vital. Breech provided one example of the extensive internal network that had to be laid after Henry Ford's death. In describing the new formal performance-reporting channels integral to his financial-control system (discussed earlier in this chapter), Breech explained that each year "every manager of a profit center" submitted "a profit plan . . . based upon a standard volume (rate of operations)." The company's top executives reviewed these forecasts "with each manager and his principal assistants." A thorough annual review was also made of expenses not linked to "profit centers, such as those covering administrative and commercial expenses, engineering, and so on." Four-month profit forecasts were submitted each month by the operational units along with "monthly and year-to-date profit statements" (Hickerson, 1968:141).

SYNCHRONIZING THE FIRM AND ENVIRONMENT

Henry Ford's lack of concern with establishing an internal communication network for FM was mirrored by an equal disregard for exchanging information with the firm's external environment. Although incoming information from its environs was vital to FM's continued success, Ford failed to provide for it. His approach toward outgoing information was haphazard as well. In spite of the fact that his firm operated in an extremely competitive industry (and the one which spent the most on national advertising), Ford's promotional activities were quite sporadic. He would fitfully permit ad campaigns to run, then have them withdrawn because he held that advertising was an economic waste.

While often unwilling to promote his cars, Ford himself did not avoid publicity. With his phenomenal early success and his growing stature as an American folk hero, Ford became more and more captivated with the idea of issuing (often offensive) ad hoc statements to the press, which widely aired them. In addition, Ford launched a propaganda barrage to vent his numerous prejudices. For FM this adverse publicity created a considerable negative reaction against the firm's products.

On the output side, Ford failed to keep abreast of short-run demand fluctuations so production could be synchronized with dealers' sales. "All the companies except Ford were members of the National Automobile Chamber of Commerce and rendered to it monthly reports of their sales which were accessible to members" (Soule, 1947:167). But Ford did not approve of associations (Rae, 1965:39). Since he also ignored R. L. Polk's sales registration statistics, he had no way to judge current consumer demand and simply forced the immense inventories generated by this lack of foresight on FM's dealers. Up until 1938 some 52 percent of these dealers had complained about overstocking, compared with 23 percent for Chrysler and 28 percent for GM (USFTC, 1939:194).

Ford's most serious environmental communication mistake was the failure to set up channels to monitor the varying tastes of the automobile-buying public. Although Ford often is considered to have neglected marketing because of his preoccupation with production, he was well aware of the consumer's crucial role. As early as 1922 he stated emphatically:

> Ordinarily business is conceived as starting with a manufacturing process and ending with a consumer. If that consumer does not want to buy what you have to sell him and has not the money to buy it, then the manufacturer blames the consumer and says that business is bad and thus, hitching the cart before the horse, he goes on his way lamenting. . . .
> But what business ever started with the manufacturer and ended with the consumer? Where does the money to make the wheels go 'round come from? From the consumer, of course. He is the only man who counts. And success in manufacture is based solely upon an ability to serve that consumer to his liking. (Ford, 1922a:261)

Four years later in 1926, when his Model T was hopelessly mismatched to the automobile environment, Ford again claimed to realize that "the manufacturing side may so predominate that the resulting article, while cheap and easy to make, does not exactly fill the public need and therefore cannot be sold" (Ford, 1926a:38).

Nevertheless, Ford did not establish any communication channels that could provide him with information on what the consumers desired. In reality, he behaved in a way diametrically opposed to his pronouncements. Couzens (1921) explained the Ford marketing strategy in the following way: "It was thought that selling started with the customer and worked back to the factory, that the factory existed to supply what the customer asked for. What the Ford Company really did . . . was to reverse the process" (p. 264).

Ford's approach worked at first because the Model T was synchronized exactly with environmental conditions. As a farm boy through-and-through, he knew precisely what people then needed and, more importantly, wanted: cheap, durable transportation capable of negotiating the country's deeply rutted roads. But as the country shifted from a rural toward an urban society with paved streets and highways, Ford failed to keep pace.

Paradoxically, Henry Ford with his Model T had helped start America off on a period of rapid change. (See, for example, Carson, 1965:279–294, and Perrett, 1982:298.) Yet "Ford never understood the implications of the revolution he himself had created. He could not grasp . . . that the day was ending when there was one car for the masses and a very different car for the classes" ("Selling to an Age of Plenty," *Business Week*, May 5, 1956:124). As a further irony, the immense success of the Model T played a major role in isolating Ford from the changing market conditions. Great success led to arrogance, which in turn produced a closed mind.

Convinced that he more than anyone else had his finger on the country's pulse, Ford actually cut many of the incoming communication channels that had been built when Couzens and then Hawkins ran FM's marketing. During the middle 1920s crisis, with the Model T faltering precariously, Ford reacted by trimming his marketing force drastically. "To permit a reduction in branch clerical forces, the company ordered the wholesale elimination of record-keeping forms" (Nevins and Hill, 1957:426).

Even when environmental information managed to penetrate the FM organization, Ford usually refused to consider it. He failed, for example, to accept the warnings coming in from his dealers about changing consumer tastes. When Henry Ford had a rare meeting with a group of Ford dealers in the early 1920s, the dealers all told him that changes were imperative in the engine, styling, color, etc. of the Model T; Ford responded that the only thing he need worry about was the best way to make more cars (Nevins and Hill, 1957:389), that is, more cars just like the already obsolescent Model T. Thus, "the manufacturer neither read the signs nor cared what other people thought about his conviction that the T was as perfect and timeless as Pike's Peak itself. Self-centered and bullheaded, he spurned the counsel of his dealers when they began to beg him to modernize his product" (Sward, 1948:197-198). In general, "the dealer had little opportunity to be heard" and, more importantly, "he was taught, often by bitter experience, to keep quiet about his difficulties, even when they were also the difficulties of the Ford Motor Company" (Nevins and Hill, 1957:259).

During the 1926 marketing crisis Ford displayed a similar intolerance for environmental intelligence information regarding competitors' activities. Ford claimed that the R. L. Polk Company's annual summary of automobile registration statistics was "rigged" and that the Polk Company was under GM's influence (Jardim, 1970:217). More generally, Ford "held that time given to the study of competition is just that much time lost from one's own business" (Ford, 1928:135). Accordingly, Ford chided his executives whenever they mentioned the competitors' efforts or products.

In 1933, a major *Fortune* article entitled "Mr. Ford Doesn't Care" quoted Ford as saying: "I don't know how many cars Chevrolet sold last year. I don't know how many they're selling this year. I don't know how many they may sell next year. And—I don't care" (p. 63). To Ford's comment the *Fortune* editors added the observation: "There are too many people who will do what Mr. Ford thinks and there are not enough people who can influence what Mr. Ford thinks. This keeps [FM] out of touch with the world, with its markets" (p. 133).

In addition, Ford lagged in building a test track capable of evaluating his cars against competitive makes. Although GM had already established its Proving

Ground by 1926, Ford would not hear of one. Instead, FM used "the plant yards and public roads to test new vehicles. It was another ten years [1937] before the growing congestion of highways, the laws governing motor traffic, and the possibility of needless accidents forced him into modern practice" (Nevins and Hill, 1957:253). But even when Ford came to build a test facility, it was inadequate, for he simply converted an unused FM airport. Henry Ford II rebuilt the proving ground in the late 1940s.

Along with his hostility to generating information on the current state of auto development, Ford failed to research future design possibilities. "At the time Ford had no research or engineering organization worthy of the name. His laboratories, all agree, were an empty shell partly because Ford was suspicious of college-trained men" (Galbraith, 1960:161). Ford simply wasted the lavish research investment in his Laboratories Building, opened in 1924. Never appointing a research director, Ford himself "decided what important experimental work should be undertaken, and . . . dealt directly with an engineer or a workman, assigning tasks and checking on progress" (Nevins and Hill, 1957:252). His engineers had desks without chairs, were not encouraged to maintain records, and worked with a "startling lack of dynamometers" (p. 252)—the essential test instrument for engine and drive-train development. Ford "launched a number of projects which wasted time and manpower" (p. 252) and "his captiousness also interfered with effective design or development" (p. 253).

Without adequate testing and research facilities, FM could not avail itself of a steady stream of product improvements. To Ford this shortcoming may not have seemed serious, for he had no desire to improve the Model T, nor later the Model A, nor even after that the Ford V-8.

In addition to ignoring or restricting vital incoming information during the competitive 1920s and 1930s, Ford pursued an erratic policy toward outgoing communications. FM's advertising department, idle during 1917 and 1918 because of the war, remained out of commission for five years. Ford stuck with his nonadvertising policy even in poor sales periods such as the 1920–1921 recession, and FM "bought not a line of space (except in behalf of tractors and Lincoln cars) until 1923" (Lewis, 1976:126).

In Ford's defense, it should be noted that while he did forgo advertising FM's products, he garnered considerable publicity through the distribution of Ford films and via his own newsworthiness. Reporters pursued him endlessly (p. 129); and when amenable, he was adept at giving them something to print, usually on the front page. Hence, he felt he saved the cost of advertising (p. 126). But publicizing himself or even FM did not focus the customer's attention directly on the Ford product.

Still Ford felt directed campaigns were unnecessary. While he conceded the need to advertise a new product, an established item, like the Model T, needed no such support (Lewis, 1976:126, 201)—regardless of the competitors' efforts.

"By eliminating 'useless advertising,' he argued, the manufacturer reduced his overhead and passed the saving on to the customer" (Nevins and Hill, 1957:263).

In reality, advertising was needed to move the FM product. Ford had only managed to shift its cost and management onto the firm's dealers. "Those in the larger cities had 'clubbed together' to conduct newspaper and billboard campaigns, spending an estimated $3,000,000 annually. This system, however, did not permit national campaigns and left much to be desired in . . . distribution and uniformity of sales appeals" (Lewis, 1976:190).

Edsel managed to have FM's Advertising Department reestablished in 1923, and for the next three years, "Ford was one of the nation's biggest advertisers" (p. 191). The money to support the advertising budget came from an assessment on the dealers. "However, Henry Ford's basic dislike of advertising reasserted itself in June 1926, when he ordered the dealer assessment discontinued and told the dealers to arrange for their own advertising" (p. 191). Consequently, "in 1926, Ford, still [by] a shade the largest producer of cars, was outspent in magazine advertising by seven other makes of cars" ("Selling to an Age of Plenty," *Business Week*, May 5, 1956:124). This policy reversal came at an inopportune time, for sales of the increasingly outmoded Model T began to weaken. So by 1927, *Printers' Ink* could make this (somewhat self-serving) claim: "As the Ford advertising passed out of the picture, Chevrolet's gained both in volume and momentum. Meanwhile, the Ford plants began running on part time. . . . The business it got came from the combined momentum of its past reputation and history and the individual efforts of its dealer organization" ("Ford to Resume Advertising," *Printers' Ink*, June 2, 1927:33–34). FM finally resumed advertising with the introduction of the Model A.

Although he deeply resented the need to advertise his cars even when they met serious sales resistance, Ford was not averse to publicizing his personal prejudices. Thus, in several campaigns over the years, Ford authorized his weekly magazine, *The Dearborn Independent*, to carry attacks on various groups he disliked. Most notably, in the early 1920s the *Independent* created "an unpleasant distinction for itself as the leading anti-Semitic publication in the world" (Perrett, 1982:258). "Starting in May 1920," for instance, "in a series of ninety-two articles sanctioned by Ford, the *Independent* blamed Jews for many of the world's problems" (Lewis, 1976:116). A boycott of Ford products ensued, but sales were not appreciably affected. "Ford, however, was to learn during his second anti-Semitic campaign, waged during the mid-1920s, that the aging Model T no longer could afford him the luxury of complacency about boycotts" (p. 141). As a result of his anti-Semitism, there was a "virtual boycott" of Ford cars in Europe (Jardim, 1970:209). The director of the company's European operations, Warren C. Anderson, was reported to have "made repeated entreaties, appeals, and finally demands that the Jewish attacks cease," but "instead the attacks became even more bitter" (p. 209). (As a reward for trying to improve

FM's projected image, Anderson was fired.) Similarly, in the United States, "many Americans of Jewish origin still refused to buy a Ford car. Hollywood organizations hesitated to use one in a motion picture or for carrying equipment or personnel" (Nevins and Hill, 1963:63). In sum, "the *Independent*'s attacks on Jews . . . cost the Ford Company untold millions of dollars worth of sales" and became "a source of embarrassment to the firm and one of its leading public relations problems for several decades" (Lewis, 1976:135).

Ford's "outbursts against tobacco" and "denunciations of liquor" (Nevins and Hill, 1963:63) generated additional enemies. Probably most damaging to FM in the turbulent 1930s was Ford's protracted battle with labor. Long after GM had settled its labor disputes, Ford continued the fight and reaped much adverse publicity. With the elder Ford's death, Henry Ford II pursued a very different advertising and public-relations policy. He became noted for his conciliatory stance towards conflicts, especially labor-relations problems.

Under Henry Ford II, also, FM's designers established departments to survey both the input and output markets that were vital to its performance. According to Albert J. Browning (1947), new vice president and director of purchases: "We organized a Commodity Research Department to provide both short-term and long-range information on basic commodities such as steel, copper, lead, etc. A Purchase Analysis Department was included to help our buying departments on product and price analysis" (p. 24). The Purchase Analysis Department compiled and analyzed price and cost trend data and obtained information about suppliers' financial positions, earnings, products, and manufacturing processes "with a view to aiding buyers and suppliers in reducing their costs without reducing suppliers' profits below a reasonable level" (p. 20). That Department also analyzed parts in order "to simplify and substitute less critical materials" (p. 24). FM created a Central Follow-up Department to track suppliers' shipments to branch plants. Further, "to handle the flow of papers and materials from other company units [and] to act as a clearing point for exchange of information," the Procurement Services Department was formed (p. 24).

On FM's output side, a GM-style National Dealers Council was used to develop an "idea of local reactions to Ford policies, merchandising, and advertising" ("For Future Sales," *Business Week*, December 13, 1947:52). As an FM executive observed, " 'The dealers eventually soak up, through their pores, a composite portrait of tomorrow's car purchaser—what he will be wanting, more important, expecting, in next year's automobiles' " (p. 52). Thus several of the more significant gaps in FM's environmental communications network were filled.

On the consumer front of the environment, a research department was instituted to compile data on "registration figures, prices and specifications, types and number of financed new-car transactions, and geographical analysis" ("Ford Asks Some Questions," *Business Week*, May 29, 1948:76). The geographical analysis covered car sales in every county of the United States from 1932 to 1941. That analysis was launched to help determine variations in automobile sales cycles, to forecast sales and fix sales quotas accurately, and to reveal areas "that can stand new dealerships—where there's a high degree of customer concentration—or areas where dealer territories overlap" (p. 76). The new FM market research department also conducted public-opinion polls, contacted dealers regarding customer reaction to FM products, and interviewed purchasers of competitive automobiles.

Still another unit—composed of financial analysts under the then FM controller, Robert McNamara—collected data on market and price trends, analyzed capital requirements, and checked purchase prices and production costs in relation to those of competitors (Stryker, 1952:158).

EVALUATING THE PERFORMANCE

Finally, Henry Ford did little to evaluate the performance of his firm, nor was there much he could do, having never defined a performance measure with which to judge FM's state of well-being. He appeared unconcerned about return rates, market shares, and even profit figures. How well FM compared to GM or any other competitor did not interest him.

Ford held such a cavalier attitude simply because he recognized no authority that could hold him accountable for the long-run performance of his firm. He scorned his Board of Directors. Board meetings, as Bennett (1951) recalled, "had no purpose other than to comply with the law. When Mr. Ford failed to show up, it was pretty funny, because all the directors dared do was" to ratify "what Mr. Ford had already done, or . . . what they knew he'd approve." None ever dared take any "initiative, with the sole exception of Sorensen, who sometimes did this," to the others' "great discomfort." But Bennett found that the "meetings reached their peak of humor . . . when Mr. Ford did show up." He "would come in, walk around, shake hands with everyone, and then say, 'Come on, Harry, let's get the hell out of here. We'll probably change everything they do, anyway'" (p. 167). Since FM was not then a public company, Ford did not have to bother with inquisitive stockholders or scrutinizing securities analysts. And he went to great lengths to avoid becoming dependent on the banking community, which would have carefully monitored FM's methods and results.

Ford also did much to isolate himself from society in general. He refused to read or even discuss almost all of his correspondence and surrounded himself with a series of toadies whose job it was to protect him from the outside world. Among these men was Ernest Liebold, who "acted as a buffer between Ford and the numerous individuals who wanted to see him" (Nevins and Hill, 1957:273). Moreover, for many years, Liebold directed the *Dearborn Independent* and "was completely in charge of the manufacturer's public relations, including his political activities and his contacts with the press" (p. 273). Newsmen came to dislike Liebold because he rarely granted them access to his boss. "Ford, for his part, was aware of the way Liebold treated the press and approved of it" (Lewis, 1976:130). Later, Harry Bennett and his henchmen served to isolate Ford from the country's social and political environment.

As FM's impact on the surrounding society increased, Ford's isolation became more and more a problem. External groups and authorities began to look askance at FM policies and practices. Even during the laissez-faire 1920s the social and political problems resulting from Ford's poor performance-control methods gained the attention of politicians and community leaders. During the unnecessarily lengthy shutdown for the Model A retooling, for example, FM's hard-pressed dealers were left without merchandise to sell. Similarly, 40,000 FM workers were thrown out of work for months. "Unemployment in Detroit rose by more than 60,000," for many others "depended on the wages paid to Ford workers. It was a changeover without advance planning" (Perrett, 1982:298–299) and with serious repercussions. "Thomas A. Dolan, then Detroit's commissioner of public welfare, reported that the Ford shutdown was responsible for 45 per cent of the city's relief load in 1927. . . . The taxpayers of Detroit in 1927 were compelled to increase expenditures for relief, over the previous year, by more than a million dollars" (Sward, 1948:201).

With the tremendous upheaval associated with the 1930s depression, Ford grew even more out-of-phase with the country's abruptly altered social and political environment. For instance, during the labor strife of the middle and late 1930s Chrysler and GM signed the code of the National Industrial Recovery Act of 1933; Ford refused. Thus, Ford's competitors "were now receiving the stickers, placards, and other materials supplied by [the] NRA, and had the right to display the Blue Eagle. Chrysler, in a full-page advertisement, boasted that it flew from 'every flagstaff' of his factories. 'Proud and glad to do our part,' ran the Chevrolet slogan" (Nevins and Hill, 1963:20).

Even though Chevrolet "fanned the anti-Ford sentiment" (Lewis, 1976:243) with its ads, Sloan and his colleagues were undoubtedly as much opposed to the New Deal legislation as Ford was. But Sloan and Brown were not about to tarnish GM's public image in a hopeless battle. When long-stable patterns shifted, it was better to accept the change and conduct business under the new circumstances than to hold out for the return of antiquated relationships. Thus, "the opposition to the rising power of the state in the thirties, like opposition to the rising power of the unions, was led not by the mature corporations but by the surviving entrepreneurs" (Galbraith, 1967:302) like Ford. "General Motors ... and other mature corporations were much more inclined to accept such innovations as NRA, to be somewhat more philosophical about Roosevelt and otherwise to accommodate themselves to the New Deal" (p. 302). So only FM, of the big three auto makers, lagged in adjusting to the permanently changed political situation. "On this occasion Ford's opposition to ... the Roosevelt Administration made him a storm center of controversy and brought down on his head, for the first time in his career, vociferous criticism ... from a sizable portion of the nation's press and the American public" (Lewis, 1976:241). The Recovery Administration head, General Hugh S. Johnson, extremely disturbed by Ford's recalcitrance, decreed that the government would purchase no Ford products and called for a public boycott as well: "I think the American people will crack down on him when the Blue Eagle is on other cars. . . . No corporation is rich enough and no group strong enough to block this nation" (Nevins and Hill, 1963:20). Though praised by avid anti-New Dealers (Lewis, 1976:243) and eventually supported by the Supreme Court, Ford's stance cost his company vital depression-era sales as "rabid New Dealers shunned Ford cars" (Nevins and Hill, 1963:63).

Undaunted by this skirmish with government, Ford escalated his fight with labor in 1937. This time the battle would be conducted not with pronouncements and decrees but with fists and clubs. Besides using the shabby spy tactics employed by GM, Ford added a frightening dimension of violence to the labor struggle. Probably the worst incident was described as the "Battle of the Overpass," in which Ford thugs under Harry Bennett's direction beat Walter Reuther and several other union organizers on a street overpass leading to the River Rouge Plant. "Company men also attacked the press corps" (Lewis, 1976:250). Naturally, "the reporters and their incensed editors outdid themselves in describing the viciousness of the attack" (p. 250). As a result, the National Labor Relations Board filed a complaint that accused the company of committing virtually "every unfair labor practice defined by the Wagner Act" (p. 250). Still Henry Ford could not bring himself to accept a new working relationship with labor as GM had. So the physical violence, legal actions, and work stoppages continued at FM well into the 1940s. Not surprisingly, such "bitter opposition to unionization hurt the company's popularity and to some extent its sales" (Rae, 1965:117).

On still another front, FM again fell short in meeting the demands of the larger society. When mounting dealer complaints led the Federal Trade Commission to investigate dealer-manufacturer relations, Ford's treatment of his dealers was by far the worst among the major manufacturers.

FM's approach to evaluating its performance, like its actions related to the other design phases, altered dramatically after Henry Ford's death. Under Henry Ford II, for instance, the Board of Directors became deeply involved in monitoring FM's policies and performances. After 1956, when FM went public, stockholders would demand from directors a management capable of sustaining profits and dividends. "Henceforth assets, earnings, salaries of officers or directors, and other facts long secret or clouded would be known to all;" competitors could review "data on blunders and successes; labor leaders could measure worker-returns against owner-returns; and . . . Federal departments and commissions could detect any disregard of the public interest" (Nevins and Hill, 1963:424–425). Under Henry Ford II, FM would have much less disregard for the public interest than it had under his grandfather's reign. FM's relationships with customers, dealers, labor, and the government all would improve. FM would be run as a modern public corporation rather than as a purely private enterprise.

Thus it is not surprising that GM had far outpaced FM by the late 1930s. FM's position, in fact, had so deteriorated that Chrysler had assumed second place in the industry's sales battle, leaving FM in third. Similarly, GM's return rates on invested capital came to outdistance those generated by FM.

PART IV

The Results

14

Comparing the Performances

What results were generated by the disparate approaches of GM and FM toward performance-control systems? To answer this question, the two firms' automotive-market shares and capital-investment returns are compared during three time periods. The first period, 1918 to 1923, juxtaposes GM's performance still under Durant's influence with FM's achievements still influenced by the planning efforts of Couzens, Klingensmith, Hawkins, Kanzler, and Edsel Ford. The second period, 1924 to 1929, contrasts the Sloan-Brown design accomplishments with the Ford antiplanning results. The third analytical period is 1930 to 1938, a time with economic conditions conducive to high performance at FM (because of its low-price emphasis) and to low performance at GM (because of its mid- to high-price concentration). Via this third assessment it becomes clear that the boom of the 1920s alone did not fuel GM's surge and retard FM's potential.

GM–FM MARKET-SHARE PERFORMANCES

The 1918–1923 Period

GM, in 1918, even with Durant's proliferation of mid- to high-price product offerings, held only about 20 percent of the motor-vehicle-industry sales, as seen in Figure 14-1. Although GM posted a slight market-share increase in the postwar boom year of 1919, by late 1920 the postwar recession had begun, and GM's market share dropped noticeably. In 1921, the Sloan-Brown group took GM's helm. An increase in market share was registered in 1922 and again in 1923—but a drop occurred in 1924, another recession year. The market-share increases captured by the still very young Sloan-Brown administration do not exceed past fluctuations, and no substantial performance-improvement claims can be made for the GM designers in this early period.

While GM made no significant improvements in share-of-market results before 1924, FM attained new heights. In 1921, for instance, FM achieved the highest share of market ever attained by a single manufacturer (about 59 percent), while GM registered a considerable loss. Yet then FM began its descent. By 1924 that descent was accelerating rapidly.

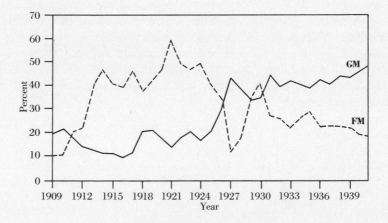

Figure 14-1. Share of market percent, 1909–1941. *Source:* Based on data presented in General Motors Corporation, *The Automobile Industry: A Case Study of Competition,* 1968:7.

The 1924–1929 Period

GM's big jump in market share came between 1924 and 1929. GM advanced from about 16 percent of the market in 1924 to more than 43 percent in 1927, when FM abruptly halted production of the outdated Model T and sluggishly switched to the Model A. While GM did not reach the heights FM had attained, it did establish a strong grip on the market. In 1929, when FM's Model A approached the zenith of its popularity and availability, GM descended only to the 33 percent level, from which it quickly bounced back.

The favorable conditions of the middle and late 1920s, coupled with the Sloan-Brown performance-control system, generated peak GM performances that doubled those achieved under Durant. Even the weak GM performances during this period were all one-and-a-half times what they had been before. Thus the Sloan-Brown planners had succeeded in moving their firm's market-share performances to a considerably higher level.

Much of GM's success between 1924 and 1929 must be attributed to FM's failure to capitalize on the growing prosperity of the period. After 1924 the Model T lost customers rapidly; and with the extended retooling shutdown for the Model A, FM permanently lost its once solid market footing. The introduction of the Model A late in 1927 allowed FM to make a temporary recovery that lasted only until 1930.

The 1930–1938 Period

During the 1930–1938 period, the Sloan-Brown team achieved a lasting domination over both FM and the other domestic automobile manufacturers. More specifically, GM advanced "from 32.8% of all passenger car sales in 1929 to 34.5% in '30 and 43.3% in '31." In 1932 "the ratio declined to 41.5%," but it was "again on the upturn" (McClary, 1933:277). And after 1932 GM showed a steady upward trend in market share. Notably, GM assumed sales leadership under circumstances more favorable to FM than to itself: a major depression was not helpful to any durable-good manufacturer and especially not kind to one whose products had low-cost substitutes, like the Ford Model A.

In a period when the lowest price class in the automobile market climbed from 83.2 percent (1926) to 98.1 percent (1939) of total industry sales (Ross, 1941:700), GM prospered by shifting most of its product offerings to the lowest price class. Furthermore, GM offered consumers a highly desirable product line. In a survey of buyers' attitudes toward car appearance conducted in 1938 by the trade journal *Automotive Industries*, purchasers rated *all* GM models more highly than the two Ford models (MacGowan, 1938:258). GM's Chevrolet Division played the decisive role in the defeat of FM, for it met Ford on his own ground—the high-volume low-price field—and overwhelmed him. In 1937, 1938, and 1939, more Chevrolet cars were sold than any other make in the United States. "It was in 1927, the year Ford abandoned the immortal Model T, that Chevrolet first passed Ford in total new passenger-car registrations" ("General Motors II," *Fortune,* 1939:37). This performance Chevy repeated in eight of the next eleven years. How? By making Chevrolet "the greatest common denominator of what the American public thinks a good car ought to be. The Chevrolet is a Mass Man's car" (p. 38).

By 1938 Sloan had brought GM's actual market-share performance almost into complete correspondence with his previously mentioned "bogie" of 45 percent ("General Motors," *Fortune,* 1938:158). This showing during the depression period presents considerable evidence that GM's improved performance resulted, not from favorable economic conditions, but instead from the Sloan-Brown performance-control system—especially from those procedures designed to synchronize GM's prices and product offerings with its customers' desires.

GM–FM INVESTMENT-RETURN PERFORMANCES

The high market shares GM captured after 1924 were vitally important to its (subsequent) consistently high rates of return, while FM's dwindling market share adversely affected its ROI performances. GM's investment return also increased greatly because of the tight controls the planners imposed over other critical profit-influencing variables such as cash balances, productive- and finished-inventory turnover rates, material and supply costs, production and marketing costs, and capital-investment levels.

The 1918–1923 Period

To establish a reference point once again, one should note that Durant competed quite unsuccessfully with the Ford organization in terms of investment-return rates from 1918 to 1923; Figure 14-2 graphs Durant's difficulties. Excessive inventory expenditures and overexpansion marked this period, for GM held a "capacity to produce 750,000 cars and trucks a year," yet it "sold only 457,000 in 1922" (Sloan, 1964:196). Not until 1923 would this excess capacity be utilized. The improvements realized by 1923, however, did not reveal any great contribution by the Sloan-Brown design group.

Throughout the 1918–1923 period FM exhibited substantially higher return rates than did GM. But the performance trends of the two firms turned in 1921 when FM came entirely under Henry Ford's capricious direction and GM came under Sloan and Brown's planning influence. FM return rates first leveled and then moved downward while GM investment returns climbed.

The 1924–1929 Period

GM's design team demonstrated its full potential during the 1924 to 1929 period when the firm rapidly overtook, then greatly outdistanced FM in return-rate performance. In short, earlier trends solidified into significant differences.

Even though the recession year of 1924 visited a temporary setback on the Sloan-Brown group, further improvements in product acceptance and manufacturing efficiency in 1925 reversed GM's rate-of-return position vis-à-vis FM's, in spite of FM's larger market share. "Both in 1925 and in 1926," GM's "automobile output and financial volume . . . reached new high levels; but its most striking gains took place in . . . profits. . . . In 1927, when the aggregate American motor-vehicle output suffered a decline of 20 per cent, General Motors scored further substantial gains" (Seltzer, 1928:215).

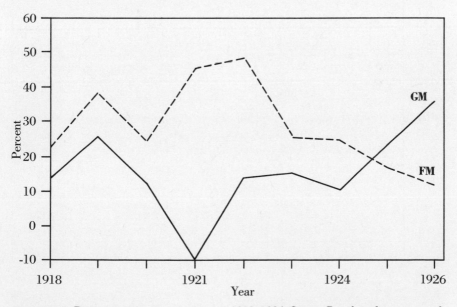

Figure 14-2. Return on investment, percent, 1918–1926. *Source:* Based on data presented in Seltzer, *The Financial History of the American Automobile Industry,* 1928:129, 232.

How did the profit performance improve? Via the increases in volume, of course, and the implementation of controls. GM, for example, "reduced its [cash] 'float' from ten million to approximately four" (Swayne, 1924:23); increased its productive inventory turnover from a little more than twice in 1921 to eleven times in 1927 (Reeves, 1927:36); and increased its total inventory turnover from less than five times in 1923 to eight times by 1926 (Bradley, 1927:427).

From Figure 14-3 it can be seen that the years 1927 and 1928 produced great performance differences between GM and FM. GM's improved performance controls now were coupled with its substantial market share (volume) increases—obtained almost exclusively from FM.

Ford and his company thus were stalled with a massive capital investment left almost completely idle. FM's rates of return for both 1927 and 1928 were negative, reflecting the firm's substantial losses on its capital equipment. In 1927 GM outstripped FM by almost 55 points on the ROI scale and in 1928 by almost 60 points. The year 1929—when the FM Model A was having its only truly successful year—was the last year in which FM was able to earn a two-digit return rate. Its performance was to grow still worse.

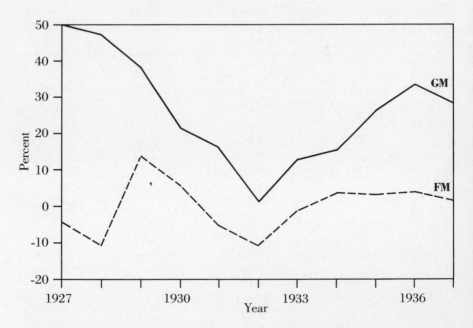

Figure 14-3. Return on investment, percent, 1927–1937. *Source:* Based on data presented by the United States Federal Trade Commission, *Report on the Motor Vehicle Industry,* 1939:487, 671.

The 1930–1938 Period

With the great depression, the performance-control system established at GM by the Sloan-Brown designers faced its most difficult test. The industry's productive capacity in 1932 was much higher than it had been in the 1921 recession; manufacturers like GM and FM held massive capital investments with little sales volume to support them. Since the market shrank to minuscule levels in relation to the industry's total productive capacity, a high market share of a shrunken market was not enough. Yet, "with a flexibility that is still, even in retrospect, almost unbelievable," GM "came through the depression without registering a loss or omitting a dividend in a single year" although its sales had dropped over 70 percent. Meanwhile, "its somewhat smaller and therefore presumably more flexible competitors were losing money: Ford, $132,000,000; Chrysler, $11,000,000" ("Alfred P. Sloan Jr.: Chairman," *Fortune*, 1938:75).

In contrast, FM's performance during the 1930s depression gave "the impression . . . that the survival of the Ford Motor Company was more a matter of habit than of any conscious planning." This result was not surprising, of course, as Henry Ford "despised systematic organization" (Rae, 1965:115).

Thus GM stood alone in the automobile industry, with an assessed value equal to FM, Chrysler, and Studebaker combined, "with a few of the lesser shops thrown in for good measure. . . . It challenged the invincible Ford and licked him" (Romer, 1937:96).

The Sloan-Brown planners' achievements were long-lasting: years after the period under study here GM was still earning its comparatively high returns. Between 1946 and 1967 White (1971:249) found the average return rates were 9.85 for FM, 7.59 for Chrysler, and 14.67 for GM. He concluded that "General Motors' superior profitability must be attributed to its superior management: a superior ability to utilize efficiently the resources at hand and to make decisions under conditions of uncertainty" (p. 265).

Shortly after the publication of White's (1971) study—with its glowing assessment of GM—the American automobile industry encountered a quantity of novel conditions not seen since the late 1920s and the early 1930s. Neither GM nor any other American manufacturer responded well, initially at least, to the new conditions of uncertainty. Indeed, many industry observers wondered if GM and the American automobile industry would ever regain the leadership position.

15

Epilogue: An Overview of GM's Subsequent Decline

With the Sloan-Brown performance-control system operating flawlessly, GM had not only passed FM but far outdistanced it by 1938. Despite the many virtues of GM's performance-control system, it provided the seedbed for several serious long-run problems.

Specifically, Sloan kept GM's management narrowly focused on stockholders' interest at the expense of labor, consumers, and the general public. Over the years, he placed too much stress on styling and on larger mid-price automobiles. Throughout his tenure, he leaned toward centralization and coordination to the extent that the divisions were in constant danger of losing what little remaining latitude they had for environmental adaptation. Because of Sloan's conservative product strategy and his centralized organizational structure, Chevrolet eventually started to gravitate toward Cadillac and thereby lost the ability to produce a functionally engineered, low-cost, small-size car, while Cadillac began to drift toward Chevrolet and thereby surrendered the ability to build a technologically sophisticated, high-margin, mid-size vehicle. And, lastly, Sloan's emphasis on rate-of-return performance would eventually thrust too many financially oriented executives into GM's top corporate decision-making positions, men often without knowledge of automobile design, production, and marketing. Under their myopic cost-cutting and profit-maximizing direction, GM's divisions built a line of cars that looked startlingly similar, exhibited very poor quality, and cost far too much.

So long as Sloan and his cohorts remained active in GM's affairs, none of these proclivities had gotten out of hand or become serious problems, for they were either compensated for or counterbalanced by the early planners. By the middle 1950s, however, the original designers' influence had diminished substantially. And without their constant attention, tendencies quickly grew into trends. So with each passing year, GM found itself in increasing difficulty as a narrowly focused, inadequately diversified, overcentralized, financially dominated manufacturing firm. In 1980, ironically, GM had a product line more muddled, overlapped, and ineffective than the one Pierre duPont and Sloan had inherited from Durant.

In terms of Ashby's Law of Requisite Variety, GM's difficulties in the 1980s are related to a failure to retain adequate internal variety. Throughout the 1940s, 1950s, and early 1960s, GM (as well as FM and Chrysler) faced little environmental variety that it was not equipped to counter quickly or to exploit advantageously. But political, social, economic, and worldwide competitive conditions began to change, slowly at first and then with startling swiftness and dramatic proportion. GM's relationships with government, labor, consumers, and its foreign competition had been placed on a whole new footing as the decade of the 1980s opened.

The federal government had enacted restrictive—but much needed—legislation on safety, pollution, and fuel economy. These strictures would become tighter and tighter as the 1980s progressed.

An increasingly alienated work force had repeatedly refused to cooperate in raising productivity or even maintaining quality, yet demanded wage increases that gave foreign competitors a decided cost advantage. "The average GM blue-collar employee received $35,919 in wages and fringe benefits in 1980" (Ross, 1982:40). While organized labor would become slightly more cooperative in the depressed early 1980s, previous precedent-setting contracts could not be altered quickly or extensively.

Neither would the deterioration of GM's quality image be reversed easily. It would go on as an unvarying parameter despite GM's advertising campaigns, for the image had been built up through years of dismal labor relations, insufficient domestic competition, and shortsighted cost cutting.

With suspicions growing about GM's product quality, a series of fuel shortages caused consumers to question long-held enthusiasms for GM's overlarge, grossly inefficient products. And a long inflationary spiral, fueled first by the Vietnam War and then by skyrocketing energy prices, forced "sticker-shocked" purchasers to be even more cautious. The protracted recession and high interest rates of the early 1980s further accentuated the trend. Gone were the days when demand could be stimulated by styling changes and new accessories. Consumers now wanted functional value that GM—along with FM and Chrysler— could not deliver immediately. A long history of technological stagnation stood in their way. From the 1920s, the major American automobile firms had emphasized mass production and concentrated on productivity gains that only came with "attendant losses in innovative capability" (Abernathy, 1978:3-4). "More than 80 percent of all cars produced in the United States in 1970," for instance, "had V-8 engines; all ran on gasoline; all had rear-wheel drive; none used fuel injection; and all had longitudinally mounted engines" (Abernathy, Clark, and Kantrow, 1983:105). Unfortunately for the domestic auto industry, progressive firms in Europe and Japan offered a much more varied array of products. Moreover, because Detroit had avoided the low-cost small car and the high-cost luxury-car markets for so long, foreign manufacturers had finally established the extensive dealer networks so necessary to compete in the vast American marketplace. Accordingly, "the share of the American market captured by imports (especially Japanese) jumped from 15% in 1972 to 27% at the end of 1981" (Lawrence and Dyer, 1983:19).

In the face of this variety onslaught coming from government, labor, consumers, and foreign competition, the American auto firms staggered into the 1980s. Chrysler had requested and received massive government-guaranteed loans to weather the transition. FM depended heavily on its foreign operations to sustain itself while it (and the UAW) lobbied for government-imposed import quotas on foreign competition. Even GM, with its vast economic resources and vaunted performance-control system, shuddered under a variety load it had not prepared itself to counter. Accordingly, GM's market-share and return-rate performances, long held steady at enviable levels, dipped dangerously. For instance, in 1980, for the first time since the Sloan-Brown planners assumed control of GM in 1921, the firm failed to earn a profit. Indeed, with each passing year the evidence mounted that a permanent adjustment in domestic social and political relationships as well as a permanent realignment of international economic relationships had occurred for which GM was initially unprepared.

By no means can most of GM's difficulties in the 1980s be blamed on the performance-control system designed by Sloan, Brown, and their colleagues. Obviously, much time has passed, and, more importantly, subsequent GM managements have made many alterations not in keeping with the early planners' policies.

In the early 1950s, for example, the then GM president, Harlow Curtice, adopted a Durant-style management based on dynamic personal leadership. "Like Durant, he toured factories and instantly dispensed millions of dollars for expansion without so much as a by-your-leave to the finance committee. The committee system of which Alfred Sloan and Pierre duPont were so proud gradually slipped into disuse" (Cray, 1980:353).

More significantly, Curtice disregarded the 45 percent upper limit Sloan had placed on GM's market share to avoid government antitrust actions. "Under his lash, G.M.'s share of the car market rose from 42 per cent in 1952 to 47 per cent in 1953, to 50.7 per cent in 1954" (Sheenan, 1956:8). Whether from the need to make his own mark or from the lure to compete, Curtice intensified the sales battle between FM and GM, in particular between the Chevrolet and Ford makes. Nevertheless, under its newly remodeled management, FM too increased its share of the pie (p. 8). Hence only the few remaining independent manufacturers really suffered.

Chevrolet's significantly updated 1955 model played a key role in Curtice's competitive strategy. This car for the first time offered an optional V-8 engine. Over the ensuing years, Chevrolet would sell fewer and fewer of its reliable, easily rebuilt, economical, but stodgy "Blue Flame Sixes." Moreover, to satisfy the youth market, the high-compression overhead-valve V-8 would have its power boosted with displacement and compression increases, dual exhausts, four-barrel carburetors, high-lift/long-duration camshafts, etc. Detroit's horsepower race had accelerated.

Car size and accessory loads kept pace with the power increases. Throughout the 1950s and 1960s GM added high-profit items like automatic transmissions, power steering, power brakes, power windows, air conditioners, and numerous less notable gadgets.

During this period GM's stylists resorted to superfluous gimmicks such as Cadillac's tail fins. "The American cars of the era" before long "were the embodiment of the tasteless exploitation of style at the expense of functional vitality" (Yates, 1983:186). What had begun under Harley Earl 25 or 30 years earlier as a noteworthy effort to streamline and beautify was pushed to exhaustion. Clean lines gave way to distracting novelties and chrome expanses; tired designs kept breathing only with the help of sales stimulants like the portholes Curtice had stamped in the Buick front fenders when he managed that division. That Buick's marketers needed to christen these functionless chrome-ringed holes with the technical name "ventiports" only points to the superficiality of GM's latest strategy.

Curtice then helped end an era of economical, functional transportation and started two decades of what in retrospect can only be seen as nonfunctional motoring. So "of all the corporation's presidents—Durant, duPont, Sloan, Knudsen, and Wilson before him—Curtice was to be the most influential in shaping Automobile America and the least recognized for that influence" (Cray, 1980:352).

With Curtice relentlessly pressing GM's divisions for increased sales, the firm's dealers found themselves under severe pressure. The "factory forcing" that Sloan had done so much to minimize in the 1920s and 1930s again ran rampant. Unresponsive dealers had their franchise agreements cancelled. Cray (1980) reported that there were 4600 such "voluntary terminations" (p. 400) between 1951 and 1955. "By the mid-1950s the extensive dealer network set up so carefully by R. H. Grant two decades before was a shambles" (p. 401).

The meaningless battles over market shares during the 1950s also started an erosion of the relationship between dealers and consumers. Since the early

1930s, the GM planners had known that this linkage in the distribution chain was vital to continued sales success. In fact, the consumer's satisfaction with dealers' sales and service efforts played a far bigger role in selling cars than did advertising ("G.M. III," *Fortune*, 1939:105). But when GM's divisions pressured their franchises for increased volume, the dealers in turn started to use high-pressure sales pitches on their customers. An era of shoddy and short-sighted selling tactics followed. (For a brief though vivid description of some of these scams, see Mandel, 1982:267, 271, 274, 276.) At the same time, the industry's push for sales volume began to diminish product quality. Moreover, high-volume dealers who skimped on customer service were not admonished by the factories to correct their ways. Consumers started to feel abused, but as yet had nowhere to turn.

Still, as did "everybody else in General Motors, Curtice believe[d] implicitly that the best interests of G.M. and of the national well-being [were] coterminous" (Sheenan, 1956:11). More than ever before, however, influential outsiders dismissed this purported identity of interest.

Not surprisingly, the federal government again focused its attention on the automobile industry. Repeating the scenario of the 1930s, the dealers' lobby petitioned the federal government for protection against the manufacturers. Before one committee chaired by Senator O'Mahoney, "dealer after dealer appeared . . . blaming Mr. Curtice for the way they were being treated by the field representatives of various car divisions" (Ellis, 1968:107). The government also became concerned about consumer treatment—by both the dealers and the manufacturers. Unwittingly, Curtice had succeeded in keeping the spotlight of government attention focused directly on GM. (See, for example, *Business Week*, June 17, 1961, "GM's Size Is Again an Issue"; April 22, 1961, "GM Takes Antitrust Flank Attack"; and December 1, 1962, "Antitrusters Try to Pin Monopoly Tag on GM.")

In response to GM's growing market strength and the weakening of the few remaining independent manufacturers, the government's administrative branch launched a series of antitrust actions. The most notable of these was the Justice Department's divestiture suit against DuPont's ownership of GM stock. When the Supreme Court finally ruled on the case, it directed DuPont to sell its GM holdings. The effect of this outcome was ironic. Unexpectedly, it had given Curtice and his successors a freer hand in conducting GM's affairs, for no longer could aging board members "Sloan, Pratt, and Mott . . . muster a controlling interest in the company's stock in order to enforce their decisions" (Cray, 1980:387). Thus, it hastened the time when GM, instead of being guided by long-held proprietary interests, became dominated by professional managers (Sheenan, 1965:51). Unfortunately for all involved, both the board's and management's time perspectives shortened from long-term well-being to near-term profit performance. GM's horizons, rather than expanding as they should have, further narrowed and constricted.

For his part, Sloan had aggravated the lack of accountability for GM's subsequent managements by refusing "to clutter up the Board of Directors with outside members," restricting it instead largely to current or retired executives of the company. So constituted, "the Board of Directors [became] usable as part of the executive organization," but useless as a source of outside opinion. "Awareness of the problem, even of its existence," as Drucker cautioned in 1946, "is by no means general, and it is, of course, least understood where the isolation is greatest" (p. 92).

The less GM's board kept the firm's professional managers in check, the more the federal government became involved in overseeing the firm's operations. In addition to the Justice Department's various antitrust investigations and suits, the Senate conducted numerous hearings into GM's activities. In 1957, for instance, Senator Estes Kefauver's Antitrust and Monopoly Subcommittee questioned Curtice and Frederic Donner, the chairman of GM's Financial Policy Committee, about the firm's failure to build "medium-sized cars" when there appeared to be a strong demand for them (Cray, 1980:394). Kefauver's committee also heard testimony from George Romney, then the president of American Motors. Romney called for the breakup of GM and FM and castigated them for failing to offer consumers reasonably sized functional transportation. Though no legislation resulted, GM had been put on official notice that it was not satisfying the consumers' needs.

Similar information began to come in from the marketplace directly. "A number of European producers introduced small cars, and the market share of imports rose from 1 percent in 1955 to 10 percent in 1959" (Abernathy, 1978:38). Disturbingly, many of these sales were achieved by a car as Spartan as Ford's Model T: Volkswagen's "Beetle." Even more distressing, most of the people buying these little cars were affluent, as two-thirds "had incomes about 40 percent higher than the average new-car buyer" and "two of three owned more than one car" (Cray, 1980:402). To stem this unwelcome tide, Curtice decided to import GM's German-made Opel and have it marketed through the Buick Division's dealerships.

Since the sales strength of the imports did not subside, Curtice next planned to offer an American car to counter the Volkswagen. This Chevrolet car would be called the Corvair, and like the Beetle, it would be driven by an air-cooled rear engine.

Shortly after authorizing the Corvair's development, Curtice retired as GM's president and chief executive officer. Simultaneously, Albert Bradley—who had assumed the board chairmanship from Sloan—retired. Frederic Donner ascended to the chairmanship of the board and in addition acquired the role of GM's chief executive officer. Not since Sloan had become both chairman and chief executive officer in 1937 had one person held GM's top two positions. But in stark contrast to Sloan, Donner arrived at GM's pinnacle via finance, knowing relatively little about domestic operations. (This deficiency is discussed more fully later in the chapter.) As John DeLorean recognized, "The modern General Motors Corporation took a dramatic change in its course" (Wright, 1979:227).

Shortly thereafter, GM's first domestic small-car venture turned into a debacle. The most vocal of many critics, Ralph Nader (1972) called the Corvair affair "one of the greatest acts of industrial irresponsibility in the present century" (pp. 4–5). Eventually, the Corvair's bad publicity opened the floodgates holding back a torrent of federal legislation onerous to the domestic automobile industry.

As part of a myopic cost-cutting program, Chevrolet's division manager and chassis engineers proposed a cheap swing-axle design for the Corvair's rear suspension. Worse yet, they did not install a stabilizer bar, which would have cost "less than four dollars a car" (Nader, 1972:lxiv). The result, critics charged, was that the car's rear wheels "tucked under," leaving it to careen uncontrollably, spin out, or roll over. After test-Corvairs flipped at GM's Proving Ground, several prominent corporate engineers fought hard to eliminate the car's defects (Wright, 1979:65). They lost the battle, and the Corvair entered production unaltered. Before the corporate headquarters finally relented and allowed Chevrolet's new division manager to correct the 1964 models, accidents involving the Corvair killed and maimed a large number of people. As Nader (1972) put it, "The tragedy was overwhelmingly the fault of cutting corners to shave costs" (p. 18). GM now calculated its cost estimates to "the fourth and fifth decimal place," and a difference of two cents per car often added up to $10,000 over a lengthy production run (p. 40). A rising tide of costly lawsuits soon rendered such shortsighted accounting exercises meaningless.

While Donner's administration must take immediate blame for the Corvair affair, it was following a long-established, though obviously unstated, General Motors policy. Going back to Sloan's early reluctance to offer safety glass, GM often has been negative toward automobile safety. Of course, "auto safety was not seen by them [GM, FM, and Chrysler] as a profitable venture, and that is why they were slow" (White, 1971:260). Yet, by far, GM took the most negative view of consumer protection. In 1955 FM and Chrysler agreed to help finance an expanding study of crash-injury protection at Cornell. However, GM's engineering vice president, Charles Chayne, declined to cooperate, and both Sloan and Curtice refused to reverse his decision (Cordtz, 1966:207). A year later Robert McNamara, then in charge of FM's Ford Division, focused his sales campaign on impact-absorbing steering wheels, safety door latches, padded instrument panels, seat belts, and the like. In response, GM's executives exploded, according to a *Fortune* report, and "many an outraged telephone call went out from the G.M. Building in Detroit to Ford headquarters in nearby

Dearborn" (p. 207). "GM said stop and Ford literally screeched to a halt," as Nader (1972) so graphically put it, "with . . . ex-GM Ford executives getting McNamara to switch gears to an advertising campaign that emphasized styling and performance rather than safety" (p. xiii).

Probably GM's greatest failure in the realm of public service, then, has been its laggard performance in the safety arena. As Donner himself later conceded, "We have got a tradition in General Motors of maybe too much sticking with our business problems" (Cordtz, 1966:117).

In the middle 1960's an activist administration and Congress demonstrated that if GM would not be responsible on its own then the government would intervene. Up until that time the only significant legislation specific to the industry was the Automobile Information Disclosure Act of 1958. It required that manufacturers affix to each new car a sticker describing make, model, equipment, prices, etc. In 1965 Senator Abraham Ribicoff started the legislative ball rolling again by using his Subcommittee on Executive Reorganization to hold hearings on automobile safety. Donner and then GM president James Roche testified about GM's meager safety-related research activities. "Even viewed charitably, the corporation's appearance before the Ribicoff subcommittee was a public embarrassment. General Motors no longer could pretend it held the best interests of the motoring public foremost" (Cray, 1980:421). But in spite of these disclosures, Senator Ribicoff could not push his auto-safety legislation past the Senate.

The aftermath of the cost-cutting efforts on the Corvair, however, would soon free the legislative logjam. On learning of Ralph Nader's forthcoming book, *Unsafe at Any Speed,* GM's lawyers placed Nader under surveillance in the hope of discrediting his well-documented criticisms of the Corvair. Caught in the act by a Senate guard, GM attracted much adverse publicity and catapulted Nader and his book into instant fame. GM's behavior toward Nader helped pass a whole series of legislative acts that GM's adroit lobbyists — without the adverse publicity — could have killed behind the scenes. Here was a public-relations blunder that matched Henry Ford's worst.

The most immediate result of GM's gumshoe blunder was the passage of the National Highway Traffic Safety Act in 1966 (Cordtz, 1966:210). "It was the first in a series of legislative acts that brought Washington squarely into Detroit boardrooms and, some said, brought Detroit design studios inside Washington regulatory agencies" (Mandel, 1982:340). Accordingly, variables once under the exclusive control of the automobile firms now became parameters or "givens" dictated by legislative acts, administrative edicts, and judicial rulings.

Not since the middle 1930s with their spate of labor legislation had GM seen such inroads made into its once exclusive decision-making domain. As previously, GM's management tried unceasingly to thwart the government's efforts. (See, for example, Nader, 1972:xi–xciii.) Perhaps had GM's management taken a conciliatory stance toward the government's safety legislation they could have turned it toward vanquishing the lightweight and questionably safe Volkswagen.

Besides limiting the manufacturers' design freedoms by requiring dual brake systems, seat belts, collapsible steering columns, standardized bumper heights, and crashworthiness, etc., an amendment to the Highway Safety Act required that the purchasers of defectively designed or produced cars be notified. A steady stream of recall notices thus served to keep consumers reminded of Detroit's latest failures. Probably the most publicized recall case took place "when, in 1971, the corporation called back 6.7 million 1965–69 Chevrolet cars to repair defective motor mounts" (Wright, 1979:68). Again, GM had responded tardily and only after much adverse media attention.

With such incidents, GM bestowed on Nader the status of a national folk hero. His asceticism and idealism allowed him to use this press exposure to become a spokesman for a generation of American youth disenchanted with their society after President Kennedy's assassination and President Johnson's cynical post-election reversal on Vietnam intervention. Nader and other critics like him told a generation of World War II babies about the American auto industry's wasteful practices. This adverse publicity presented Detroit with a very serious problem. These youngsters "constituted a huge age cohort just beginning to move as adults through the population, about to become the 'new consumers,' at the leading edge of the demographic buying group that formed the juicy target for marketers of hard goods—particularly hard goods with four wheels and an engine" (Mandel, 1982:329). As GM's research scientist Kettering had warned almost forty years earlier, "So long as we have younger generations, . . . we will have changes. Their views are new, their tastes are new, their likes are new—and emphatic" (O'Shea, 1928:359). Unfortunately for GM, many in this latest generation—for which the firm held high hopes because of its size and affluence—no longer wanted to be saddled with pushing GM's large cars through life. Whether or not GM noticed it, a dangerous trend had started to accelerate.

Ironically, Kettering had been an early supporter of the smaller cars Nader's disciples gravitated towards. Kettering's early research into improved gasolines and high-compression engines stemmed from the gas shortages of the early 1920s. And "while the glut conditions of the late 1920s and 1930s diverted most automotive engineers in other directions, Kettering never lost sight of fuel economy as a primary goal of the automobile designer" (Leslie, 1983:175). As early as 1925 he proposed a fuel-efficient auto "with light weight, a short wheel base, . . . streamlining, . . . a factory-set carburetor and spark advance automatically corrected for load and speed, . . . low-drag brakes, a four-speed transmission with overdrive, and an economical rear end" (p. 176). "Kettering returned to his 'light car' design in 1932" to address what he perceived as "a major weakness in the contemporary automobile market—the lack of a true 'economy car'" (p. 176). Kettering kept refining his lightweight, economy car until World War II interrupted the work. In 1944 (a year when many of Nader's supporters were born), Kettering, Mott, and C. E. Wilson, then the GM president, again considered a small car for the corporation (Cray, 1980:323). "The concept squared with the corporation's own market research as well as a survey conducted by the Society of Automotive Engineers just before the end of the war. According to the SAE, prospective automobile purchasers in New York, Chicago, New Orleans, and San Francisco wanted smaller, less expensive, and more functional automobiles" (p. 323). Supposedly, Sloan and GM's financial people vetoed Wilson's proposals on the ground that small cars commanded small profit margins. As Sloan undoubtedly knew, this decision would have no immediate impact on GM because of the tremendous pent-up demand for cars after the war.

But Sloan's opposition to the small-car philosophy would make it difficult for successive generations of automotive men to gain approval for their small-car projects from the financial staff. Once Sloan opted for the more lucrative larger cars, Kettering "salvaged the [light-car] project's best ideas for new high-compression and high-performance [V-8] engines" (Leslie, 1983:179). These engines first appeared in the Cadillacs and Oldsmobiles of the late 1940s, but quickly spread to GM's lower-priced, high-volume lines. This adaptation, unfortunately, "led ultimately to a more rapid depletion of petroleum and the gradual disappearance from the industry of Kettering's original vision of conservation" (p. 180). And eventually, when gas supplies dwindled, Sloan's past emphasis on the comparatively high returns of the large cars would severely jeopardize GM's long-run well-being. To make money on small cars required constant development—especially in engines. But after GM first introduced its high-performance V-8 engines, their further development focused not on costly efficiency improvements but rather on cheap power gains attained largely through displacement increases. GM had taken a wrong turn, and it increasingly lagged behind the German and Japanese manufacturers in the development of small-displacement, high-efficiency engines. Even Ford and Chrysler gained an edge on GM in this increasingly vital race (Yates, 1983:208-209). Sloan's successors, much to their dismay, would soon learn that despite its resources GM could not compete successfully in the small-car field. It remained by far the most difficult market segment to conquer. In fact, unless forced into it by depressed economic conditions, Sloan had always astutely skirted this battleground, as Henry Ford was the only domestic manufacturer ever to achieve any noteworthy financial success there. But with the dawning of a "small is beautiful" era, GM could no longer hide, just as Ford had been unable to hide in a "bigger is better" era.

Another growing public concern championed by Nader was automobile-generated pollution. Throughout the country, but especially in the Los Angeles

Basin, smog problems grew more severe as the horsepower race shifted into high gear in the middle 1960s. For the 1964 model year John DeLorean at GM's Pontiac Division, for example, introduced the firm's first "muscle car": the Pontiac GTO (Wright, 1979:109–110). But excessive engine sizes and heavy cars produced more pollutants, and the new freeways, which the industry had lobbied for so effectively, only aggravated the problem. Again the federal government responded to protect the general well-being and mandated strict pollution requirements. Via the Clean Air Act of 1970, government had begun to set engine-design parameters that GM had to satisfy. Another set of variables had been expropriated from GM's decision-making domain.

In addition, the American public received another publicity barrage about the domestic automobile manufacturers' inability to keep their own house clean. The evidence mounted constantly that consumers and the general public were not being well served.

The American manufacturers did nothing to alleviate this impression when they decided to meet the new regulations by expediency rather than by thorough redesign. They simply added a series of efficiency-robbing emission controls to their already inefficient overlarge, power-tuned engines. Smoothness, or "driveability," suffered; performance declined; and fuel economy plummeted to an all-time low. On the theory that America's unsophisticated consumers did not want "overengineered" cars, United States automakers left it to the Japanese and Europeans to develop designs for combustion chambers, valve trains, fuel-injection devices, and ignition components that not only met pollution standards but improved performance and lowered fuel consumption. Honda, for instance, went as far as modifying a Chevrolet engine with its own CVCC (compound vortex controlled combustion) cylinder head. Not only did performance improve, but the engine met the "emissions standards, regulations [supposedly] so rigid that General Motors was protesting loudly in the press and in the halls of Washington that they could not be attained with the current technology" (Yates, 1983:38).

In their fight against safety and pollution legislation, the American auto manufacturers gained an unexpected ally in the UAW. Union representatives in Washington lobbied to block these reforms, believing that they would raise costs and thereby eliminate jobs. Thus while the federal government attempted to aid auto consumers and the general public, organized labor too ignored its broader responsibilities, knowing that the exclusive collective-bargaining rights granted by the 1930s labor legislation protected its position. To some extent, the UAW even benefited from GM's big-car policy, for higher prices allowed higher wages to be extracted. And despite its rhetoric about enriching jobs or making manufacturers responsible, the UAW invariably opted for *more* money at the bargaining table. Accordingly, the UAW negotiated wage rates that were probably the highest in the world for semiskilled labor.

With the Vietnam-induced inflation spiraling, the UAW demanded even higher wages in a vain attempt to match a seemingly endless round of price increases. After a lengthy strike in 1970, GM's negotiators acquiesced again and signed another inflationary labor contract with the UAW. First, wages would be increased by $1.40 per hour over the three-year contract term (Cray, 1980:467). More seriously, GM agreed to give the UAW's members a cost-of-living adjustment (COLA) that was not tied to productivity increases. "The settlement also provided a model for contracts in other industries over the next six months; the post office, the copper and aluminum industry, aircraft and communication workers, and steelworkers successively received 50-cents-per-hour wage boosts and/or cost-of-living clauses" (pp. 467–468). So GM would not only be paying higher wage bills but would also be paying more for its materials and services. The inflationary spiral continued unabated.

If one speculates on why GM would negotiate such a dangerously expensive contract, four reasons come to mind immediately. First, and most obvious, the UAW strike was generating serious losses. Second, the GM management probably thought that it need not worry about being at a competitive disadvantage since the contract negotiated with one domestic manufacturer was always adopted by the other two.

Third, the firm may have expected to buy itself greater industrial peace. Since the middle 1930s, GM had countered worker militancy with more lucrative contracts: "GM pioneered the wage increase linked to the cost of living back in 1948" (Burck, 1981:56). Besides, such liberal wage policies had been viewed as a business stimulant since the day Henry Ford first announced FM's $5-a-day pay rate. Workers had to be well paid if they were to boost demand by purchasing the cars they produced. Sloan (1941), too, believed that human progress sprung from endless cycles of higher wages, greater demand, increased volume, and heightened productivity.

Only management, however, could make sure that productivity rose, for Sloan (1941) did not place much faith in the "common" production worker's ability to improve GM's productive processes. In his industrial scenario, workers did exactly what management told them, and the union representatives helped enforce discipline. Sloan's antipathy toward labor and labor's representatives was a "legacy of the 1937 sitdown strikes, which more than anything else prevent[ed] either side from approaching a common problem in a spirit of understanding and sympathy" (Drucker, 1946:177). Unfortunately by 1970 little had changed. Because of a pattern established in 1937, "far too many people in management prefer even today to escape into the belief that workers are a race apart and almost subhuman . . . rather than face a difficult and dangerous problem." Since many of GM's early workers had been poorly educated immigrants from Europe and migrants from the South, there existed "a tendency, especially among the lower ranks of management, to feel superior to the worker, or at least to see him as an alien."

With such attitudes permeating management's thinking at all levels and solidifying over time, it took more and more money to keep each successive generation of better-educated workers on the job. But even with more money the workers cooperated little in management's efforts to raise productivity, in spite of the union leadership's expected effort to deliver a compliant work force. In sum, "the U.S. auto industry's high wages and fringe benefits have failed to win it the affection, the loyalty, or even the consistent attendance of its workers" (Burck, 1981:56). But in 1970 GM's management had tried again, for a high-wage policy was its only response to a disgruntled work force. "Business was so good that it was easier to pay off, essentially in the form of industrial blackmail, a work force more interested in creating an opposing power structure than in joining a campaign to increase efficiency" (Yates, 1983:172).

Fourth, whether this tactic succeeded or failed did not make much difference, for GM's management undoubtedly expected to be able to pass its increased costs on to the consumers. GM, after all, enjoyed the role of industry price leader. Ford and Chrysler had no choice but to follow, and the consumers would have to pay (Galbraith, 1967:30). Indeed, whenever two economic monoliths like GM and the UAW got together to divide the spoils reaped in the oligopolistically controlled automobile industry, the unorganized mass of consumers picked up the tab. They had to: the modern American transportation system—with the help of the auto, highway, and union lobbies—depended on the car. Automobiles were no longer luxuries but necessities. One could barely exist without them.

But with prices escalating, more consumers sought out transportation that was more reasonably priced. Facing a growing demand for small and inexpensive cars along with prodding from the Johnson administration, GM announced that it would develop the XP 887 car, later marketed as the Chevrolet Vega. GM intended the Vega to be a model project. With the help of automatic (i.e., robot) welding machines, for instance, the assembly line would run at the phenomenal rate of slightly more than 100 cars per hour. Turning out a car every 36 seconds meant minimum production costs and strong profit margins. To head this effort to beat back the low-cost imports, GM looked to its fastest-rising star. After his very successful experience in tapping the performance-oriented youth market for Pontiac, John DeLorean was made the Chevrolet Division's manager and charged with developing the Vega. His problems began immediately. Chevrolet's engineers were unhappy with corporate design for the car. Worse yet, "after eight miles . . . on the test track, the front end of the Vega literally fell off. It would be only the first of many problems" (Cray, 1980:473).

To avoid labor trouble, GM chose Lordstown, Ohio, as the plant site to produce the Vega. By locating in a rural setting, the firm hoped to avert the labor strife and Black unrest that so permeated the large industrial cities of the east and midwest.

Indeed, during the late 1960s and early 1970s GM faced an environment on its labor front far more tumultuous than the one it then encountered in the marketplace. With the widespread social unrest and economic prosperity caused by the Vietnam War, a new generation of auto workers became restive. They seemed to feel that there was more to life than the mindless drudgery of the casting foundry, stamping plant, or assembly line. They wanted more than just the high wages that had satisfied their fathers and mothers and that GM's management and the UAW establishment routinely granted them. And if they could not get any human respect or job satisfaction they would minimize the time spent working. The day had dawned when production employees worked four days a week only because their wage rates would not allow them to make a living in three days. "At General Motors, for instance, as [the] new generation of workers entered the plants, absenteeism rose 50% between 1965 and 1969, and ran as high as 20% on Mondays and Fridays. Over the same period employee turnover climbed 72%, the number of grievances lodged against supervisory personnel rose 38%, and disciplinary dismissals increased 44%" (Lawrence and Dyer, 1983:43).

Into this dangerously flammable atmosphere GM's corporate management introduced its hottest spark: General Motors Assembly Division (GMAD). Donner had created this unit in 1965 to specialize in assembling *all* the corporation's cars. GMAD, noted for its severe cost cutting, would run the Lordstown Plant after October 1, 1971, in an effort to improve the small car's return rate. Almost immediately, "seven hundred workers, including the bulk of those assigned to quality control, were summarily dismissed as super-numeraries" (Cray, 1980:474). Spurred on by GMAD's myopic cost-cutting and speed-up efforts, even the supposedly cooperative workers at rural plants like Lordstown readopted the militancy of the middle 1930s. Frustrated and infuriated, they struck the plant for three weeks in early 1972. (For the workers' side of the story, the reader is referred to Rothschild, 1973). Thereafter, "for six months the 'Lordstown Syndrome' was a staple of newspaper editorials and television reports. Striking back at a corporation they considered ruthless, the workers were quick to point out to reporters defects in the cars" (Cray, 1980:474). Worse yet, "workers claimed that supervisors authorized shipment of defective cars; the company claimed that workers attacked [that is, sabotaged] the paint, body, upholstery, and controls of the Vegas" (Rothschild, 1973:101). In Rothschild's view, "Even business opinion came close to criticizing GM for a lack of humanity in its plant management" (p. 17). After the Corvair affair with Ralph Nader, GM could hardly afford to tarnish its image further with such damaging publicity.

Though it is hard to imagine, GM's model project encountered still more difficulties. The Vega's "cooling system proved inadequate and the engine block warped; the front disc brakes wore rapidly, or, worse, gave out suddenly; the combination of a faulty carburetor and ruptured muffler could lead to the automobile catching fire; the accelerator could jam open; and, most disconcerting of all, the rear wheels were liable to drop off due to an error in production that left the axle a fraction of an inch too short" (Cray, 1980:473).

Lastly, the Vega failed to give GM an entry in the low-price field. "A foot longer and $300 more expensive, the Vega really did not compete with the lighter Volkswagen, Datsun, and Toyota" (Cray, 1980:473). And by loading many Vegas with options like air conditioning, GM further increased the car's initial cost as well as its operating expense. The firm's small-car program now had two strikes against it, with the second coming at a most inopportune time.

Another blow came in late 1973, just when GM introduced its newly restyled large cars. After the Yom Kippur War between Egypt and Israel and the accompanying Arab oil embargo, many people realized that the domestic auto makers had led the country on a dangerous binge of big cars, large V-8 engines, highway building, and suburbanization. America was now over a barrel. "If any company appeared threatened by the oil embargo, it was the largest, the most heavily reliant on big-car sales, the corporation that had waxed rich on cheap energy. General Motors depended so heavily on large cars that its corporate average fuel economy was the worst in the industry, just 12.2 miles per gallon, substantially below the 14 miles per gallon average of all cars" (Cray, 1980:485). Even with GM's offerings dragging it down, the average for all domestically built 1974 models reached 12.8 (Yergin, 1979:148).

Once again the federal government reached into the automakers' realm and removed more decision variables from their control: Congress passed the Energy Policy and Conservation Act of 1975 (Yergin, 1979:150). This law established "corporate average fuel economy" (CAFE) standards starting after 1977. All the cars sold by a given manufacturer would have to average 18 miles per gallon in 1978, 19 in 1979, 20 in 1980, and 27.5 in 1985.

GM's strategists now found themselves confronting extremely tight—perhaps even conflicting—constraints. The federal government's safety legislation necessitated weight additions to satisfy "crashworthiness" standards, but heavier cars sapped fuel economy. Similarly, the American manufacturers' approach to pollution control further drained gasoline tanks. Only one strategy would solve these constraints simultaneously: the small car.

To fill the strong demand for fuel-efficient vehicles and to meet the most immediate CAFE standards, the automobile manufacturers desperately needed smaller models to sell. If they sold enough gas-saving cars, the government would permit them to continue marketing their high-profit behemoths without incurring fines. Over the longer run, however, the latter vehicles would have to be "down-sized." In the future all cars would be relatively small whether they were inexpensive subcompacts or luxury sedans. American manufacturers would not only have to reduce engine capacity but also car size and weight especially. Improved engine efficiency and body aerodynamics would be necessary too. Engineers, not stylists, would be designing future cars if a company was to remain competitive.

Despite a renewed emphasis on engineering, GM encountered problems selling enough small cars to offset its still overly inefficient large cars. The firm thus would fail to meet the 1983 CAFE standards, and with the increasing demand for large cars in 1984, it faced the prospect of enormous fines ("Detroit's Feverish Maneuvering to Sell Small Cars," *Business Week,* January 9, 1984:26). Japanese and European auto manufacturers were already well-positioned for these new environmental conditions. Consumers in their countries had long paid realistic fuel prices and thus had demanded efficient, reasonably sized cars.

GM and the other domestic manufacturers had difficulty meeting the government's fuel-economy standards because they had lost their almost exclusive grip on technological leadership. The problem was widespread: America had lost technological leadership, not just in automobiles but in a whole range of basic support industries. The United States—preoccupied with Vietnam, political protests, racial strife, Watergate, and double-digit inflation—temporarily turned its back on the world, only to find when it looked outward again that the world had changed immensely. Hence, industrial firms in Europe and Japan after years of quiet but steady technological development (usually in close cooperation with their respective governments) burst onto the international economic scene as the technological leaders in such traditional industries as railroads, steel, machine tools, construction equipment, consumer electronics, and automobiles. They also had gained a lead in such new areas as robotic assembly and welding equipment. Along with their technological leadership, many had cooperative labor forces that accepted or even suggested innovations and received considerably less for their services than their American counterparts. And long in tune with producing for quality-conscious, value-oriented consumers, they penetrated deeply into the American market with an array of sophisticated, reasonably sized cars ranging from the cheapest to the most expensive.

The American manufacturers, on the other hand, were ill prepared. GM especially had not made a truly successful effort to diversify into the low-price small-car market. Since the days of challenging Ford's Model T, GM tended to stay up-market from the lowest-price segment. Sloan and his cohorts well knew that the return rates were significantly higher there than in Ford's barebones utility market. Nevertheless, as was proved during the 1930s, Sloan and GM could compete most effectively in the low-price field. Subsequent GM managements, however, had moved too far away from the low-price strategy that saved GM during the depression. By the late 1960s and the early 1970s, GM had firmly entrenched itself in the large-car market. True, the firm tried the Corvair and Vega, but both had become public-relations disasters because of efforts to eke GM's accustomed return rate out of the cars.

It should be pointed out that FM fared no better with its low-cost Pinto. In some collisions gasoline "spilled out and frequently ignited" (Iacocca, 1984:161) with terrible consequences. Denying the problem (probably for liability and public relations reasons), Ford delayed in recalling more than a million Pintos (p. 162).

Foreign car manufacturers like Volkswagen, Toyota, Honda, and Datsun held strong positions in the low-cost small-car market. The Japanese, in particular, "had tested this market as early as 1959 with an execrable product, withdrawn for redesign and study of American methods and then returned with an increasingly attractive group of cars" (Mandel, 1983:405). The Japanese auto makers had done their homework well; no longer did their lightweight low-cost cars collapse under the sustained loads of long-distance high-speed driving sometimes encountered in the United States. In general, they held up far better than the Vega and Pinto. Moreover, the Japanese used the 1960s and 1970s to develop extensive distribution facilities and dealer networks. They could not thereafter be dislodged easily, especially by manufacturers burdened with questionable product images.

At GM, management's "big" problem stemmed from its big cars. The loss of technological leadership hurt most in the middle and expensive price ranges. By the mid-1970s, GM no longer set the standard to which the American consumer aspired. The market share for makes like the Saab, Volvo, Peugeot, Porsche-Audi, BMW, and especially Mercedes-Benz may not have been large, but more than ever these cars epitomized the qualities affluent Americans demanded when purchasing their personal transportation. More and more, such vehicles were seen in parking lots at university campuses, law offices, medical buildings, executive headquarters, and exclusive shopping malls. Energy-efficient, soundly engineered for safety and durability, well-built, and functionally styled, they offered value; and, more importantly, they retained it. Though admittedly expensive for their outward size, they represented stable investments, very much unlike GM's product offerings, which depreciated immensely as soon as the purchaser drove them from the dealer's lot. To GM's chagrin, it meant little that few Americans really understood or appreciated the engineering components that made these cars America's newest status symbols. Nor did it matter that America's straight highways, suburban drivers, and recently lowered speed limits did not always demand such technical excellence. If GM could project affluence with vinyl roofs, opera windows, concealed windshield wipers, hidden radio antennas, and rectangular headlights, why not achieve the same end with overhead camshafts; electronic fuel injection; hemispherical combustion chambers; removable wet cylinder liners; turbo chargers; transitorized ignitions; five-speed overdrive transmissions; lightweight aluminum cylinder heads, intake manifolds, and transmission housings; controlled-crush unitized bodies; rack-and-pinion steering gear; four-wheel independent suspensions with disc brakes; radial tires; and ergonomically designed interiors? Here, at least, the consumer got more value because the cars were generally safer, performed better, and lasted longer than Detroit's.

Perhaps even more distressing to GM was its belated realization that technologically innovative vehicles commanded high prices. Such cars were luxury goods where the high prices added to the status, resale value, and demand. By ignoring technological development and concentrating instead on valueless styling changes, GM had quite literally missed a golden opportunity to earn the high profit margins it so coveted. Worse yet, GM's knack for developing and introducing meaningful technical improvements had atrophied. Catching up would take considerable time.

Though it took many years to prove, Kettering was wrong when in 1927 he wrote to Sloan that European automotive engineers too often ignored the commercial side of the business. "Sophisticated engineering for its own sake was, in his view, a luxury the American automobile industry would have to forgo" (Leslie, 1983:198). It was this attitude that allowed GM's once strong engineering arm to atrophy, except for such convenience devices as air conditioners and automatic transmissions. Cadillac, noted as a technological innovator and renowned as a standard of excellence in the 1920s, now shared its engineering with the commonplace Chevrolet. Chevrolet's engineering tradition, unfortunately, focused on low-cost production, not on technological advancement and high quality.

Probably more than anything else, years of double-digit inflation shifted the values of American auto buyers. Having discovered just how dependent the industrialized world had become on its petroleum supplies, the oil cartel of the Organization of Petroleum Exporting Countries (OPEC) began escalating the price for their crude oil. These energy price increases fed the inflationary spiral begun with President Johnson's escalation of the Vietnam War. Suddenly, GM found the great mass of consumers becoming value conscious. Repeated price increases of sizable proportions in the middle 1970s gave less affluent new car purchasers "sticker shock" whenever they walked into dealers' showrooms. For instance, in 1974 alone GM increased its prices by 20 percent or $1000. On top of this tab, the overburdened consumer had to fuel the car with much more expensive gasoline. To ease the pain, lenders lengthened installment contracts from 36 to 48 months—a move that only made consumers even more value conscious. If they needed to keep their cars a full four years, they wanted them both to last longer and also to operate more efficiently than their trade-ins. They demanded economy, reliability, and quality. "Tangible criteria" became the "common denominator" to consumers when "making purchasing decisions" (Abernathy, 1978:39).

Here, again, the American auto manufacturers stumbled badly. Having only each other and the still fairly weak foreign manufacturers as competition for many years, they let quality slip. *Consumer Reports* first noticed the slide during the Curtice-initiated sales wars of the middle 1950's. Donner's shortsighted cost-cutting pressures did not improve the situation. Thus, in addressing "Quality Control, Warranties, and a Crisis in Confidence" (March, 1965), the magazine reported that the condition of the 1965 cars it purchased was "about the worst, so far as sloppiness in production goes, in the whole 10-year stretch of deterioration that began in 1955" (p. 173). More specifically, owners of 1964 cars responding to the magazine questionnaires ranked GM quality below that of Volkswagen, Chrysler, Ford, and American Motors (p. 174). Still, *Consumer Reports* had seen nothing yet, for Donner had just established the cost-cutting GMAD in 1965.

Until the middle 1970s quality had not really mattered very much, since many new-car purchasers traded their cars fairly quickly. But as they held onto their vehicles longer, their demands shifted rather abruptly. Customers required to spend close to $10,000 for a car did not easily forget "lemons" or high repair bills, especially when they suddenly had high-quality, good-value substitutes available to them. Not surprisingly, a Chrysler study discovered a startling reversal in consumer preferences between 1975 and 1979 (Ceppos, 1981:43). Brand loyalty plummeted "from first in importance to sixteenth." Styling fell "from second to ninth." Even more ominously for the domestic manufacturers, quality soared from sixteenth to second. And " 'value' . . . jumped from eighth to the number-one slot."

After the middle 1970s, then, GM faced trouble along its entire front of offerings. Japanese manufacturers threatened the low-price segment with their high-quality small cars. European firms like Mercedes-Benz threatened the high-price range with their advanced engineering and classic styling. More seriously, both these groups could be expected to advance into GM's lucrative middle-price bracket. Not only did they have an edge in engineering and quality, but in addition their cars were already properly sized.

As Whyte (1956) put it, "Even the largest corporations must respond to changes in the environment; a settled company may have its very existence threatened by technological advances unless it makes a bold shift to a new type of market. What, then, of the pruning and molding that adapted it so beautifully to its original environment? The dinosaur was a formidable animal" (p. 216). Having squandered its lead time by denying, ignoring, or fighting the market's changes, GM needed to adapt and quickly.

With the evidence now unmistakable, GM's management finally initiated a series of quick adjustments. First, for the 1976 model year GM "pulled together a crash program to replace the hard-luck Vega with the subcompact Chevette" (Burck, 1981:50). To save time and money, "the Chevette would be built, using component designs from Opel and other overseas divisions, mainly Brazil" (Burck, 1978:89). As GM's then president, Elliot (Pete) Estes, explained to an interviewer from *Car and Driver*, the Chevette was introduced for "the kid out of college who hasn't got very much money and is infatuated with small cars, and fuel economy, and ecology, he's the guy we have to get, because he's going into that damn Volkswagen or Toyota" (Quinter, 1976). Nevertheless, the particular Chevette that was highlighted in the Estes interview had about $1550 worth of accessories added to the car's base price of roughly $3100. These options included a $424 air-conditioner that reduced fuel economy by adding weight and absorbing power. GM's addiction to high profit margins would not end just because the market forced it to get serious about small cars.

Second, to fend off the high-priced imports, the Cadillac Division introduced the Seville for the 1976 model year. The Seville's primary target, of course, was Mercedes. "From about 22,500 deliveries in the United States in 1968, the marque had doubled sales to about 44,000 by 1973—versus a forty-three percent increase for luxury cars as a whole. Moreover, Mercedes was attracting a different kind of U.S. buyer . . . younger, female, better educated, and 'more upscale' in occupation than the average domestic car buyer" (Hendry, 1983:333). The latter problem was the most serious, for Cadillac ownership was no longer seen as a status symbol for a very influential segment of the market.

GM planned the Seville as "a luxury Cadillac fitted to a reinforced frame from a compact [Chevrolet] Nova" (Cray, 1980:491). Thus, the Seville would be considerably smaller than the typical Cadillac. To give it the now coveted European luxury look, GM emulated the boxy, squared-off, functional styling of the Mercedes-Benz four-door sedans.

Third, to improve the fuel economy of its models in the middle price range, GM's management accelerated the down-sizing of its big cars. "Originally, the shrinkage was to have begun with a modest scaling-down of the 1977 full-sized cars, but after the embargo the company more than doubled the planned weight reductions, from 400 to 1,000 pounds" (Burck, 1981:50).

For the 1979 model year GM planned an even more dramatic adjustment to the new marketing environment. For its intermediate-size X-car, GM "plotted a complete and systematic break with the past" (Burck, 1981:50). Most significantly, these so-called "high-technology" cars featured front-wheel drive. Besides reducing weight this arrangement increased the usable space within the passenger compartment by eliminating the driveshaft tunnel. "Virtually everything about the car was new, and its major components, such as transmission, axles, and brake system—along with their manufacturing facilities—would be the basis for many new models to follow, both bigger and smaller" (p. 50).

The firm's immense capital resources thus enabled management to revamp its product offering fairly quickly. On this round, the timing was perfect. Once the Shah of Iran had been overthrown in 1979, the supply of Iranian oil was cut. A second oil crisis erupted and the long gas lines returned. GM "was launching a new line of fuel-efficient cars, the X-cars, just as a fuel-conscious public was clamoring for them" (*Business Week,* October 4, 1982:75). The future looked bright as positive trends began to appear; an "exultant management" watched "the company's share of the total auto market climb from 42% in 1974 to nearly 48% in 1978, while profits that year reached a record $3.5 billion on sales of $63.2 billion" (p. 75). In contrast, FM and Chrysler staggered under the latest blow. Ford seemed by far in the worst shape since it had retained many of its dated product designs to save money. So "Ford, in a state of shock, had to tear up its product plans and scramble desperately. GM simply accelerated projects already in place" (Burck, 1981:50).

Shortly thereafter, however, several significant problems surfaced that severely damaged GM's abortive recovery. One problem dealt with the increasingly deteriorating quality of GM's cars, and the other resulted from the confusing and overlapping product lines GM offered to consumers.

With respect to the quality issue, "the X-car, which GM promised would deliver unprecedented reliability, has suffered from an awful lot of defects" (p. 56). More specifically, "the X-car was recalled seven times for safety-related problems during its first year" (*Consumer Reports,* January, 1981:12). Somewhat later, *Consumer Reports* concluded, after receiving about 19,000 owner responses on all four nameplates, that: "X-cars have serious trouble in many areas: manual transmission and clutch, electrical system, fuel system, and paint" (February, 1982:92). So "after nearly three years' experience with the X-car design, one would think that GM would be able to turn out a debugged, reliable car. But that is not the case" (p. 95). Finally, in August, 1983, the Department of Justice sought $4 million in penalties and the recall of 1.1 million 1980 X-cars, alleging that GM tried to conceal, not correct, an infamous brake problem.

Still more damaging for GM's management and disheartening to its consumers, the X-car quality problem was not isolated. For instance, "the failure of thousands of Oldsmobile diesel engines, for reasons ranging from water damage to their injectors to broken crankshafts, has left customers wondering whether they are being used to debug new GM products" (Burck, 1981:56). In a similar vein, *Consumer Reports* issued "some early warnings" on the J-cars' anticipated reliability (January, 1982:11). J-cars were a smaller descendant of the X-cars. The problems encountered by the magazine's staff and a smattering of consumers included the usual fit-and-finish defects, but also several more serious engine and transmission malfunctions (p. 11). Still other maladies afflicted GM's front-wheel-drive A-cars, which represented the larger version of the X-cars. For A-cars with four-cylinder engines, *Consumer Reports* predicted a "worse than average" repair incidence. "Trouble spots include the engine and brakes—not a comforting prospect. The V6 versions are even worse; we predict a much-worse-than-average repair record, with problems in the engine, automatic transmission, brakes, cooling system, fuel system, and other areas" (February, 1983:77). Quality problems surfaced even in rear-drive models, despite GM's decades of experience in building them. "The 1979–80 [Buick] *Regals* were among the more reliable of the domestic models. The 1981 and 1982 *Regals*, however, have been a disaster, with a much-worse-than-average predicted repair incidence. The other GM G-cars share a similarly unhappy prognosis" (p. 78). In yet another critical article *Consumer Reports* noted that "the large *Chevrolet*, like many other GM models, has shown a decline in reliability in recent years" (March, 1983:130). Similarly, "the reliability of *Cadillacs* has slipped in recent years, according to our Annual Questionnaires. . . . The new 4.1-liter aluminum V8 has been especially troublesome" (p. 129).

In sum, GM's reliability problems seemingly mounted in the late 1970s and the early 1980s. This apparent decline in quality came at a time when astronomical interest rates made an already sticker-shocked, value-conscious consumer even more cautious about buying GM's products. Not since the early 1920s — when Sloan doggedly pursued the Chevrolet and Oakland Divisions to improve their cars — had GM products suffered under such a deleterious reputation for quality.

GM, however, still insisted that its quality problems affected only superficial fit-and-finish characteristics and that some American consumers had jumped to an erroneous conclusion about the lack of mechanical reliability in its products. According to *Business Week* (October 4, 1982), "GM has voluminous studies that say domestic car-makers lead in durability of such components as engines, brakes, and transmission" (p. 76). It remained an open question, then, why GM's statistics were not corroborated by *Consumer Reports'* survey data or reflected in the GM cars purchased by that organization for testing. Furthermore, one might ask why GM could not assemble even its expensive cars with fits and finishes comparable to low-cost Japanese automobiles. GM, after all had prided itself since the 1920s on the "appearance" of its products — particularly, in the quality of its paints and interior finishing materials — and knew full well that technically unsophisticated, tire-kicking American car buyers focused on such superficialities.

The other problem confronting GM at this time was the muddled images that the divisions' products projected to the consumers. Under constant pressure from the corporate accountants to cut costs, the divisions' cars came to look more and more alike. (See, for example, the Chevrolet, Oldsmobile, Buick, and Pontiac clones on the cover of the August 23, 1983, *Fortune*). While Sloan had started this trend in the 1930s with his common-body program, he always worked hard (as was shown in Chapter 6) to maintain GM's product diversity. But "before the end of the 1960s, the Chevrolet Monza, Pontiac Sunbird, and Buick Skyhawk; the Buick Special and Pontiac Tempest; and the Chevrolet

Camaro and Pontiac Firebird were strikingly similar" (Cray, 1980:449). Similarly, "the front-drive X-cars [GM] introduced through four divisions in 1979 all looked about the same" ("General Motors: The Next Move Is to Restore Each Division's Identity," *Business Week*, October 4, 1982:76). Given the strong demand for fuel-efficient vehicles after the 1979 oil crisis, this lack of diversity did not hurt the early X-car sales. Encouraged by the initial success of the almost identical X-cars, GM's financially oriented corporate headquarters moved further from Sloan's (1964) dictum, "Each line of General Motors cars produced should preserve a distinction of appearance, so that one knows on sight a Chevrolet, a Pontiac, an Oldsmobile, a Buick, or a Cadillac" (p. 265). GM's new subcompact J-cars thus appeared far too much alike. A similar problem afflicted the sales of GM's newly designed mid-size products. "So, while GM argues that its new cars have plenty of unique sheet metal from one division to the next, critics say the cars still look too much alike" (*Business Week*, October 4, 1982:76). The critics seemed to be right, for "the company's clutter of new, look-alike models has been confusing customers and blurring the traditional marketing distinctions among its five automobile divisions" (p. 75).

GM's relentless effort to generate big-car profits from its small car sales added to the consumers' bewilderment. Many small cars loaded with expensive accessories landed in dealers' showrooms with sticker prices equal to or in excess of the firm's big cars. The fact that practically every division offered a variant of the X-car, A-car, G-car, etc. further blurred the divisional distinctions. Thus, "GM destroyed the sharp differences that once made it easy for buyers to know if they should be shopping for a Chevrolet or a Buick" (p. 78). General Motors offered a product line even more redundant than the one Sloan and Pierre duPont had inherited from Durant in 1920–1921. Durant, after all, left the divisions alone.

As was noted in Chapter 1, GM's management of the early 1980s was attempting to reestablish distinct images for the five car divisions by readopting the marketing strategy Sloan set down in the 1920s (pp. 75–76). Key elements in this new (or old) approach were to be a graduated price structure with little overlap among makes as well as increased styling and product differentiation.

GM's product differentiation also suffered in less obvious ways. When the firm's engineers developed the new J-car, for instance, the usual emphasis on return rates forced them to use many key chassis and drive-train parts from the heavier X-cars (Yates, 1983:29). But one simply could not build the J-car using major X-car components and still finish with a significantly lighter, more maneuverable vehicle. Sharing parts from one generation of cars to the next, moreover, inhibited innovation. So when marketed, GM's "revolutionary" J-car resembled an underpowered X-car. "In design, they had little that seemed new and sales were sluggish" (Dorfman, 1983:101).

The attempt to apply Sloan's interchangeability philosophy in this case was questionable, for interchangeability only yields acceptable results where cars are quite similar in size and weight. Furthermore, it is risky during periods of rapid technological change. Strong foreign competitors like the Japanese, for example, prefer to engineer completely new vehicles rather than accept the compromises and hidden costs of fitting old parts.

Much of the blame for the design defects, quality decline, and diminished differentiation in GM's products ultimately is attributable to the firm's increasingly centralized structure. "GM began centralizing product engineering in the 1950s," for instance, and "centralization reached its peak during the 1970s" (Burck, 1983:98). To a large extent, even GM's failure to compete successfully in the 1960s and 1970s small-car market resulted from the firm's over-centralization. Moreover, such problems permeated the entire corporation because GM's decision-making was dominated by a single entity: the corporate headquarters. Over the years, the tremendous number of variables set by corporate headquarters along with the stringent review procedures imposed on the divisions served to stifle divisional adaptability. GM no longer could expect its divisions to adapt for it, as the divisions were tightly constrained to follow corporate dictates. Yet, unfortunately for all involved, the corporate headquarters may have lost its ability to adapt, the original Sloan-Brown designers having long passed from the scene.

As was shown previously (especially in Chapter 6), Sloan began centralizing GM's decision making in the 1920s. This trend accelerated in the 1930s because the depression forced an increase in the consolidation of and coordination among the divisions. With the precedent well established by Sloan, subsequent corporate management teams found it comparatively easy to make further inroads into the less and less decentralized divisional domains. Harlow Curtice, for instance, "made an increasing number of decisions himself" (Dale, 1960:107). Frederic Donner's administration probably did the most to constrain the divisions' decision-making freedom. Donner's creation of the GMAD, over "the objections from the automotive divisions" (Cray, 1980:448), for instance, stripped the automobile assembly function from the product divisions. But even this move was in keeping with some old GM traditions originated by Sloan. As GM's diversification strategy had always remained quite restricted, there was a strong tendency for this essentially one-product firm to gravitate toward a functional structure where one subunit focused exclusively on assembling finished automobiles. In fact, the first forerunner of the GMAD was Sloan's old Interdivisional Committee for Works Managers.

Besides creating GMAD, "Donner progressively trimmed the authority of the automotive divisions to design their own cars, insisting instead that preliminary design work be done at the corporate level and that the divisions share as many parts as possible" (Cray, 1980:406). These reorganizations exemplify Donner's strong disposition toward centralization. Generally "intolerant of dissent, Donner's insistence upon executive loyalty above all else was to eventually gut the automotive divisions of their independence and what little creativity remained in the corporate world dominated by professional managers" (p. 406).

Though Donner implemented his centralization moves with an abrupt and heavy hand hardly in keeping with Sloan's patient, subtle ways, Donner still did not stray far from Sloan's strong predisposition toward autocracy. How is it, then, that Sloan's movements toward centralization helped GM prosper while Donner's damaged its prospects? Much of the difference lies in the type of executives chosen to fill the top corporate posts. As the firm grew more centralized over time, it—more than ever before—needed corporate executives with entrepreneurial initiative, long-run vision, and automotive expertise. But men with such qualities no longer ascended to the top of the hierarchical ladder.

When Sloan ran GM, however, the corporate offices had such executives. First, during GM's formative days men like Pierre duPont, Raskob, Sloan, Mott, and Kettering guided GM's destiny; for "in a young organization, independent, entrepreneurial types come to the top." However, with success, these early risk-takers grow conservative, even intolerant of the "behavior that enabled them to succeed." The aspiring manager then must choose among leaving, cloaking "initiative under a facade of conformity, or acquiescing to puppetry." This leadership style rewards conformity (Myers and Myers, 1974:13). Even Sloan (1964) himself eventually acknowledged that "General Motors is not the appropriate organization for purely intuitive executives, but it provides a favorable environment for capable and rational men" (p. 433). When he left the firm, DeLorean thought the organization "so rigid" that "an innovative thinker like Alfred P. Sloan, Jr." himself would not "qualify for a job in the upper ranks of General Motors" (Wright, 1979:280).

Besides exerting considerable entrepreneurial initiative, the owner-managers who ran GM under Sloan took a long-run viewpoint. As DeLorean put it, "Their decisions were biased as much in favor of the long-term growth and health of the company as they were in favor of the short-term profit statement" (Wright, 1979:259). Not only did men like Donner overemphasize short-run performance with their suboptimal cost-cutting campaigns, but they did not have "the sense of history and understanding of General Motors system necessary to manage the business effectively" (p. 250).

Third, during Sloan's tenure GM's corporate executives were broadly based, financially oriented automotive engineers, production experts, and marketing strategists. Sloan's own career established the prototype since he had gained considerable experience in all these areas. Under Donner, on the other hand, narrowly focused financial specialists—lacking almost any knowledge regarding automotive design, production, and marketing—dominated the company. Donner perceived no problem in such a lack of industry expertise, for he too "had spent his entire career in finance; he knew comparatively little about operations" (Cray, 1980:405–406). In short, he "was not a man well-schooled and experienced in the operations of the corporation" (Wright, 1970:227). So even though Sloan approved of Donner's short-run performance achievements, DeLorean concluded that "the delicate balance at the top of the world's largest industrial corporation was starting to tip toward the financial side of the business, and in the process, the substance of the organizational system so thoroughly thought out by Sloan was beginning to dissipate" (p. 227). For instance, "skipping over a half-dozen men from the automobile side, the Donner-manipulated board of directors named the corporation's treasurer, Thomas A. Murphy, as vice-president in charge of the auto and truck divisions" (Cray, 1980:480). Unbelievably, the bookkeeper Murphy knew *nothing* about operating an assembly plant, meeting a production schedule, or even the functioning of an automobile (pp. 496–497). He needed to acquire a little quick operating experience, however, as Donner had slated him to become the chairman of GM's Board of Directors. Murphy held GM's top position from 1974 to 1980. Before retiring Murphy maneuvered the selection of his successor, Roger Smith, another financial executive (Kanter, 1983:344). Evidently, "operating executives with a strong product and people orientation" dissented "because of their concern that the financial side had too long dominated the company and that 'the corporation uses figures too much.' But the 'financial votes' prevailed" (p. 344).

By the time Murphy retired in 1980, GM was struggling to adapt to an environment that had undergone a complete metamorphosis. The shift had not been sudden except at the climax around 1980, for the various factors that contributed to the change had started at different times on several fronts and progressed at different rates. Government regulation grew, labor dissatisfaction and compensation increased, energy shortages developed and costs rose, foreign competition arrived, inflation raged, interest rates skyrocketed, and consumer tastes shifted first incrementally but then dramatically. Pollution problems and energy shortages convinced Americans that an age of apparently limitless growth had ended. High inflation and the concomitant interest rates eroded their purchasing power, giving them no choice but to accept a level or reduced standard of living. Lastly, a declining economy, led by the domestic automobile industry with its massive unemployment, ushered the United States fully into an age of diminished prospects. In such an era, function superseded form.

Ironically, Sloan's GM had passed Ford's FM in the dissimilarly expansive 1920s by offering style or form first and placing function second. And now, in the contracting 1970s and 1980s, foreign manufacturers gained on GM by providing consumers function over form. The tide had turned. Indeed, "manufacturers . . . like Volkswagen and Mercedes-Benz, who did not visibly change their annual product, discovered that this refusal to change itself became a valued 'special feature' with a sales appeal that was quite novel" (Boorstin, 1973:554).

By far the most unmistakable early warning of the changing tide had surfaced during the early 1950s with the introduction of the Volkswagen Beetle to the United States market. Sales took off in 1955, and more and more Volkswagens were seen on American roads. As "the first foreign passenger car to sell enormously well in postwar America, the Beetle began to rid us of the notion that big is best. In fact, it introduced the reverse, and as a trendy cultural phenomenon at that" (Mandel, 1982:417). The success of the Beetle, in turn, offered much-needed encouragement to the Japanese as well as important guidelines for mastering the American market (Sobel, 1984:29, 46).

Consumers, particularly the young college-educated, so prized the small, inexpensive, and durable VW Bug that it even outsold the 15,000,000-unit record held by Ford's similarly functional Model T. Though markedly different in outward appearance, the similarities between these two cars went quite deep. After all, Ferdinand Porsche had designed the Volkswagen to embody the Model T's characteristics and to employ its production processes. An ardent admirer of Ford, Porsche had even visited FM factories in the middle 1930s when the VW design was evolving. Like Ford's Model T, then, Porsche's Volkswagen was simple in design, well made of excellent materials, light in weight, short in wheelbase, and easy to service. Similarly, when production began after World War II under Heinz Nordoff (a former executive and director of GM's Opel subsidiary), Ford's static-model policy was followed religiously. Few technical changes were made, the biggest being the early adoption of hydraulic brakes and synchromesh transmission. More importantly, Nordoff eschewed major styling "advances" and thus permitted only minor changes such as making the rear window one piece (Sobel, 1984:37). Function had revived sufficiently to challenge form.

Still, throughout all of the prosperous 1940s, 1950s, 1960s, and even the early 1970s, GM assumed the almost unassailable position Ford had held before the middle 1920s. No manufacturer—either domestic or foreign—could hope to challenge it seriously in the enormously lucrative American market. GM alone set the standard. Consumers accepted the tastes of GM's engineers and stylists as the norm, even when it meant buying cars with gas-guzzling V-8 engines or with outlandish tailfins inspired by a World War II fighter plane. "Even the small Mercedes-Benz 230 sedan sported vestigial tail fins until the late 1960's" (Yates, 1983:42). Gas was cheap, times were good, and GM had the country's pulse. All foreign products—whether from the low-cost Japanese manufacturers or even the high-price European firms—were laughable (except to a few automotive purists). Neither GM nor the great mass of American consumers took them seriously. These quaint foreign cars had funny little engines almost like sewing machines, looked like stacked boxes or inverted bathtubs, and accelerated like snails.

All their manufacturers could hope for was that the market conditions would change, that they would be exactly in step with these shifts, and that GM would not keep pace. Thus, the Japanese and European auto manufacturers *welcomed* the altered economic environment because change to them presented an otherwise unavailable opportunity to crack the high-volume American market wide open.

GM, along with FM and Chrysler, in contrast had little to gain from such altered circumstances. To those on top, change offered mostly discontinuity, uncertainty, and risk. Therefore it is not altogether surprising that America's automotive executives procrastinated.

Also, until the very end, the signs of change were never clear-cut—without the aid of considerable hindsight. For instance, energy flows dropped to a trickle, then surged into a glut. Similarly, sales of small foreign cars soared, only to dive abruptly and leave the importers' docks jammed with immovable inventories.

At GM, another factor could have blinded the corporate executives to the forthcoming competitive challenge: The men in power in the 1960s, 1970s, and early 1980s had never known their firm as anything but the top automobile company. To them, success was natural, they were always right, consumers did not want reasonably sized, high-quality, fuel-efficient cars. Unlike Sloan and his cohorts, they had not seen GM when it survived on the market's leavings that the then dominant FM organization neglected to gather. They were too secure! And as GM's inveterate innovator, Charles Kettering (1929) warned: "Nations and industries that have become satisfied with themselves and their ways of doing things, don't last. While they are sitting back and admiring themselves other nations and concerns have forgotten the looking-glass and have been moving ahead" (p. 31).

Perhaps such a limited sense of history even explains why GM's supposedly sophisticated corporate managers failed to monitor the market's shifts, thereby repeating the mistake the once successful Henry Ford had made 50 years earlier. As DeLorean complained bitterly in a draft memo to his "cloistered" corporate superiors: "Contrary to Mr. Sloan's teachings we do not use the best information available to make decisions. None of the modern marketing tools are used regularly or extensively. When they are used the results are generally disregarded" (Wright, 1979:261). Furthermore, the market-test information that did manage to filter through to the corporate decision makers could be incomplete and therefore distorted (Yates, 1983:52–53).

Maybe, like Ford in the 1920s, GM's top executives thought they could use their firm's size to force a halt in progress and thereby maintain a place of prominence. But whatever the cause, GM's modern managements jeopardized, possibly even lost, the firm's leadership position by repeating several of the more blatant mistakes Henry Ford had committed during the 1920s and 1930s.

Most notably, they overlooked the possibility that the developing foreign manufacturers could profit from the strategy Sloan used to pass Ford back when FM was supposedly unchallengeable. In those days, FM alone set the standard. Even Ford's strongest competitor, GM, could only hope for a shift in consumer preferences from basic function to stylized form and be ready to exploit the flux at Ford's expense. In essence, here was Sloan's strategy.

As Sloan (1964) himself explained, "Now it so happened—luckily for us—that during the first part of the 1920s, and especially in the years 1924 to 1926, certain changes took place in the nature of the automobile market which transformed it into something different from what it had been all the years up to that time" (p. 149). A "parameter shift occurred" (Thomas, 1973:121) favoring the large manufacturer—especially a high-variety firm like GM. Indeed, the 1920s market shift rivaled in importance Ford's momentous introduction of his low-cost, high-value Model T in 1908.

"As the economy, led by the automobile industry, rose to a new high level in the twenties, a complex of new elements came into existence to transform the market once again and create the watershed which divides the present from the past" (Sloan, 1964:150). The more obvious factors that so revolutionized the automobile market in the 1920s included: "installment selling, the used-car trade-in, the closed body, and the annual model," along with "improved roads" and then increased urbanization.

Yet, in reality, the change was more profound, for the 1920s marked America's full emergence into the twentieth century from its "Victorian/Puritan/Frontier past" (Perrett, 1982:10). Throughout American life "a deep break in continuity" occurred, "with the sense of release that liberation brings, along with all the anxiety occasioned by the unknown" (pp. 10-11).

This shift, in turn, sparked a heightened taste for style (see, for example, Bush's *The Streamlined Decade* and Meikle's *Twentieth Century Limited*). For the most part, an impressionable America embraced the Art Deco style that originated with the 1925 Exposition Internationale des Arts Decoratifs et Industriels Modernes at Paris. Soon thereafter, Art Deco themes appeared all over the United States "in the design of beauty parlors, ocean liners, handbags, shoes, lampposts, book covers, cigarette lighters, ashtrays, dresses, ties, hats, hotels, furniture, hip flasks and factories" (Perrett, 1982:421). Of more importance to the automobile industry, the Art Deco movement marked a "complete acceptance of the Machine Age," and "featured new materials, such as Bakelite and stainless steel." It aimed "to break down the traditional divisions between art and industry, craft production and mass production." Its "colors were brilliant. Fashionable people who had lived their entire lives with subdued walls suddenly decided to paint them scarlet, or yellow, or jade green, or, best of all, orange." Style superseded form.

A new age had dawned. But whereas Ford stubbornly resisted the shift, Sloan and his colleagues eagerly embraced it "because as a challenger to the then established position of Ford we were favored by change. We had no stake in the old ways of the automobile business; for us, change meant opportunity. We were glad to bend our efforts to go with it and make the most of it. We were prepared, too, with the various business concepts" (Sloan, 1964:149-150). In sum, "the change would not be sudden except at its climax" (p. 151), which came with the Model T's demise in May of 1927. So "if a date can be put on it, May, 1927, marks a great divide in modern times. It can be used handily to date the transition from the Age of Production to the Age of Distribution" ("Selling to an Age of Plenty," *Business Week*, May 5, 1956:121).

Hence, Sloan receives both too much credit as well as too much criticism for making the Model T obsolete and leading America on a tawdry parade of styling changes. Sloan did not convert consumers to higher aspirations with his variety-marketing strategy; a changing society and the booming 1920s economy did so. Sloan's accomplishment was simply one of constantly measuring the direction and speed of the populace's taste variations and keeping GM's variety of offerings in step with their latest preferences. Often the firm had to push hard to maintain the pace, but simultaneously it was always at risk of getting too far ahead or of taking a wrong tack and having the crowd fail to respond. But with Sloan at the helm, GM maintained the proper course through the tumultuous 1920s and 1930s.

After the late 1930s GM's economic performances soared still higher, buoyed by an extended period of prosperity. Here, at last, were the conditions Sloan and his colleagues envisioned when they first started designing GM's performance control system in the early 1920s: a broadly prosperous America. As GM thrived, still more and more effort went into fine-tuning Sloan's original design. With Sloan's approval, his successors honed, polished, and oiled GM's internal workings. Though certainly not designers, these financial technicians excelled as "detailers" and "tuners"; they coaxed even more dollars from the corporate engine. The firm, set to exact specifications, performed even better in the economic arena. In fact, GM's fit with the economy after World War II became so perfect that it almost singlehandedly served as the cornerstone for the national prosperity. Sloan could only be gratified.

Both Sloan and his successors, however, overlooked or ignored the possibility that the Age of Distribution might yield to an Age of Limitation. Yet the transition came. And if a date can be attached to it, the middle of February, 1966, marked the shift for the American automobile manufacturers. On February 10, 1966, Ralph Nader testified before Congress about GM's Corvair. On February 17, almost as if scripted by a Grand Designer, Alfred Pritchard Sloan,

Jr., the unchallenged master of the Age of Distribution, died at ninety. His death coincided with the ending of an era of plenty and the opening of an era of constraint.

There had been a hint of this possibility as early as the 1930s when organized labor forced GM to share more of its earnings with the workers. But renewed prosperity allowed GM's management to keep its and the stockholders' portions constant by making the pie larger. When the prosperity finally began to flag somewhat in the late 1950s, both management and labor sliced into the consumers' and the public's shares. Soon restrictive government regulations ended this shortsightedness. Then the changes resulting from a dramatically rearranged world order buffeted American society with social and political unrest, disrupted its economy with oil shortages and inflation, and convulsed its automobile industry with foreign competition. Given GM's focal position, it felt the full force of these blows. The firm's finely tuned performance-control system could not cope, for it had not been designed to anticipate or counter disturbances emanating from the social and political sectors. The great ship that Sloan had righted in 1921 foundered again.

GM had not extended its environmental sensors far enough. Sloan's critical mistake occurred when he turned GM inward in the late 1930s. Thereafter, he contented himself with fine-tuning his original design, and thus he concentrated more and more on GM's internal details. When focusing on lower-level issues, as Sloan did, one indeed obtains a more detailed insight into the divisions' operations, and when concentrating on more global concerns one naturally attains a broader perspective on the corporation's responsibilities. Movement in either direction, however, requires that some previously held knowledge be brushed aside for the new perspective, lest the complexity grow overwhelming. No single individual or group can master more than a few perspectives. Sloan chose to go downward and inward rather than outward and upward. He traded breadth for detail.

Bibliography*

Abernathy, W. J. *The Productivity Dilemma: Roadblock to Innovation in the Automobile Industry.* Baltimore: Johns Hopkins University Press, 1978.

Abernathy, W. J., and K. Wayne. "Limits of the Learning Curve." *Harvard Business Review,* 1974 (September–October), *52,* 109–119.

Abernathy, W. J., K. B. Clark, and A. M. Kantrow. *Industrial Renaissance: Producing a Competitive Future for America.* New York: Basic Books, 1983.

Ackoff, R. L. *A Concept of Corporate Planning.* New York: Wiley-Interscience, 1970.

"Alfred P. Sloan, Jr.: Chairman." *Fortune,* 1938 (April), *17,* 72–77+.

Allen, A. H. "Ford's Administrative Organization Gets Another Reshuffling, Latest Change Being Another Step in Company's Decentralization Plan." *Steel,* 1949 (February 21), *124,* 79.

Ashby, W. R. *An Introduction to Cybernetics.* London: Chapman & Hall, 1956.

Ashby, W. R. *Design for a Brain: The Origin of Adaptive Behavior* (2nd ed., rev.). London: Science Paperbacks, 1966.

Ayres, E. *What's Good for GM.* Nashville: Aurora Publishers, 1970.

Ayres, L. "Maturity of Automotive Industry." *Journal of the Society of Automotive Engineers,* 1927, *20,* 372.

Bachman, B. B. "S.A.E. Standards." *Journal of the Society of Automotive Engineers,* 1921, *9,* 355–356.

Baird, D. G. "Eliminating Needless Cost and Confusion." *Industrial Management,* 1923, *65,* 334–337.

Baird, D. G. "Management Coordination Keeps GM Fleet Sales on the Up Grade throughout Depression Years." *Sales Management,* 1935, *36,* 188–189+.

---

*NOTE: Periodical articles without author credit are not included in the bibliography, except in the case of *Fortune* articles. Full citations for such articles are included in the text.

Barnes, J. K. "The Men Who 'Standardize' Automobile Parts." *World's Work,* 1921, *42,* 204-208.

Barton, A. H. "Organizations: Methods of Research." In *International Encyclopedia of the Social Sciences.* New York: Macmillan, 1968, Vol. 11, pp. 334-343.

Bassett, H. H. "Better Production and Conveying Machinery Important Factors." *Industrial Management,* 1926, *71,* 268-270.

Bauer, G. F. "Financing Automobile Sales Abroad." *Management and Administration,* 1925, *9,* 365-368 (a).

Bauer, G. F. "Profitable Export Market Analysis: Methods of American Automobile Manufacturers." *Management and Administration,* 1925, *9,* 561-564 (b).

Bayley, S. *Harley Earl and the Dream Machine.* New York: Knopf, 1983.

Beasley, N. *Knudsen: A Biography.* New York: McGraw-Hill, 1947.

Bennett, H., as told to P. Marcus. *We Never Called Him Henry.* New York: Fawcett Publications, 1951.

Benson, A. L. *The New Henry Ford.* New York: Funk & Wagnalls, 1923.

Benson, A. L. "Sloan: A Study by Allan L. Benson of the Only Man Who Ever Made Ford Take a Back Seat." *Hearst's International-Cosmopolitan,* 1928 (January), *84,* 32-33+.

Boorstin, D. J. *The Americans: The Democratic Experience.* New York: Vintage Books, 1973.

Boyd, T. *Professional Amateur: The Biography of Charles Franklin Kettering.* New York: Dutton, 1957.

Bradley, A. H. "Forecasting Stabilizes Operations: Program of General Motors Corporation Aids Formulation of Fundamental Policies." *Iron Age,* 1926, *117,* 691-693 (a).

Bradley, A. H. "General Motors Prepared to Vary Production Every 10 Days." *Automotive Industries,* 1926, *54,* 488-490 (b).

Bradley, A. H. "How General Motors Copes with the Seasonal Problem." *Printers' Ink,* 1926, *134* (March 18), 156+ (c).

Bradley, A. H. "Setting up a Forecasting Program." American Management Association, *Annual Convention Series,* 1926, *41,* 3-18 (d). Reprinted in Chandler, 1979.

Bradley, A. H. "Financial Control Policies of General Motors Corporation and Their Relationship to Cost Accounting." *N.A.C.A. Bulletin,* 1927, 7, 412-433.

Brough, J. *The Ford Dynasty: An American Story.* Garden City, N.Y.: Doubleday, 1977.

Brown, D. "Pricing Policy in Relation to Financial Control." *Management and Administration,* 1924, *7,* 195-198 (a). Reprinted in Chandler, 1979.

Brown, D. "Pricing Policy in Relation to Financial Control." *Management and Administration,* 1924, *7,* 283-286 (b). Reprinted in Chandler, 1979.

Brown, D. "Pricing Policy Applied to Financial Control." *Management and Administration,* 1924, 7, 417-422 (c). Reprinted in Chandler, 1979.

Brown, D. "Centralized Control with Decentralized Responsibilities." American Management Association, *Annual Convention Series,* 1927, *57,* 3-24. Reprinted in Chandler, 1979.

Brown, D. "Forecasting and Planning." *Survey,* 1929, *68,* 34-35 (a).

Brown, D. "Forecasting and Planning as a Factor in Stabilizing Industry." *Sales Management and Advertisers' Weekly,* 1929, *17,* 181-183+ (b).

Brown, D. "Forecasting and Planning as a Factor in Stabilizing Industry (Part 2)." *Sales Management and Advertisers' Weekly*, 1929, *17*, 258-259+ (c).

Brown, D. "Industrial Management as a National Resource." *Conference Board Management Record*, 1943, *5*, 142-148.

Brown, D. *Some Reminiscences of an Industrialist*. Port Deposit, Md.: Author, 1957. Reprinted: Ann Arbor, Michigan: University Microfilms, 1981.

Brown, D. "Coordinated Control." In E. Dale (Ed.), *Readings in Management*. New York: McGraw-Hill, 1965, pp. 332-333.

Browning, A. J. "Ford Modernizes Its Buying Methods." *Automotive Industries*, 1947 (November 1), *97*, 24-27+.

Buick Motor Company. *The Factory behind the Car*. Flint, Mich.: Buick Motor Company, 1925.

Burck, C. G. "How G.M. Turned Itself Around." *Fortune*, 1978 (January 16), *97*, 87-89+.

Burck, C. G. "How GM Stays Ahead." *Fortune*, 1981 (March 9), *103*, 48-56.

Burck, C. G. "Can Detroit Catch Up?" *Fortune*, 1982 (February 8), *105*, 34-39.

Burck, C. G. "Will Success Spoil General Motors?" *Fortune*, 1983 (August 22), *108*, 94-100+.

Burlingame, R. *Henry Ford: A Great Life in Brief*. New York: Knopf, 1957.

Bush, D. J. *The Streamlined Decade*. New York: George Braziller, 1975.

Butler, R. N., and M. I. Lewis. *Aging and Mental Health*. St. Louis, Missouri: C. V. Mosby, 1982.

Cadillac Motor Car Company. *The New Cadillac: In Which the Modern Trend of the Motor Car Is Embodied in a Rich Variety of Luxurious, Distinguished Models*. Detroit: Evan-Winter-Hebb, 1927.

Carson, G. *The Old Country Store*. New York: Dutton, 1965.

Ceppos, R. "Quality." *Car and Driver*, 1981 (March), *26*, 43+.

Chandler, A. D., Jr. *Giant Enterprise: Ford, General Motors, and the Automobile Industry*. New York: Harcourt Brace Jovanovich, 1964.

Chandler, A. D., Jr. *Strategy and Structure*. New York: Anchor Books, 1966.

Chandler, A. D., Jr. *The Visible Hand: The Managerial Revolution in American Business*. Cambridge, Mass.: Belknap Press, 1977.

Chandler, A. D., Jr. (Ed.) *Managerial Innovation of General Motors*. New York: Arno Press, 1979.

Chandler, A. D., Jr., and S. Salsbury. *Pierre S. duPont and the Making of the Modern Corporation*. New York: Harper & Row, 1971.

Chrysler, W. P. *Life of an American Workman*. New York: Dodd, Mead & Co., 1937.

Churchman, C. W. *The Systems Approach*. New York: Dell, 1968.

Churchman, C. W. *The Design of Inquiring Systems: Basic Concepts of Systems and Organization*. New York: Basic Books, 1971.

Clark, W. W., Jr. "Buick Saves $3,500,000 in Nine Months." *Manufacturing Industries*, 1926, *11*, 85-88.

Cole, R. E., and T. Yakushiji. *The American and Japanese Auto Industries in Transition*. Ann Arbor: University of Michigan, Center for Japanese Studies, 1984.

Colston, R. "The World's Most Profitable Business Enterprise." *Magazine of Wall Street*, 1939, *64*, 606-609+.

Condit, K. H. "Getting Facts for Automobile Designers." *American Machinist*, 1926, *64*, 9-12.

Cordtz, D. "The Face in the Mirror at General Motors." *Fortune*, 1966 (August), *74*, 117-119+.

"Corporate Management." *Fortune*, 1933 (June), *7*, 47-51+.

Couzens, J. "What I Learned about Business from Ford." *System*, 1921, *40*, 261-264+.

Crabb, R. *Birth of a Giant*. New York: Chilton, 1969.

Cray, E. *Chrome Colossus: General Motors and Its Times*. New York: McGraw-Hill, 1980.

Crow, C. *The City of Flint Grows Up*. New York: Harper & Brothers, 1945.

Crowther, S. "John J. Raskob, and the World's Largest Business." *World's Work*, 1920, *40*(10), 612-617.

Customer Research Staff. *Customer Research*. Detroit: General Motors Corporation, 1937.

Dale, E. "Contributions to Administration by Alfred P. Sloan, Jr., and GM." *Administrative Science Quarterly*, 1956, *1*, 30-61.

Dale, E. *The Great Organizers*. New York: McGraw-Hill, 1960.

Dale, E. (Ed.). *Readings in Management*. New York: McGraw-Hill, 1965.

Dale, E., and L. F. Urwick, *Staff in Organization*. New York: McGraw-Hill, 1960.

Department of Organization and Management, General Motors Institute. *Industrial Management*. Flint, Mich.: General Motors Institute, 1951.

Department of Public Relations, General Motors. *The Responsibility of Management*. Detroit: General Motors Corporation, 1946.

Dibble, L. C. "New General Motors College to Offer Wide Range of Courses." *Automotive Industries*, 1926, *55*, 180-181.

Donner, F. G. *The World-wide Industrial Enterprise*. New York: McGraw-Hill, 1967.

Dorfman, J. "Make or Break." *Forbes*, 1983 (April 25), *131*, 98-102+.

Douglass, P. F. *Six upon the World: Toward an American Culture for an Industrial Age*. Boston: Little, Brown & Co., 1954.

Drucker, P. F. *Concept of the Corporation*. New York: John Day, 1946.

Drucker, P. F. *The Practice of Management*. New York: Harper & Brothers, 1954.

Drucker, P. F. *Concept of the Corporation* (rev. ed.). New York: John Day, 1972.

Drucker, P. F. *Management*. New York: Harper & Row, 1973.

Durant, W. C. "The True Story of General Motors." In E. Dale (Ed.), *Readings in Management*. New York: McGraw-Hill, 1965, pp. 14-19.

Durham, C. B. "We Make 1400% More Cars with 10% More Men." *Magazine of Business*, 1927, *52*, 29-31+.

Ellis, J. *Billboards to Buicks*. New York: Abelard-Schuman, 1968.

Epstein, R. C. "The Rise and Fall of Firms in the Automobile Industry." *Harvard Business Review*, 1927, *5*, 157-173 (a).

Epstein, R. C. "Leadership in the Automobile Industry, 1903-1924." *Harvard Business Review*, 1927, *5*, 281-292 (b).

Epstein, R. C. *The Automobile Industry*. Chicago: A. W. Shaw, 1928. Reprinted: New York: Arno Press, 1972.

Fanning, L. M. *Titans of Business*. Philadelphia and New York: Lippincott, 1964.

Farnham, D. T. "The Vertical Trust." *Industrial Management*, 1924, *67*, 257-263.

Farnham, D. T. "Types of Consolidations and Mergers in America and Europe." American Management Association, *General Management Series*, 1929, No. 88.

Faulkner, H. U. *From Versailles to the New Deal: A Chronicle of the Harding-Coolidge-Hoover Era.* New Haven: Yale University Press, 1950.

Faurote, F. L. "Single-purpose Manufacturing." *Factory and Industrial Management*, 1928, *75*, 769-773.

Faurote, F. L. "Specific Training in the General Motors Institute of Technology." *Iron Age*, 1930, *126*, 425-429.

Filipetti, G. "The Dealer Study. Appendix 5." In E. R. A. Seligman, *The Economics of Installment Selling.* New York: Harper & Brothers, 1927, pp. 395-472.

Fine, S. *The Automobile under the Blue Eagle.* Ann Arbor: University of Michigan Press, 1963.

Fine, S. *Sit-down: The General Motors Strike of 1936-1937.* Ann Arbor: University of Michigan Press, 1969.

Finn, D. *The Corporate Oligarch.* New York: Simon & Schuster, 1969.

Forbes, B. C. " 'We Face the Future without Fear, with Faith'—Sloan." *Forbes,* 1924 (March 29), *13*, 759+.

Forbes, B. C. "Ford Loses Motor Leadership: Amazing Facts." *Forbes,* 1927 (April 15), *19*, 9-10+ (a).

Forbes, B. C. "Why Our Articles *on* Henry Ford Are Being Published." *Forbes,* 1927 (May 15), *19*, 16 (b).

Forbes, B. C. "How Ford Dealers Are Treated as Described by One of Them." *Forbes,* 1927 (May 15), *19*, 17-19 (c).

Forbes, B. C. "Can Ford Troubles Be Cured?" *Forbes,* 1927 (June 1), *19*, 12-14+ (d).

Forbes, B. C. *America's Fifty Foremost Business Leaders.* New York: Forbes & Sons, 1948.

Forbes, B. C., and O. D. Foster. *Automotive Giants of America.* New York: Forbes, 1926.

Forbes, J. D. *Stettinius, Sr.: Portrait of a Morgan Partner.* Charlottesville: University Press of Virginia, 1974.

Ford, H., in an interview with S. Crowther. "My Rule for Making Steady Profits." *System,* 1922, *41*, 259-263 (a).

Ford, H., in collaboration with S. Crowther. *My Life and Work.* New York: Doubleday, Page & Co., 1922 (b).

Ford, H., in collaboration with S. Crowther. "What I Have Learned about Management in the Last 25 Years." *System,* 1926, *49*, 37-40+ (a).

Ford, H., in collaboration with S. Crowther. *Today and Tomorrow.* New York: Doubleday, Page & Co., 1926 (b).

Ford, H., in collaboration with S. Crowther. "When Is a Business Worth While?" *Magazine of Business,* 1928, *54*, 133-136.

Ford, H., and S. Crowther. "Management and Size." *Saturday Evening Post,* 1930 (September 20), *203*, 24-25+.

Ford, H., an interview by F. L. Faurote. *My Philosophy of Industry.* New York: Coward-McCann, 1929.

Ford Motor Company. *The Ford Industries: Facts about the Ford Motor Company and Its Subsidiaries.* Detroit: Ford Motor Company, 1924.

Ford Motor Company. *Ford at Fifty.* New York: Simon & Schuster, 1953.

Fordham, T. B., and E. H. Tingley. "Control through Organization and Budgets." *Management and Administration,* 1923, *6,* 719-724. Reprinted in Chandler, 1979.

Fordham, T. B., and E. H. Tingley. "The Compilation of a Budget: 'Control through Organization and Budgets'—Article II." *Management and Administration,* 1924, *7,* 57-62 (a). Reprinted in Chandler, 1979.

Fordham, T. B., and E. H. Tingley. "Applying the Budget to Industrial Operations: 'Control through Organization and Budgets'—Article III." *Management and Administration,* 1924, *7,* 205-208 (b). Reprinted in Chandler, 1979.

Fordham, T. B., and E. H. Tingley. "Operating Factory Divisions by Budget: 'Control through Organization and Budgets'—Article IV." *Management and Administration,* 1924, *7,* 291-294 (c). Reprinted in Chandler, 1979.

"Ford's New Managers." *Fortune,* 1953 (May), *47,* 142-143+.

Galbraith, J. K. *The Liberal Hour.* Boston: Houghton Mifflin, 1960.

Galbraith, J. K. *The New Industrial State.* Boston: Houghton Mifflin, 1967.

Gates, W. "General Motors Heads for New Records." *Magazine of Wall Street,* 1936, *58,* 463-465+.

Gelderman, C. *Henry Ford: The Wayward Capitalist.* New York: Dial Press, 1981.

General Motors Corporation. *Annual Reports.*

General Motors Corporation. *The Proving Ground for the Products of General Motors.* New York: General Motors Corporation, 1926.

General Motors Corporation. *Before Buying Another Car See What General Motors Offers.* New York: General Motors Corporation, 1929 (a).

General Motors Corporation. *General Motors Acceptance Corporation: The Sales Financing Organization of General Motors.* New York: General Motors Corporation, 1929 (b).

General Motors Corporation. *General Motors Institute of Technology.* New York: General Motors Corporation, 1929 (c).

General Motors Corporation. *The New Devices Committee of General Motors.* New York: General Motors Corporation, 1929 (d).

General Motors Corporation. *The Story of Knee Action.* Detroit: General Motors Corporation, 1935.

General Motors Corporation. *The Dynamics of Automobile Demand.* New York: General Motors Corporation, 1939.

General Motors Corporation. *Research, Science, and the Motor Car.* Detroit: General Motors Corporation, 1941.

General Motors Corporation. *The Automobile Industry: A Case Study of Competition.* Detroit: General Motors Corporation, 1968.

General Motors Corporation. *The Locomotive Industry and General Motors.* Detroit: General Motors Corporation, 1973.

General Motors Sales Corporation. *General Motors Dealers Standard Accounting System Manual.* Detroit: General Motors Sales Corporation, 1940.

"General Motors: Part I of a Study in Bigness." *Fortune,* 1938 (December), *18,* 40-47+.

"General Motors II: Chevrolet." *Fortune,* 1939 (January), *19,* 36-46+.

"G.M. III: How to Sell Automobiles." *Fortune,* 1939 (February), *19,* 70-78+.

"General Motors IV: A Unit in Society." *Fortune,* 1939 (March), *19,* 44-52+.

General Motors: The First 75 Years. New York: Crown Publishers, 1983.

Getz, C. H. "Financing the Purchase of Automobiles." *Bankers Magazine,* 1924, *108,* 869–872.

Gillette, H. P. "Ford's Business Philosophy." *Engineering and Contracting,* 1928, *67,* 137–140.

Golden, L. L. "The Sloan Touch." *Saturday Review,* 1966 (April 9), 78.

Grant, R. H. "Better Retail Credit Risks." *American Bankers Association Journal,* 1931, *24,* 197+.

Green, P. M. "General Motors Corporation; A Study in Corporate Consolidation and Finance." Unpublished doctoral dissertation, University of Illinois, Urbana, 1933.

Greenleaf, William. *From These Beginnings: The Early Philanthropies of Henry and Edsel Ford, 1911–1936.* Detroit: Wayne State University Press, 1964.

Griffin, C. E. "The Evolution of the Automobile Market." *Harvard Business Review,* 1926, *5,* 407–416.

Grimes, W. A. *Financing Automobile Sales by the Time-Payment Plan.* New York: A. W. Shaw, 1926.

Gronseth, H. E. " 'Alma Mater' at Flint." *Automotive Industries,* 1936, *75,* 404–407+.

Gustin, L. R. *Billy Durant: Creator of General Motors.* Grand Rapids, Mich.: William B. Eerdmans, 1973.

Hagemann, G. E. "Testing Products from the User's Angle: General Motors Makes Exhaustive Studies of Motor Vehicles." *Manufacturing Industries,* 1926, *11,* 431–434.

Harris, W. B. "Ford's Fight for First." *Fortune,* 1954 (September), *50,* 123–127+.

Hearings Before the Subcommittee on Antitrust and Monopoly of the Senate Committee on the Judiciary, 84th Cong., 1st Sess., pursuant to S. Res. 61, Parts 6, 7, and 8 (1955).

Hearings Before Subcommittees of the Senate Select Committee on Small Business, 90th Cong., 2nd Sess., on The Question: Are Planning and Regulation Replacing Competition in the American Economy? (The Automobile Industry as a Case Study) (1968).

Hendry, M. D. *Cadillac: The Complete History.* New York: Bonanza Books, 1983.

Hickerson, J. M. *Ernie Breech.* New York: Meredith Press, 1968.

Hounshell, D. A. *From the American System to Mass Production: The Development of Manufacturing Technology in the United States.* Baltimore: Johns Hopkins University Press, 1984.

Iacocca, L., with W. Novak. *Iacocca: An Autobiography.* New York: Bantam Books, 1984.

"The Industrial War." *Fortune,* 1937 (November), *16,* 104–110+.

Jardim, A. *The First Henry Ford: A Study in Personality and Business Leadership.* Cambridge, Mass.: M.I.T. Press, 1970.

John, W. A. P. "That Man Durant." *Motor,* 1924 (January), 70–71+.

Kanter, R. M. *The Change Masters: Innovation and Entrepreneurship in the American Corporation.* New York: Simon & Schuster, 1983.

Kennedy, E. *The Automobile Industry.* New York: Reynal and Hitchcock, 1941.

Kettering, C. F. "Research, Horse-sense and Profits." *Factory and Industrial Management,* 1928, *75,* 735–739.

Kettering, C. F. "Keep the Customer Dissatisfied." *Nation's Business,* 1929 (January), *17,* 30-31+.

Kettering, C. F., and A. Orth. *The New Necessity.* Baltimore: Williams & Wilkins, 1932.

Kidder, T. *The Soul of a New Machine.* Boston: Little, Brown & Co., 1981.

Knudsen, W. S. " 'For Economical Transportation': How the Chevrolet Motor Company Applies Its Own Slogan to Production." *Industrial Management,* 1927, *74,* 65-68.

Koskey, G. H. "The Relation of Cost Control to the Evolution of System." *Industrial Management,* 1926, *72,* 34-38.

Kraar, L. "Japan's Automakers Shift Strategies." *Fortune,* 1980 (August 11), *102,* 106-111.

Kreusser, O. T. "Automotive Proving Grounds of the General Motors Corporation." *Journal of the Society of Automotive Engineers,* 1926, *18,* 27-32.

Kuhn, A. J. "An Application of a System Control Model to Business History: The General Motors Corporation under Alfred P. Sloan, 1920 to 1935." Unpublished doctoral dissertation, University of California, Berkeley, 1972.

Kuhn, A. J. *Organizational Cybernetics and Business Policy: System Design for Performance Control.* University Park: Pennsylvania State University Press, 1986.

Lawrence, P. R., and D. Dyer. *Renewing American Industry.* New York: Free Press, 1983.

Leonard, J. N. *The Tragedy of Henry Ford.* New York: G. P. Putnam's Sons, 1932.

Leslie, S. W. *Boss Kettering: Wizard of General Motors.* New York: Columbia University Press, 1983.

Lewis, D. *The Public Image of Henry Ford.* Detroit: Wayne State University Press, 1976.

"Lincoln-Mercury Moves Up." *Fortune,* 1952 (March), *45,* 96-99+.

Livingston, J. A. *The American Stockholder.* Philadelphia and New York: J. B. Lippincott, 1958.

Lundberg, F. *The Rich and the Super-rich.* New York: Lyle Stuart, 1968.

Macaulay, S. *Law and the Balance of Power.* New York: Russell Sage Foundation, 1966.

MacGowan, T. G. "How Do Buyers Feel about Today's Cars . . . Today? Article III." *Automotive Industries,* 1938, *79,* 254-260.

MacManus, T., and N. Beasley. *Men, Money, and Motors.* New York: Harper & Brothers, 1930.

McClary, T. C. "In the Battle of Motor Giants." *Magazine of Wall Street,* 1933, *52,* 276-278+.

McDonald, J. *The Game of Business.* Garden City, N.Y.: Anchor Books, 1977.

Mandel, L. *American Cars.* New York: Stewart, Tabori & Chang, 1982.

Marquis, S. S. *Henry Ford: An Interpretation.* Boston: Little, Brown & Co., 1923.

Meikle, J. L. *Twentieth Century Limited.* Philadelphia: Temple University Press, 1979.

Meyer, S. *The Five Dollar Day.* Albany: State University of New York Press, 1981.

Mooney, J. D. "Selling the Automobile Overseas: How General Motors' Export Business Is Handled." *Management and Administration,* 1924, *8,* 27-32.

Mooney, J. D., and A. C. Reiley. *Onward Industry.* New York: Harper & Brothers, 1931.

Mooney, J. D., and A. C. Reiley, *The Principles of Organization.* New York: Harper & Brothers, 1939.

Moritz, M. *The Little Kingdom: The Private Story of Apple Computer.* New York: William Morrow and Co., 1984.

Mott, C. S. "Organizing a Great Industrial." *Management and Administration*, 1924, 7, 523-527. Reprinted in Chandler, 1979.

"Mr. Ford Doesn't Care . . . What Chevrolet Does." *Fortune*, 1933 (December), 8, 62-69+.

Myers, M. S., and S. S. Myers. "Toward Understanding the Changing Work Ethic." *California Management Review*, 1974, 16 (3), 7-19.

Nader, R. *Unsafe at Any Speed.* New York: Simon & Schuster, 1965.

Nader, R. *Unsafe at Any Speed* (2nd ed.). New York: Grossman Publishers, 1972.

National Industrial Conference Board. *Industrial Standardization.* New York: National Industrial Conference Board, 1929.

Nevins, A., and F. E. Hill. *Ford: The Times, the Man, the Company.* New York: Scribner's, 1954.

Nevins, A., and F. E. Hill. *Ford: Expansion and Challenge, 1915-1933.* New York: Scribner's, 1957.

Nevins, A., and F. E. Hill. *Ford: Decline and Rebirth, 1933-1962.* New York: Scribner's, 1963.

Nichols, G. A. "General Motors Places Its Cars on Non-Competitive Basis." *Printers' Ink*, 1923 (December 13), 125, 17-19.

Nichols, G. A. "What Will Take Place of Advertising in Ford's Marketing Scheme?" *Printers' Ink*, 1926 (June 17), 135, 17-20.

Nielsen, W. A. *The Big Foundations.* New York: Columbia University Press, 1972.

Noble, D. F. *America by Design.* New York: Knopf, 1977.

O'Shea, P. F. "General Motors Budgets for Change: an Interview with Charles F. Kettering." *Magazine of Business*, 1928, 54, 359-361+.

Perrett, G. *America in the Twenties: A History.* New York: Simon & Schuster, 1982.

Pound, A. *The Turning Wheel: the Story of General Motors through Twenty-Five Years, 1908-1933.* Garden City, N.Y.: Doubleday, Doran & Co., 1934.

Pound, A. "Precision and Perspective: General Motors' Philosophy of Industrial Progress." *Atlantic Monthly*, 1935, 155 (1), 123-130.

Quinter, D. "The Man Who Made Detroit's First Minicar." *Car and Driver*, 1976 (January), 22, 37-39.

Rae, J. B. "The Fabulous Billy Durant." *Business History Review*, 1958, 32, 255-271.

Rae, J. B. *American Automobile Manufacturers: The First Forty Years.* Chicago: University of Chicago Press, 1959.

Rae, J. B. *The American Automobile: A Brief History.* Chicago: University of Chicago Press, 1965.

Rae, J. B. (Ed.) *Henry Ford.* Englewood Cliffs, N.J.: Prentice-Hall, 1969.

Raskob, J. J. "Management the Major Factor in All Industry." *Industrial Management*, 1927, 74, 129-135.

Raskob, J. J. "Our New Industrial Set-Up." *Magazine of Business*, 1928, 34, 23-26.

"The Rebirth of Ford." *Fortune*, 1947 (May), 35, 81-89+.

Reck, F. M. *On Time: The History of Electro-Motive Division of General Motors Corporation.* New York: Electro-Motive Division of General Motors Corporation, 1948.

Reeves, E. "The Secret of General Motors' Amazing Earnings." *American Bankers Association Journal*, 1927, 20, 36+.

Rickard, E. B. "A Study in Decentralization: Controllership in a Divisional Organization." *N.A.C.A. Bulletin,* 1950, *31,* 567-578.

Romer, S. "Profile of General Motors." *Nation,* 1937 (January 23), *144,* 96-98.

Ross, I. "The New UAW Contract: A *Fortune* Proposal." *Fortune,* 1982 (February 8), *105,* 40-45.

Ross, T. J. "The Automobile Industry." In J. G. Glover and W. B. Cornell (Eds.) *The Development of American Industries* (rev. ed.). New York: Prentice-Hall, 1941, pp. 679-705.

Rothschild, E. *Paradise Lost: The Decline of the Auto-industrial Age.* New York: Random House, 1973.

Rukeyser, M. S. "General Motors and Ford: A Race for Leadership." *American Review of Reviews,* 1927, *76,* 372-379.

Schnapp, J., and J. Cassettari. *Corporate Strategies of the Automotive Manufacturers.* Lexington, Mass.: Lexington Books, 1979.

Schneider, R. A. *Sixteen Cylinder Motorcars.* Arcadia, California: Heritage House, 1974.

Seligman, E. R. A. *The Economics of Instalment Selling.* (Vol. 1). New York: Harper & Brothers, 1927 (a).

Seligman, E. R. A. *The Economics of Instalment Selling.* (Vol. 2). New York: Harper & Brothers, 1927 (b).

Seltzer, L. *Financial History of American Automobile Industry.* New York: Houghton Mifflin, 1928.

Sheenan, R. "How Harlow Curtice Earns His $750,000." In Editors of *Fortune, The Art of Success.* Philadelphia and New York: J. B. Lippincott, 1956, pp. 3-20.

Sheenan, R. "G.M.'s Remodeled Management." In M. Carter, W. Weintraub, and C. Ray (Eds.), *Management: Challenge and Response.* New York: Holt, Rinehart & Winston, 1965, pp. 46-58.

Shidle, N. G. "Ford's New Prices—What Do They Portend?" *Automotive Industries,* 1926, *55,* 10-11.

Shidle, N. G. "Trend toward More Car Models Helping Parts Makers." *Automotive Industries,* 1927, *57,* 145-147 (a).

Shidle, N. G. "Ford's New Model—Will It Place Him on Top Again?" *Automotive Industries,* 1927, *57,* 253-256 (b).

Shidle, N. G. "Competition Sharpens Ford Dilemma: New Models or Lower Prices Looming." *Automotive Industries,* 1931, *68,* 939-941+.

Shidle, N. G. "What Does the Public Want?" *Automotive Industries,* 1933, *68,* 666-669+.

Simon, H. A. *Administrative Behavior* (2nd ed.). New York: Free Press, 1965.

Sinsabaugh, C. G. *Who Me? Forty Years of Automobile History.* Detroit: Arnold-Powers, 1940.

Sloan, A. P., Jr. "The Most Important Thing I Ever Learned about Management." *System,* 1924, *46,* 137-141.

Sloan, A. P., Jr. "Make Smaller Profits Pay." *Factory,* 1926, *37,* 993-997+.

Sloan, A. P., Jr. " 'Getting the Facts' Is Keystone of General Motors Success." *Automotive Industries,* 1927, *57,* 550-551 (a).

Sloan, A. P., Jr. "The Coming Competition between Ford and General Motors." *Printers' Ink,* 1927 (October 6), *141,* 164+ (b).

Sloan, A. P., Jr. *The Principles and Policies behind General Motors.* New York: General Motors Corporation, 1927 (c).

Sloan, A. P., Jr. "Sloan of General Motors Predicts a Revolution in Distribution." *Printers' Ink,* 1929 (February 7), *146,* 88+.

Sloan, A. P., Jr. *How Members of the General Motors Family Are Made Partners in General Motors.* New York: General Motors Corporation, 1930.

Sloan, A. P., Jr. "Proving Ground of Public Opinion: General Motors' Consumer Research Staff Is Part of an Operating Philosophy." *Printers' Ink,* 1933 (September 21), *164,* 92-93.

Sloan, A. P., Jr. "Sloan Urges Industry to Stand and, If Necessary, to Fight for a Square Deal." *Automotive Industries,* 1934, *70,* 522-523+.

Sloan, A. P., Jr. "The Broadened Responsibilities of Industries Executives." In J. G. Frederick (Ed.), *For Top Executives Only.* New York: Business Bourse, 1936, pp. 351-371.

Sloan, A. P., Jr., in collaboration with B. Sparks. *Adventures of a White Collar Man.* New York: Doubleday, Doran & Co., 1941.

Sloan, A. P., Jr. "Fact-finding for Management." In L. Greendlinger (Ed.), *Modern Business Lectures.* New York: Alexander Hamilton Institute, 1944, pp. 5-18.

Sloan, A. P., Jr. *My Years with General Motors.* Garden City, N.Y.: Doubleday & Co., 1964.

Sloan, A. P., Jr. "General Motors Corporation Study of Organization." In E. Dale (Ed.), *Readings in Management.* New York: McGraw-Hill, 1965, pp. 215-218.

Smelser, N. J. *Essays in Sociological Explanation.* Englewood Cliffs, N.J.: Prentice-Hall, 1968.

Snell, B. "Annual Style Change in the Automobile Industry as an Unfair Method of Competition." *Antitrust Law and Economics Review,* 1970 (a), *4*(1), 67-94; 1970 (b), *4*(2), 55-92.

Sobel, R. *Car Wars.* New York: E.P. Dutton, 1984.

Sobey, A. "The General Motors Institute of Technology." *Mechanical Engineering,* 1927, *49*(5a), 553-557 (a).

Sobey, A. "The New General Motors Institute of Technology." *Journal of the Society of Automotive Engineers,* 1927, *20,* 239-244 (b).

Sorensen, C. E., in collaboration with S. T. Williamson. *My Forty Years with Ford.* New York: W. W. Norton, 1956.

Soule, G. *Prosperity Decade: From War to Depression, 1917-1929.* New York: Rinehart, 1947.

Sprague, J. R. (Ed.). "Confessions of a Ford Dealer." *Harper's,* 1927 (June), *155,* 26-35.

Stern, P. *Tin Lizzie.* New York: Simon & Schuster, 1955.

Stillman, K. W. "Hand-to-mouth Buying: What Has It Done—and What Will It Do?" *Automotive Industries,* 1927, *57,* 1-4.

Stinchcombe, A. L. *Constructing Social Theories.* New York: Harcourt Brace Jovanovich, 1968.

Stryker, P. "Planning and Control for Profit." *Fortune,* 1952 (April), *45,* 128+.

Sward, K. *The Legend of Henry Ford.* New York: Rinehart, 1948.

Swayne, A. H. "Mobilization of Cash Reserves." *Management and Administration*, 1924, *7*, 21-23. Reprinted in Chandler, 1979.
Swayne, A. H. "Sound Credit Selling." *Credit Monthly*, 1929, *31* (9), 5-6.
Terkel, S. *Hard Times*. New York: Pantheon Books, 1970.
Thomas, R. P. "Style Change and the Automobile Industry During the Roaring Twenties." In L. P. Cain and P. J. Uselding (Eds.), *Business Enterprise and Economic Change.* Kent, Ohio: Kent State University Press, 1973.
Thompson, G. V. "Intercompany Technical Standardization in the Early American Automobile Industry." *Journal of Economic History*, 1954, *14*, 1-20.
Tingley, E. H. "Visualizing Budgetary Control." *Management and Administration*, 1924, *8*, 383-386.
United States Federal Trade Commission. *Report on Motor Vehicle Industry.* Washington, D.C.: U.S. Government Printing Office, 1939.
United States Senate, 84th Congress, 2d Session, 1956, Committee on the Judiciary, Subcommittee on Antitrust and Monopoly. "A Study of the Antitrust Laws," Report No. 1879. Washington, D.C.: U.S. Government Printing Office.
United States v. *DuPont & Co. et al.* (1956) 353 U.S. 586, U.S. Supreme Court Records, Briefs.
Warner, E. L., Jr. "Inter-office Communication Systems." *Automotive Industries*, 1940, *83*, 50-54+.
Weaver, H. G. "General Motors 'Purchasing Power' Index." *Manufacturing Industries*, 1926, *11*, 291-292.
Weaver, W. *U.S. Philanthropic Foundations: Their History, Structure, Management, and Record.* New York: Harper & Row, 1967.
Weisberger, B. *The Dream Maker: William C. Durant, Founder of General Motors.* Boston: Little, Brown & Company, 1979.
Wennerlund, E. K. "Quantity Control of Inventories: Physical Regulation Contrasted with Mere Financial Information." *Management and Administration*, 1924, *7*, 677-682. Reprinted in Chandler, 1979.
White, L. J. *The Automobile Industry since 1945.* Cambridge, Mass.: Harvard University Press, 1971.
White, T. M. *Famous Leaders of Industry.* Freeport, N.Y.: Books for Libraries Press, 1931.
Whyte, W. H., Jr. *The Organization Man.* Garden City, N.Y.: Doubleday, 1956.
Wiener, N. *The Human Use of Human Beings.* Garden City, N.Y.: Doubleday, 1954.
Wik, R. M. *Henry Ford and Grass-roots America.* Ann Arbor: University of Michigan Press, 1972.
Williamson, O. E. *Corporate Control and Business Behavior.* Englewood Cliffs, N.J.: Prentice-Hall, 1970.
Wolff, H. A. "The Great GM Mystery." *Harvard Business Review*, 1964 (September-October), *42*, 164-166+.
Wright, J. P. *On a Clear Day You Can See General Motors.* New York: Avon Books, 1979.
Yates, B. *The Decline and Fall of the American Automobile Industry.* New York: Empire Books, 1983.
Yergin, D. "Conservation: The Key Energy Source." In R. Stogbaugh and D. Yergin (Eds.) *Energy Future.* New York: Random House, 1979, pp. 136-182.

Young, C. H., and W. A. Quinn. *Foundation for Living: The Story of Charles Stewart Mott and Flint.* New York: McGraw-Hill, 1963.

Zurcher, A. J. *The Management of American Foundations: Administration, Policies, and Social Role.* New York: New York University Press, 1972.

Index of Names

Only the names of participants are included in the Index of Names

Index of Topics